On Distributed and Cooperative Control Design for Networks of Dynamical Systems

Von der Fakultät Konstruktions-, Produktions- und Fahrzeugtechnik
und dem Stuttgart Research Centre for Simulation Technology
der Universität Stuttgart zur Erlangung der Würde eines
Doktor-Ingenieurs (Dr.-Ing.) genehmigte Abhandlung

Vorgelegt von

Georg S. Seyboth

aus Stuttgart

Hauptberichter: Prof. Dr.-Ing. Frank Allgöwer
Mitberichter: Assoc. Prof. Dimos V. Dimarogonas, PhD
Asst. Prof. Paolo Frasca, PhD

Tag der mündlichen Prüfung: 15. Januar 2016

Institut für Systemtheorie und Regelungstechnik

Universität Stuttgart

2016

Bibliografische Information der Deutschen Nationalbibliothek

Die Deutsche Nationalbibliothek verzeichnet diese Publikation in der Deutschen Nationalbibliografie; detaillierte bibliografische Daten sind im Internet über http://dnb.d-nb.de abrufbar.

D 93

ISBN 978-3-8325-4259-7

Logos Verlag Berlin GmbH
Comeniushof, Gubener Str. 47,
10243 Berlin
Tel.: +49 (0)30 42 85 10 90
Fax: +49 (0)30 42 85 10 92
INTERNET: http://www.logos-verlag.de

Acknowledgements

The results presented in this thesis are the outcome of my research activity at the Institute for Systems Theory and Automatic Control (IST) at the University of Stuttgart. There are numerous people who supported me and who made my time as a graduate student very enjoyable.

First of all I want to express my gratitude to my advisor Prof. Frank Allgöwer for the opportunity to pursue a doctoral degree at his institute, for giving me a lot of freedom, and for his guidance and support. I want to thank Prof. Dimos Dimarogonas and Prof. Paolo Frasca for their interest in my work, for the valuable comments, and for being members of my doctoral exam committee. Moreover I am particularly happy that Prof. Arnold Kistner acted as chairman of the committee.

I want to thank Prof. Karl Henrik Johansson and Prof. Dimos Dimarogonas for hosting me as an exchange student at the Royal Institute of Technology (KTH) in Stockholm where I entered the world of research in control theory and decided to pursue a doctoral degree. I want to thank as well Prof. Brad Yu and Prof. Wei Ren for hosting me as guest researcher at the Australian National University (ANU) in Canberra and at the University of California Riverside (UCR), respectively. All these visits were truly inspiring and I greatly appreciated our fruitful collaborations.

Special thanks go to all my colleagues and fellow PhD students at the IST. You supported me not only in my research activities with several collaborations and countless discussions but you also created a very pleasant and happy atmosphere at the institute. It was great working with you, sharing an office, growing chili plants, running, biking, hiking, barbecuing, playing games, watching movies, and having fun with you. Particular thanks go to Gregor Goebel, Jingbo Wu, Jan-Maximilian Montenbruck, Rainer Blind, and Steffen Linsenmayer for reading the first draft of this thesis.

There are many more people to whom I would like to express my gratitude – former colleagues, students, guests and researchers from all over the world. Thank you very much, and my apologies that I cannot name all of you here.

Last but not least I would like to wholeheartedly thank my family and my loving girlfriend for their patience, positivity, and continuous support which finally lead me to this success.

Stuttgart, February 2016
Georg S. Seyboth

Table of Contents

Notation

The mathematical notation and frequently used symbols and acronyms of this thesis are listed below. The notation and symbols are additionally introduced at their first appearance. Specific symbols and notation used in particular passages are only introduced in the respective passage.

Acronyms

LMI	Linear Matrix Inequalities
LPV	Linear Parameter-Varying
LTI	Linear Time-Invariant
SISO	Single-Input Single-Output
MIMO	Multi-Input Multi-Output
SDP	Semi-Definite Programming
iSCC	Independent Strongly Connected Component

Real and Complex Numbers

$\mathbb{N}$	Integers
$\mathbb{R}$	Real numbers
$\mathbb{R}^+$	Positive real numbers
$\mathbb{R}_0^+$	Non-negative real numbers
$\mathbb{R}^-$	Negative real numbers
$[a,b]$	Closed interval
$]a,b[$	Open interval
$\mathbb{C}$	Complex numbers
$\mathbb{C}^-$	Open left-half complex plane
$\mathbb{C}^0$	Imaginary axis
$\mathbb{C}^+$	Open right-half complex plane
$\mathbf{Re}(z)$	Real part of $z \in \mathbb{C}$
$\mathbf{Im}(z)$	Imaginary part of $z \in \mathbb{C}$
$\arg(z)$	Argument of $z \in \mathbb{C}$
$\lvert z \rvert$	Absolute value of $z \in \mathbb{C}$
$\bar{z}$	Complex conjugate of $z \in \mathbb{C}$
$\mathbf{i}$	Imaginary unit, i.e., $\mathbf{i}^2 = -1$

Vectors, Matrices and Norms

$\mathbf{0}$	Vector of zeros
$\mathbf{1}$	Vector of ones
I_N	Identity matrix of dimension N
M^T	Transpose of matrix M
M^H	Hermitian adjoint of matrix M
$\mathrm{diag}(v)$	Diagonal matrix with the elements v_i, $i = 1,...,n$, of $v \in \mathbb{R}^n$ on the diagonal
$\mathrm{diag}(v_i)$	Diagonal matrix with the elements v_i, $i = 1,...,n$, of $v \in \mathbb{R}^n$ on the diagonal
$\mathrm{diag}(M_i)$	Block diagonal matrix with matrices M_i, $i = 1,...,n$, on the diagonal
$\mathrm{stack}(M_i)$	Stacked matrix consisting of matrices M_i, $i = 1,...,n$
$\sigma(M)$	Spectrum of the matrix $M \in \mathbb{R}^{n\times n}$ (respecting multiplicities)
$M > 0$	Matrix M is positive definite
$M \geq 0$	Matrix M is positive semi-definite
$M < 0$	Matrix M is negative definite
$M \leq 0$	Matrix M is negative semi-definite
$\lVert v \rVert$	2-norm of the vector $v \in \mathbb{R}^n$
$\lVert M \rVert$	Induced 2-norm of the matrix $M \in \mathbb{R}^{p\times q}$
$\star$	Placeholder for matrix entries not of interest
$\lVert G \rVert_\infty$	$\mathcal{H}_\infty$ norm of the transfer function matrix G
$\otimes$	Kronecker product

Subspaces

$\ker(M)$	Null space of M, i.e., $\ker(M) = \{x \in \mathbb{R}^q : Mx = \mathbf{0}\}$ for $M \in \mathbb{R}^{p\times q}$
$\mathrm{im}(M)$	Image of M, i.e., $\mathrm{im}(M) = \{y \in \mathbb{R}^p : y = Mx, x \in \mathbb{R}^q\}$ for $M \in \mathbb{R}^{p\times q}$
$\dim(\mathcal{U})$	Dimension of subspace $\mathcal{U}$
$\mathcal{U}^\perp$	Orthogonal complement of subspace $\mathcal{U}$

Graphs

$\mathcal{G}$	Graph $\mathcal{G} = (\mathcal{V}, \mathcal{E})$ or weighted graph $\mathcal{G} = (\mathcal{V}, \mathcal{E}, A_\mathcal{G})$
$\mathcal{V}$	Vertex (or node) set $\mathcal{V} = \{v_1, ..., v_N\}$
$\mathcal{E}$	Edge (or link) set $\mathcal{E} \subset \mathcal{V} \times \mathcal{V}$
$\mathcal{N}$	Index set $\mathcal{N} = \{1, ..., N\}$
$\mathcal{N}_k$	Neighbor set $\mathcal{N}_k = \{j \in \mathcal{N} : (v_j, v_k) \in \mathcal{E}\}$ of vertex v_k in $\mathcal{G}$
$A_\mathcal{G}$	Adjacency matrix of $\mathcal{G}$
$D_\mathcal{G}$	Degree matrix of $\mathcal{G}$
$L_\mathcal{G}$	Laplacian matrix of $\mathcal{G}$
$\lambda_k(L_\mathcal{G})$	The k-th eigenvalue of $L_\mathcal{G}$

Abstract

This thesis is dedicated to the study of distributed and cooperative control problems in groups of dynamical agents. In a variety of modern man-made systems, it is desirable to synthesize a cooperative behavior among individual dynamical agents by distributed control laws. Examples include multi-vehicle coordination or formation flight problems, robot cooperation in production lines, as well as power balancing in micro-grids, and many more. Even though this area has become one of the major research fields within automatic control over the past decade, there are still many open questions and problems yet to be solved.

This thesis contributes to the development of a cooperative control theory for homogeneous and heterogeneous multi-agent systems consisting of identical and non-identical dynamical agents, respectively. The goal is to explain fundamental effects of non-identical agent dynamics on the behavior of a distributed system and, primarily, to develop suitable control design methods for a wide range of multi-agent coordination problems. We investigate output synchronization problems as well as cooperative disturbance rejection and reference tracking problems in multi-agent systems. Suitable controller design methods for networks consisting of identical or non-identical linear time-invariant systems, linear parameter-varying systems, and selected classes of nonlinear systems are developed. These controller design methods provide a solution to a variety of practically relevant distributed coordination and cooperative control scenarios.

The first part of this thesis is focused on state and output synchronization as the fundamental cooperative control task for multi-agent systems. The study comprises homogeneous linear multi-agent systems consisting of identical dynamical agents, heterogeneous linear multi-agent systems consisting of non-identical dynamical agents, and networks of linear parameter-varying dynamical agents, discusses structural requirements, and presents design methods for suitable distributed control laws.

The second part of this thesis is dedicated to cooperative disturbance rejection and reference tracking tasks for multi-agent systems subject to external input signals. Novel distributed control laws with integral action are proposed in order to react to such disturbances acting on individual agents. Moreover, the distributed output regulation framework is revisited and developed further as a universal approach to distributed and cooperative control design, capturing a wide range of coordination scenarios.

In the third part of this thesis, coordination problems for groups of nonholonomic vehicles are investigated. A distributed formation control algorithm for a group of mobile robots is developed based on the distributed output regulation framework developed in part two. Moreover, motion coordination algorithms for groups of unicycle-type vehicles with constant non-identical velocities are proposed for UAV coordination purposes.

Throughout the thesis, the theoretical results are illustrated by numerical simulations for a clearer understanding and in order to demonstrate the effectiveness of the novel control methods.

Deutsche Kurzfassung

Über die verteilte und kooperative Regelung von Netzwerken dynamischer Systeme

Motivation und Fragestellung

Der rasante technologische Fortschritt über die letzten Jahrzehnte hatte einen gewaltigen Einfluss auf unsere Welt und unser Alltagsleben. Wir leben in einer Zeit brillanter Technologien und man kann durchaus sagen, dass wir in ein neues Zeitalter eintreten, welches als *Second Machine Age* (engl. zweites Maschinenzeitalter) bezeichnet wird (Brynjolfsson and McAfee (2014)). Der technologische Fortschritt beinhaltet schnelle Weiterentwicklungen im Computerwesen, in der Kommunikationstechnik und in der Messtechnik (Murray et al. (2003)). Wesentliche Errungenschaften in den Ingenieurwissenschaften, der Informatik und der Kommunikationstechnik führen zu einem rasanten Wachstum von Kommunikationsnetzwerken, zu ständig verfügbarer Rechenleistung, sowie zu einer zunehmend engen Verknüpfung von physikalischen Systemen mit Informationssystemen. Immer größere und komplexere *Cyber-Physikalische Systeme* entstehen, die aus einer Vielzahl interagierender dynamischer Systeme bestehen. Eindrucksvolle Beispiele sind intelligente Energienetze (Stichwort: Smart Grid), intelligente Fabriken (Stichwort: Industrie 4.0) und intelligente Verkehrssysteme. Eine weitere bedeutende Entwicklung ist die zunehmende *Autonomie* moderner technischer Systeme, d.h. die zunehmende Unabhängigkeit von einer äußeren Leitung. Zwei wichtige Beispiele dafür sind Flugzeuge und Automobile. Die Entwicklung der ersten Autopiloten begann bereits im frühen 20. Jahrhundert. Heutzutage werden Start, Flug, Anflug und Landung durch moderne Flight Management Systeme weitgehend automatisiert. Außerdem sind bereits unbemannte Flugzeuge (UAVs) ferngesteuert oder sogar vollständig autonom im Einsatz. Die Autonomie von Automobilen steigerte sich vom einfachen Tempomat über adaptive Tempomaten mit Abstandsregelung und kooperative Tempomaten mit Kommunikation zwischen den Fahrzeugen bis hin zu hochautomatisierten LKW (Daimler AG (2015)) und vollautomatisierten PKW ohne Fahrer (Google (2015)), siehe auch Ross (2014).

Wie Antsaklis (1999) feststellt, war das Verlangen nach Maschinen mit größerer Autonomie schon seit jeher eine treibende Kraft für die Entwicklung von Regelungssystemen. Infolgedessen und im Zuge der beschriebenen Trends trat auch die Regelungstechnik um das Jahr 2000 in ein neues Zeitalter ein (Åström and Kumar (2014)). Großskalige Regelungs- und Koordinationsprobleme entstehen, die es verlangen, ein kooperatives Verhalten von Netzwerken dynamischer Systeme zu erzeugen. Wie Murray et al. (2003) in ihrem Bericht über die Zukunft der Regelungstechnik darlegen, ist es eine wichtige Herausforderung, sich vom klassischen Verständnis eines Regelungssystems als einzelne Regelstrecke mit einem einzelnen Regler zu lösen und es vielmehr als heterogene Kombination aus physikalischen Systemen und Informationssystemen mit komplizierten Wechselwirkungen wahrzunehmen. Um Netzwerke dynamischer Systeme mit einem hohen Autonomiegrad entwickeln und realisieren zu können, werden ein entsprechender systemtheoretischer Rahmen und spezielle Regelungsmethoden dringend benötigt.

Vor diesem Hintergrund hat sich ein neues Forschungsgebiet innerhalb der System- und Regelungstheorie entwickelt, welches sich mit der *verteilten und kooperativen Regelung* von *Multi-Agenten Systemen* beschäftigt. Eine detaillierte Beschreibung der Entstehung dieses Gebiets im vergangenen Jahrzehnt erfolgt in Kapitel 2 dieser Arbeit. Im Kern steht dort die Frage, wie ein kooperatives Verhalten einer Gruppe dynamischer Agenten erzeugt werden kann, wie es sich beispielsweise bei Fisch- oder Vogelschwärmen in der Natur beobachten lässt. Nennenswerte Anwendungsbeispiele sind autonome Fahrzeugkolonnen, Formationsflug, kooperierende Roboter in Produktionslinien sowie die Lastverteilung in Energienetzen. Von besonderem Interesse sind dabei verteilte Regelgesetze, welche nur lokalen Informationsaustausch zwischen benachbarten Teilsystemen erfordern, sodass keine zentrale Datenerfassung und Verarbeitung erforderlich ist. Die wesentlichen Vorteile verteilter Regelgesetze sind die Skalierbarkeit für große Netzwerke, die Flexibilität bezüglich Hinzunahme und Entfernung von Teilsystemen sowie die Robustheit bezüglich Ausfällen einzelner Teilsysteme.

Kooperative Regelungsprobleme weisen typischerweise die folgende Struktur auf: Das Multi-Agenten System besteht aus einer Gruppe dynamischer Systeme (z.B. mobile Fahrzeuge, Roboter, UAVs, oder Erzeuger und Verbraucher eines Energienetzes), welche als Agenten bezeichnet und üblicherweise durch gewöhnliche Differentialgleichungen beschrieben werden. Das Ziel der kooperativen Regelung ist es, ein gewünschtes Verhalten des Gesamtsystems zu erzeugen, sodass eine gegebene Aufgabe erfüllt wird (z.B. Folge eines Pfades, Erreichen einer Formation, Frequenz-Synchronisation). Das kooperative Verhalten soll durch verteilte Regelgesetze erreicht werden. Jeder Agent ist mit einem lokalen Prozessor ausgestattet und kann durch Messung oder Kommunikation Informationen mit benachbarten Agenten austauschen und wertet sein Regelgesetz auf Basis dieser lokalen Informationen aus. Die Informationsstruktur eines Multi-Agenten Systems wird üblicherweise durch einen Graphen modelliert, wobei die Knoten den Agenten und die gerichteten Kanten den Informationsfluss darstellen. Diese Struktur spiegelt sich in allen in dieser Arbeit betrachteten Problemstellungen wider. Es werden jeweils die Modelle der Agenten, die Informationsstruktur sowie das gewünschte Verhalten der Gruppe spezifiziert.

Aktuelle Fragestellungen im Gebiet der verteilten und kooperativen Regelung betreffen die Analyse und den Reglerentwurf für heterogene Multi-Agenten Systeme mit komplexer und unterschiedlicher Dynamik der Agenten. Deren Bedeutung für die Anwendung liegt auf der Hand: In der Praxis werden die Agenten in der Regel nicht ganz identisch sein, beispielsweise aufgrund unterschiedlicher physikalischer Parameter oder Betriebsbedingungen. Daher ist es eine unausweichliche Aufgabe, die Auswirkung der Heterogenität auf das Verhalten des Gesamtsystems zu untersuchen und geeignete Methoden zum Entwurf verteilter Regler zu entwickeln. Zudem müssen sowohl externe Störungen als auch Referenzsignale berücksichtigt werden, um praktikable Regelungsalgorithmen zu erhalten.

Ziel der vorliegenden Arbeit ist es, zur Entwicklung der Theorie der kooperativen Regelung homogener und heterogener Multi-Agenten Systeme beizutragen, grundlegende Effekte nichtidentischer Agentendynamik zu erklären und, vor allem, geeignete Entwurfsmethoden für verteilte Regelgesetze zu entwickeln, die auf solch komplexe Multi-Agenten Systeme zugeschnitten sind.

Der erste Teil dieser Arbeit ist dem Problem der Ausgangs-Synchronisation gewidmet, welches die grundlegende kooperative Regelungsaufgabe darstellt und als Baustein für anspruchsvollere Koordinationsaufgaben dient. Dieses Problem wurde bereits viel studiert, aber die zugehörige Theorie lässt sich dennoch verfeinern und weiter ausbauen. Wir konzentrieren uns auf Reglerentwurfsmethoden für homogene lineare Netzwerke, strukturelle Anforderungen und Robustheitseigenschaften heterogener linearer Systeme sowie die Verallgemeinerung der Konzepte auf Netzwerke linear parameter-varianter Systeme.

Der zweite Teil dieser Arbeit befasst sich mit Netzwerken dynamischer Systeme, auf welche Signale von außen einwirken. Um das gewünschte kooperative Verhalten zu garantieren, ist es unabdingbar externe Störungen zu berücksichtigen, da diese in jeder praktischen Anwendung auftreten können. Auch Referenzsignale auf Netzwerkebene und auf Agentenebene werden berücksichtigt, um das Verhalten der Gruppe von außen vorgeben zu können. Um solche Probleme zu lösen, werden verteilte Regelgesetze mit Integral-Anteil entwickelt und die verteilte Ausgangsfolgeregelung als allgemeine Herangehensweise vorgestellt und weiterentwickelt.

Der dritte Teil dieser Arbeit befasst sich mit der Entwicklung von Algorithmen zu Bewegungskoordination von mobilen Fahrzeugen, die als nichtholonome Systeme beschrieben werden müssen. Die wesentliche Herausforderung für den Reglerentwurf ist die nichtlineare Dynamik der Agenten, welche spezielle Methoden erfordert. Wir konzentrieren uns auf zwei Arten von Koordinationsaufgaben und konstruieren in beiden Fällen geeignete Regelgesetze.

In dieser Arbeit wählen wir eine mathematische Herangehensweise und verwenden verschiedene regelungstheoretische Konzepte und Methoden, um die betrachteten Probleme zu lösen. Die geometrische Regelungstheorie für lineare Systeme, die Graphentheorie, die Stabilitätstheorie nach Lyapunov, sowie konvexe Optimierung spielen eine wichtige Rolle. Zahlreiche numerische Beispiele sind in der Arbeit enthalten, um das Verständnis zu erleichtern und die theoretischen Ergebnisse zu veranschaulichen.

Forschungsbeiträge und Gliederung der Arbeit

Im Folgenden werden die Gliederung dieser Arbeit vorgestellt und die wesentlichen Forschungsbeiträge der jeweiligen Kapitel zusammengefasst.

Kapitel 2: Grundlagen. In Kapitel 2 wird das Gebiet der verteilten und kooperativen Regelung von Multi-Agenten Systemen vorgestellt. In Abschnitt 2.1 wird das klassische Konsens-Problem als einführendes Beispiel und als Startpunkt für die anschließenden Untersuchungen präsentiert. Abschnitt 2.2 gibt einen Überblick über das Forschungsthema und fasst die wesentlichen Vorarbeiten auf dem Gebiet der verteilten und kooperativen Regelung zusammen.

Kapitel 3: Ausgangs-Synchronisation. In Kapitel 3 befassen wir uns mit Zustands- und Ausgangs-Synchronisationsproblemen, welche die grundlegenden kooperativen Regelungsaufgaben für Multi-Agenten Systeme darstellen.

In **Abschnitt 3.1, Homogene Lineare Netzwerke**, betrachten wir Netzwerke von identischen Agenten mit linearer zeitinvarianter (LTI) Dynamik und die Lösung von Zustands- und Ausgangs-Synchronisationsproblemen mittels statischen diffusen Kopplungen. Wir besprechen Existenzbedingungen für geeignete Kopplungen und entwickeln neue, verbesserte Entwurfsmethoden für das verteilte Regelgesetz. Teile dieses Abschnitts basieren auf Seyboth et al. (2014a, 2016). Unser wesentlicher Forschungsbeitrag in diesem Abschnitt ist:

- Wir stellen neue Entwurfsmethoden für statische diffuse Kopplungen auf Basis von linearen Matrixungleichungen (LMIs) vor, die es erlauben, Anforderungen an die Regelgüte als Polvorgabe-Nebenbedingungen für den Synchronisationsfehler zu berücksichtigen.

In **Abschnitt 3.2, Heterogene Lineare Netzwerke**, untersuchen wir heterogene Netzwerke bestehend aus nicht-identischen LTI Systemen. Wir besprechen notwendige und hinreichende Bedingungen für die Lösbarkeit des Ausgangs-Synchronisationsproblems und das Innere-Modell-Prinzip für Synchronisation. Wir untersuchen die Robustheit der Synchronisation in ausgewählten heterogenen Netzwerken und ermöglichen so ein tieferes Verständnis für die Auswirkung der Heterogenität auf das Verhalten des Gesamtsystems. Außerdem besprechen wir die Lösung mittels verteilter dynamischer Regelgesetze. Teile dieses Abschnitts basieren auf Seyboth et al. (2012a, 2015). Unsere wesentlichen Forschungsbeiträge in diesem Abschnitt sind:

- Wir diskutieren das Innere-Modell-Prinzip für Synchronisation für Netzwerke mit statischen Kopplungen, stellen eine geometrische Interpretation der Bedingungen vor und leiten eine notwendige Bedingung an die Spektren der Systemmatrizen der Agenten ab.
- Wir untersuchen die Robustheit der Synchronisation in Netzwerken gestörter Doppelintegratoren und harmonischer Oszillatoren. Für beide Netzwerktypen diskutieren wir die Implikationen des Innere-Modell-Prinzips für Synchronisation, charakterisieren das dynamische Verhalten, zeigen die Bedeutung der Netzwerktopologie auf und werten die Robustheit der Synchronisation bezüglich Parameterunsicherheiten in der Agentendynamik aus. Dadurch erweitern wir das bisherige Wissen über heterogene lineare Netzwerke.

In **Abschnitt 3.3, Netzwerke Linear Parameter-varianter Systeme**, erweitern wir die Systemklasse und betrachten das Ausgangs-Synchronisationsproblem für Netzwerke linear parameter-varianter (LPV) Systeme. Alle Agenten haben die gleiche LPV Struktur, aber die zeitabhängigen Parameter der Agenten können sich unterscheiden, sodass eine heterogene Gruppe resultiert. Wir entwickeln sowohl statische als auch dynamische, teilweise parameter-abhängige, verteilte Regelgesetze und stellen entsprechende Entwurfsverfahren vor. Teile dieses Abschnitts basieren auf Seyboth et al. (2012b,c). Unsere wesentlichen Forschungsbeiträge in diesem Abschnitt sind:

- Wir leiten strukturelle Bedingungen an die Dynamik der LPV Agenten für die Lösbarkeit des Ausgangs-Synchronisationsproblems ab und diskutieren diese.
- Wir präsentieren eine Entwurfsmethode für statische und neue parameter-abhängige verteilte Regelgesetze für homogene Netzwerke von LPV Systemen mit identischen Parametern.
- Wir präsentieren eine konstruktive, skalierbare Lösung des Ausgangs-Synchronisationsproblems für heterogene Netzwerke von LPV Systemen mit verschiedenen Parametern mittels neuer, parameter-abhängiger dynamischer diffuser Kopplungen.

Kapitel 4: Kooperative Störunterdrückung und Folgeregelung. In Kapitel 4 gehen wir einen Schritt weiter und behandeln kooperative Regelungsaufgaben für Multi-Agenten Systeme mit Eingangssignalen. Sowohl äußere Störungen als auch Referenzsignale, welche auf einzelne Agenten wirken, werden berücksichtigt. Dadurch wird die Formulierung und Lösung eines breiten Spektrums von verteilten Koordinations- und Regelungsaufgaben ermöglicht.

In **Abschnitt 4.1, Verteilte Regelung mit Integral-Anteil**, konzentrieren wir uns auf homogene lineare Multi-Agenten Systeme mit konstanten äußeren Störungen und Referenzsignalen und formulieren entsprechend ein robustes Ausgangs-Synchronisationsproblem und ein kooperatives Folgeregelungsproblem. Als wirksame Lösung übertragen wir den klassischen Ansatz der Zustandsregelung mit Integral-Anteil auf die verteilte Regelung vernetzter Systeme. Das resultierende verteilte Regelgesetz mit Integral-Anteil stellt eine wirksame Lösung für beide genannten Probleme dar. Teile dieses Abschnitts basieren auf Seyboth and Allgöwer (2015). Unsere wesentlichen Forschungsbeiträge in diesem Abschnitt sind:

- Wir entwickeln ein neuartiges verteiltes Regelgesetz mit Integral-Anteil, welches die exakte Ausgangs-Synchronisation trotz konstanter äußerer Störungen auf die Agenten garantiert. Wir leiten Bedingungen für die Existenz geeigneter Reglermatrizen her und zeigen Entwurfsmethoden auf.

- Für den Fall, dass nur Ausgangsmessungen zur Verfügung stehen, stellen wir Beobachterbasierte Implementierungen des Regelgesetzes vor, sowohl für absolute als auch für relative Ausgangsmessungen. Zudem beschreiben wird die resultierenden synchronen Trajektorien.

- Wir zeigen, dass dieses verteilte Regelgesetz mit Integral-Anteil gut geeignet ist, um das kooperative Folgeregelungsproblem zu lösen.

In **Abschnitt 4.2, Verteilte Ausgangsfolgeregelung** greifen wir das Konzept der verteilten Ausgangsfolgeregelung auf und erweitern und verallgemeinern dies. Mit diesem Konzept kann ein breites Spektrum von kooperativen Regelungsproblemen für heterogene Multi-Agenten Systeme mit äußeren Referenzsignalen und Störungen elegant formuliert und gelöst werden. Teile dieses Abschnitts basieren auf Seyboth et al. (2014a, 2016). Unsere wesentlichen Forschungsbeiträge in diesem Abschnitt sind:

- Wir führen ein strukturiertes Exosystem ein, indem wir zwischen globalen und lokalen verallgemeinerten Störungen unterscheiden. Dies führt zu einer intuitiven Zerlegung der Francis-Gleichungen und zu verteilten Reglern niedrigerer Dimension im Vergleich zu bestehenden Resultaten. Zudem formulieren wir das gesamte kooperative Regelungsproblem als ein einzelnes, zentrales Ausgangsfolgeregelungsproblem. Die Lösbarkeitsbedingungen des zentralen Problems und seine spezielle Struktur führen auf notwendige und hinreichende Lösbarkeitsbedingungen für das verteilte Regelungsproblem.

- Wir entwickeln einen neuen verteilten Ausgangsfolgeregler als Lösung des betrachteten Problems. Als wesentliche Verbesserung erweitern wir das Regelgesetz und führen eine neue Kopplung auf Agentenebene ein, welche eine kooperative Reaktion des Gesamtsystems auf lokal auftretende Störungen ermöglicht.

- Wir verallgemeinern die Problemformulierung und analysieren die Auswirkung von gekoppelten Zuständen, Messausgängen und Regelfehlern auf die Lösbarkeit des resultierenden gekoppelten verteilten Ausgangsfolgeregelungsproblems. Für jeden der drei Fälle zeigen wir geeignete Lösungsstrategien auf.

Kapitel 5: Koordination von Gruppen Nichtholonomer Fahrzeuge. In Kapitel 5 wenden wir uns einem Anwendungsgebiet der kooperativen Regelung zu und untersuchen Aufgaben der Koordination von nichtlinearen Multi-Agenten Systemen bestehend aus nichtholonomen Fahrzeugen. Zunächst betrachten wir Formationen von mobilen Robotern, anschließend koordinierte Kreisbewegungen von nichtholonomen Fahrzeugen mit unterschiedlichen konstanten Geschwindigkeiten.

In **Abschnitt 5.1, Mobile Roboter**, betrachten wir Gruppen mobiler Roboter welche sich in der Ebene bewegen. Unser wesentlicher Forschungsbeitrag in diesem Abschnitt ist:

- Wir entwickeln einen neuen verteilten Regelungsalgorithmus für die Erzeugung komplexer Formationen trotz äußerer Störungen. Der Regelungsalgorithmus verbindet die verteilte Ausgangsfolgeregelung aus Abschnitt 4.2, die Feedback-Linearisierung und nichtlineare Beobachter für die lokalen Störungen, um die gewünschte Formation zu stabilisieren

In **Abschnitt 5.2, Nichtholonome Fahrzeuge mit Konstanter Geschwindigkeit**, untersuchen wir Probleme der Bewegungskoordination für Gruppen nichtholonomer Fahrzeuge mit unterschiedlichen konstanten Geschwindigkeiten. Als grundlegendes Koordinationsproblem setzen wir den Schwerpunkt auf die Stabilisierung kreisförmiger Formationen, in denen sich die Fahrzeuge gemeinsam auf Kreisbahnen bewegen. Dieser Abschnitt basiert auf Seyboth et al. (2014b). Unsere wesentlichen Forschungsbeiträge in diesem Abschnitt sind:

- Wir diskutieren die im Fall von unterschiedlichen Geschwindigkeiten möglichen Bewegungsformen der Gruppe. Die Agenten können Kreise um einen gemeinsamen Mittelpunkt nur umlaufen, wenn entweder unterschiedliche Radien oder unterschiedliche Kreisfrequenzen erlaubt werden.

- Wir leiten neue nichtlineare Regelgesetze für verschiedene Koordinationsprobleme, einschließlich der Gleichrichtung, Stabilisierung des Gruppenmittelpunkts und Stabilisierung der oben genannten kreisförmigen Bewegungen, her. Im Gegensatz zu bestehenden Resultaten sind diese Regelgesetze anwendbar für heterogene Gruppen von Fahrzeugen mit beliebigen konstanten Geschwindigkeiten.

Kapitel 6: Fazit. Zuletzt fassen wir die wesentlichen Ergebnisse dieser Arbeit zusammen und präsentieren unsere Schlussfolgerungen. Zudem zeigen wir mögliche Anknüpfungspunkte für zukünftige Forschung auf.

Anhang A: Technischer Hintergrund. Anhang A enthält technisches Hintergrundmaterial. Eine kurze Zusammenfassung der systemtheoretischen Basis und ausgewählter Konzepte der Regelungstheorie für lineare Systeme findet sich in Abschnitt A.1. Die Graphentheorie, welche von großer Bedeutung für die Beschreibung der Informationsstruktur von Multi-Agenten Systemen ist, wird in Abschnitt A.2 eingeführt.

Anhang B: Technische Beweise. Die meisten technischen Beweise sind in den Anhang B verschoben, um den Lesefluss zu verbessern.

Chapter 1

Introduction

1.1 Motivation and Focus

The rapid technological progress over the past decades has had a tremendous impact on our world and our everyday life. We live in a time of brilliant technologies and it is fair to say that we are entering a new era, commonly referred to as the *second machine age* (Brynjolfsson and McAfee (2014)). The technological progress encompasses rapid advances in computing, communication, and sensing technology (Murray et al. (2003)). Major achievements in the engineering sciences, in computer science, and in communication technology lead to a rapid growth of communication networks, ubiquitous and distributed computational power, as well as an increasing linkage between physical systems and information systems. More and more complex and large-scale *cyber-physical systems* emerge, comprising multiple dynamical systems interacting with each other. Some of the most prominent examples are smart grids, automated highway systems, and smart factories. A further significant trend is the increase of *autonomy* in modern engineering systems, which means increasing independence of an outside supervisor. Two compelling examples are aircrafts and automobiles. The development of the first autopilots started already in the early 20th century. Nowadays, modern flight management systems automate climb, cruise, descent, approach, and landing phases. Moreover, unmanned aerial vehicles (UAVs) with remote control or completely autonomous flight capabilities are widely on duty. The autonomy of cars has increased from the simple cruise control (CC) feature via adaptive cruise control (ACC), keeping a desired inter-vehicle distance, and cooperative adaptive cruise control (CACC), allowing for inter-vehicle communication, through to completely autonomous trucks (Daimler AG (2015)) and driver-less cars (Google (2015)), cf., Ross (2014).

It was noted by Antsaklis (1999) that "the quest for machines that exhibit higher autonomy has been the driving force in the development of control systems over the centuries". As a consequence, also *control* has entered a new era around the year 2000 in the course of these developments (Åström and Kumar (2014)). Large-scale regulation and coordination problems emerge, which require the *synthesis of cooperative behavior* in *networks of dynamical systems*. In their report on "Future directions in control in an information-rich world", Murray et al. (2003) state that "the challenge to the field is to go from the traditional view of control systems as a single process with a single controller to recognizing control systems as a heterogeneous collection of physical and information systems, with intricate interconnections and interactions." In order to handle, analyze, and design networks of dynamical systems with a high degree of autonomy, a systems theoretic framework is required and appropriate control methods need to be developed.

These developments have boosted the development of a novel research field within the field of systems and control theory, denoted as *distributed and cooperative control theory* for *multi-agent systems*. A detailed account for the emergence of this field since the turn of the millennium is given in Chapter 2 of this thesis. In a variety of modern man-made systems, it is desirable to synthesize a cooperative behavior among individual dynamical agents, similarly to self-organizing bird flocks and fish schools observed in nature. Examples include multi-vehicle coordination and platooning, formation flight, robot cooperation in production lines, power balancing in micro-grids, and many more. Of particular interest are distributed control laws, which require only local information exchange between neighboring subsystems and no centralized data collection or processing entity. The main advantages of distributed control algorithms are scalability for large networks of dynamical systems, flexibility with respect to addition and removal of subsystems, and robustness with respect to failure of individual subsystems.

Cooperative control problems are typically characterized by the following basic structure: There is a group of dynamical systems (e.g., mobile vehicles, robots, UAVs, or generators and consumers in a micro-grid). These individual dynamical systems are commonly referred to as agents and described by ordinary differential equations. The entire group of agents is referred to as the multi-agent system. The goal of cooperative control is to synthesize a desired dynamical behavior of the entire group such that it achieves a given group objective (e.g., path following, formation keeping, or frequency synchronization). The cooperative behavior shall be generated by means of distributed control algorithms, based on local information. In particular, each agent is equipped with a local processor and controller that senses or receives information from neighboring agents in the group and, based on this local information, calculates its control action. The information structure of a multi-agent system is usually modeled by a graph, where nodes correspond to agents in the group and each directed edge describes information flow between the respective agents. This structure is clearly reflected in the problem formulations throughout the present thesis. In each problem formulation, the agent models, information structure, and group objective are specified individually.

Recent challenges in the field of distributed and cooperative control include the analysis and controller synthesis for heterogeneous groups of agents with complex and non-identical dynamics. The practical relevance of heterogeneous multi-agent systems is apparent: In reality, the agents in a group (such as mobile robots) will never be perfectly identical (e.g., due to different physical parameters or operating conditions). Hence it is an inevitable task to study the effects of heterogeneity in the agent dynamics on the behavior of the multi-agent system. An understanding of such effects is essential for the development of suitable distributed control laws which guarantee a desired behavior. Recent challenges include as well the treatment of external influences on networks of dynamical systems. Both disturbances and reference signals need to be taken into account in order to arrive at practicable control algorithms.

The present thesis shall contribute to the development of a cooperative control theory for homogeneous and heterogeneous multi-agent systems consisting of identical and non-identical dynamical agents. The goal is to reveal and explain fundamental effects of non-identical agent dynamics on the behavior of a distributed system and, most importantly, to develop suitable synthesis methods for distributed control algorithms tailored to complex multi-agent systems.

The first part of this thesis is dedicated to the output synchronization problem. It constitutes the most fundamental cooperative control task and serves as building block for sophisticated coordination tasks. Even though this problem has already received considerable attention in the scientific community, the corresponding theory can be refined and developed further in various aspects. Our focus is on control design methods for homogeneous linear networks, structural requirements and robustness issues in heterogeneous linear networks, and the extension of these concepts to networks of linear parameter-varying systems.

The second part of this thesis is dedicated to networks of dynamical systems with exogenous input signals. As already motivated, external disturbances need to be taken into account in order to guarantee a desirable behavior of the group since they are inevitable in almost any practical application. Reference signals are considered both on the network level and on the agent level in order to dictate a desired motion of the multi-agent system. For this purpose, distributed control laws with integral action are developed and, as a general approach to distributed cooperative control design, the distributed output regulation framework is presented and developed further in various respects.

The third part of this thesis is dedicated to the development of motion coordination algorithms for groups of mobile vehicles. Suitable for the description of nonholonomic vehicles such as mobile robots or UAVs is the unicycle model. The major challenge for the distributed control design are the nonlinear dynamics, which require tailored analysis and control synthesis techniques. We focus on two classes of coordination problems and construct suitable control algorithms in both cases.

We take a mathematical approach for the developments in this thesis and employ a variety of control theoretic tools and concepts in order solve the problems under consideration. In particular, the geometric approach to linear systems theory and tools from algebraic graph theory play a major role, as well as Lyapunov stability theory and convex optimization. In order to facilitate the understanding and illustrate the theoretic results, numerous numerical examples and simulation results are provided throughout the thesis. MATLAB and Simulink, the Multi-Parametric Toolbox (Herceg et al. (2013)), and Yalmip (Löfberg (2004)) were used for the simulations.

1.2 Contributions and Outline

In this section, we present the outline and summarize the main contributions of the present thesis.

Chapter 2: Background. Chapter 2 provides background information on the field of distributed and cooperative control for multi-agent systems. In Section 2.1, the classical single-integrator consensus problem is presented as an introductory example and as the starting point for the subsequent developments. Afterwards, a research topic overview including a summary of major achievements in the field of distributed and cooperative control is given in Section 2.2.

Chapter 3: Output Synchronization. In Chapter 3, both state and output synchronization problems are addressed as the most fundamental cooperative control tasks for multi-agent systems.

In **Section 3.1, Homogeneous Linear Networks**, we consider networks of identical agents with linear time-invariant (LTI) dynamics and the solution to the state and output synchronization problem via static diffusive couplings. After reviewing conditions on the existence of suitable coupling gains ensuring output synchronization, we develop novel improved design methods for the distributed control law. Parts of this section are based on Seyboth et al. (2014a, 2016). Our main contribution in this section is the following:

- We propose novel design methods for static diffusive couplings based on linear matrix inequality (LMI) conditions, which allow to include performance requirements conveniently in the design by means of pole placement constraints on the synchronization error.

In **Section 3.2, Heterogeneous Linear Networks**, we study heterogeneous networks of non-identical agents with LTI dynamics. We review necessary and sufficient conditions for the existence of a solution to the output synchronization problem and the internal model principle for synchronization. We study the robustness of synchronization in selected heterogeneous linear networks in order to develop a deeper understanding of the effects of heterogeneity in the agent dynamics on the behavior of the multi-agent system. Moreover, we review solutions based on dynamic diffusive couplings. Parts of this section are based on Seyboth et al. (2012a, 2015). Our main contributions in this section are the following:

- We discuss the internal model principle for synchronization for networks with static couplings, provide a geometric interpretation of its conditions, and derive a necessary condition for output synchronization based on the spectra of the agents' system matrices.
- We explore the robustness of synchronization in networks of perturbed double-integrator agents and harmonic oscillators. For both types of networks, we discuss the implications of the internal model principle for synchronization, characterize the dynamic behavior, highlight the importance of the network topology, and assess the robustness of synchronization with respect to parameter uncertainties in the agent dynamics. These results extend the current knowledge about heterogeneous linear networks.

In **Section 3.3, Networks of Linear Parameter-Varying Systems**, we address the output synchronization problem for networks of linear parameter-varying (LPV) dynamical agents, which is a significant extension of the class of agent dynamics compared to existing work. While all agents are assumed to have the same LPV structure, the time-varying parameters may be different for different agents such that the group has heterogeneous dynamics. We develop both static and dynamic, partly parameter-dependent, distributed control laws ensuring output synchronization and provide corresponding control design methods. Parts of this section are based on Seyboth et al. (2012b,c). Our main contributions in this section are the following:

- We derive and discuss structural requirements on the agent dynamics for the solvability of the output synchronization problem for networks of LPV systems.
- We present a design method for constant and novel parameter-dependent (gain-scheduled) distributed control laws for homogeneous groups of LPV systems with identical parameters.
- We present a constructive and scalable solution to the output synchronization problem for general heterogeneous groups of LPV systems with non-identical parameters based on novel gain-scheduled dynamic diffusive couplings.

Chapter 4: Cooperative Disturbance Rejection and Reference Tracking. In Chapter 4, we take a step beyond pure output synchronization and address cooperative control problems involving external input signals. Both external disturbances and reference signals acting on individual agents are taken into account in order to arrive at a more realistic problem formulation and facilitate the solution of a wide range of distributed coordination and control problems.

In **Section 4.1, Distributed Integral Action**, we focus on homogeneous linear multi-agent systems subject to constant external disturbances and reference signals and formulate corresponding robust output synchronization and cooperative reference tracking problems. As a powerful solution, we bring the classical approach of state feedback control with integral action forward to distributed control of network systems. The resulting distributed control law with integral action provides an effective solution to the aforementioned tasks. Parts of this section are based on Seyboth and Allgöwer (2015). Our main contributions in this section are the following:

- We propose a novel distributed control law with integral action ensuring offset-free output synchronization despite constant external disturbances acting on individual agents. We derive conditions for the existence of suitable control gains and provide design methods.
- We present observer-based output feedback implementations of the novel distributed control law, both for absolute and for relative output measurements, and characterize the resulting synchronous trajectories.
- We show that the distributed control law with integral action is well suited for solving a cooperative reference tracking problem which contains the distributed input averaging problem as a special case.

In **Section 4.2, Distributed Output Regulation**, we revisit, extend, and generalize the distributed output regulation framework, which allows to elegantly formulate and solve a wide range of cooperative control problems for heterogeneous multi-agent systems with both external reference signals and disturbances. Parts of this section are based on Seyboth et al. (2014a, 2016). Our main contributions in this section are the following:

- We introduce a structured exosystem distinguishing explicitly between global and local generalized disturbances, which leads to an intuitive decomposition of the regulator equations and to distributed controllers of lower dimension compared to existing work. Moreover, we formulate the overall cooperative control problem as a single centralized output regulation problem. The solvability conditions for this centralized problem and its particular structure yield necessary and sufficient solvability conditions for the distributed problem.
- We develop a novel distributed output feedback regulator, solving the distributed output regulation problem. As a major improvement, we extend the distributed regulator and introduce a novel coupling on the agent level which realizes a cooperative reaction of the group on external disturbances, quantified by the decay rate of the synchronization error of the agents' transient state components.
- We generalize the problem formulation and analyze the effects of coupled agent states, coupled measurement outputs, and coupled regulation errors on the solvability of the resulting coupled distributed output regulation problem. For each of these cases, we point out corresponding solution strategies.

Chapter 5: Motion Coordination for Groups of Nonholonomic Vehicles. In Chapter 5, we study distributed motion coordination problems for nonlinear multi-agent systems consisting of nonholonomic vehicles. First, in Section 5.1, we address formation control problems for a group of mobile robots. Then, in Section 5.2, we address circular motion coordination problems for vehicles with non-identical constant velocities motivated by UAV applications.

In **Section 5.1, Mobile Robots**, we consider groups of mobile robots moving in the plane described by nonlinear dynamical models. Our main contribution in this section is the following:

- We develop a novel distributed control algorithm tailored to complex formation control tasks under external disturbances. The control algorithm integrates the distributed regulator developed in Section 4.2, the feedback linearization technique, and nonlinear observers for the local disturbances in order to stabilize the desired formation.

In **Section 5.2, Unicycle-type Vehicles with Constant Velocities**, we study motion coordination problems for groups of unicycle-type vehicles with non-identical speed. As a fundamental coordination task, we focus on the asymptotic stabilization of circular formations in which all vehicles move collectively on circles in the plane. This section is based on Seyboth et al. (2014b). Our main contributions in this section are the following:

- We discuss the limitations caused by the heterogeneous velocities and describe realizable circular motion patterns. The agents can traverse circles around a common center point only if either different radii or different angular frequencies for each agent are permitted.
- We derive novel nonlinear control laws for a set of coordination problems including phase agreement, phase balancing, stabilization of the group center point, and finally stabilization of circular formations as described above. In contrast to prior work, these control laws are applicable to heterogeneous groups of vehicles with arbitrary constant velocities.

Chapter 6: Conclusions. Finally, we summarize the main results of this thesis and present the conclusions in Chapter 6 and indicate possible directions for future research.

Appendix A: Technical Background. Appendix A provides mathematical preliminaries and technical background material which is used throughout this thesis. A brief summary of the system theoretic basis and selected concepts from control theory for linear systems is provided in Section A.1. Second, graph theory is reviewed in Section A.2 as an essential framework for the description of the information structure of the multi-agent systems under consideration.

Appendix B: Technical Proofs. Most of the technical proofs are deferred to the Appendix B in order to improve readability.

Chapter 2

Background

This chapter presents the classical consensus problem as the initial and most basic distributed control problem. Moreover, it presents a brief history and overview on the research field of distributed and cooperative control of multi-agent systems in order to set the stage for the developments in the remainder of this thesis.

2.1 The Consensus Problem

At the heart of the distributed and cooperative control theory for networks of dynamical systems lies the consensus problem, which we review in this section. The following discussion illustrates the interplay of control theory and graph theory and allows to introduce some useful terminology. For technical background material on linear systems and control theory as well as graph theory, the reader is referred to Appendix A.

The consensus problem, also termed agreement problem, goes back to early work in computer science and physics. It has found its way into the field of systems and control over the area of distributed computation over networks (Tsitsiklis (1984); Tsitsiklis et al. (1986); Blondel et al. (2005)), as well as over the study and simulation of particle swarms and flocking phenomena in nature (Reynolds (1987); Vicsek et al. (1995); Jadbabaie et al. (2003)). These developments led to the seminal papers by Olfati-Saber and Murray (2004); Fax and Murray (2004); Ren and Beard (2005); and Moreau (2005), which provide a rigorous in-depth study of the consensus problem for groups of dynamical agents. A comprehensive treatise and further references are found in the survey papers by Ren et al. (2005a); Olfati-Saber et al. (2007); and Ren et al. (2007).

The essence of the consensus problem is that all agents in a given group shall agree on a common quantity of interest via local interaction rules. The group of dynamical agents is referred to as a multi-agent system in the following. We consider a multi-agent system consisting of N agents with continuous-time single-integrator dynamics

$$\dot{x}_k = u_k, \tag{2.1}$$

where $x_k(t) \in \mathbb{R}$ is the state, $u_k(t) \in \mathbb{R}$ is the input, and $x_k(0) = x_{0,k} \in \mathbb{R}$ is the initial condition of agent $k \in \mathcal{N}$. The goal, called group objective, is that for arbitrary initial conditions, the agents reach an agreement on a common state value, i.e., the states converge to a configuration such that $x_1 = x_2 = \cdots = x_N$.

We have to find a control law u_k for each agent which takes only local information of agent k and its neighbors into account. For this purpose, the neighboring relations between the agents are described by a graph $\mathcal{G} = (\mathcal{V}, \mathcal{E})$. In particular, each node in $\mathcal{V}$ corresponds to one agent in the group and the links $\mathcal{E}$ define the underlying communication topology, i.e., the information flow among agents. A link (v_j, v_k) can either be understood as a communication channel in a network in the sense that agent j can send packets to agent k, or as a measurement in the sense that agent k can obtain measurement information about agent j. The graph $\mathcal{G}$ is also referred to as the communication graph of the multi-agent system. We are looking for a distributed control law in the sense that agent k may take into account only its own state x_k as well as the states x_j of its neighbors $j \in \mathcal{N}_k = \{j \in \mathcal{N} : (v_j, v_k) \in \mathcal{E}\}$ for the computation of u_k. The following distributed control law is known as the consensus protocol:

$$u_k = \sum_{j=1}^{N} a_{kj}(x_j - x_k). \tag{2.2}$$

The closed loop consisting of the agents (2.1) and (2.4) can be written as

$$\dot{x} = -L_{\mathcal{G}} x \tag{2.3}$$

with $x = [x_1 \ \cdots \ x_N]^\mathsf{T}$ and where $L_{\mathcal{G}}$ is the Laplacian matrix of the underlying graph $\mathcal{G}$.

Definition 2.1 (Consensus)**.** The multi-agent system (2.3) is said to reach *consensus*, if for all $x_{0,k} \in \mathbb{R}$, $k \in \mathcal{N}$, there exists a consensus value $x^* \in \mathbb{R}$ such that $\lim_{t\to\infty} x_k(t) = x^*$ for all $k \in \mathcal{N}$.

Note that the final consensus value x^* is not defined a-priori. An equivalent formulation of the consensus objective is as follows. Let $x \in \mathbb{R}^N$ be the stack vector of the individual agent states and define the consensus subspace

$$\mathcal{C} = \{x \in \mathbb{R}^N : x_1 = x_2 = \cdots = x_N\} \subset \mathbb{R}^N. \tag{2.4}$$

The system reaches consensus if and only if $\mathcal{C}$ is asymptotically stable with respect to (2.3).

Theorem 2.1 (Consensus (Ren et al., 2005b, Thm. 2))**.** *The multi-agent system* (2.3) *reaches consensus if and only if the underlying graph* $\mathcal{G}$ *is connected.*

Theorem 2.1 is a consequence of Theorem A.25 and the fact that the eigenvector of $L_{\mathcal{G}}$ corresponding to the zero eigenvalue is $\mathbf{1}$. Here, we use geometric arguments in order to provide more insight and justify the result.

For any graph $\mathcal{G}$, the null space of $L_{\mathcal{G}}$ is invariant with respect to (2.3). Due to Theorem A.25, all non-zero eigenvalues of $L_{\mathcal{G}}$ have positive real parts. Hence, the dynamics of (2.3) restricted to the complement of $\ker(L_{\mathcal{G}})$ are asymptotically stable. Hence, the subspace $\ker(L_{\mathcal{G}})$ is asymptotically stable with respect to (2.3), independently of the connectivity properties of $\mathcal{G}$. By construction of $L_{\mathcal{G}}$, it holds that $\mathbf{1} \in \ker(L_{\mathcal{G}})$. From Theorem A.27, it follows that $\operatorname{rank}(L_{\mathcal{G}}) = N - 1$ if and only if $\mathcal{G}$ contains a spanning tree. In this case, $\dim(\ker(L_{\mathcal{G}})) = 1$ by the Rank–nullity Theorem A.13. Hence, it follows that $\ker(L_{\mathcal{G}}) = \operatorname{im}(\mathbf{1})$ if and only if the graph $\mathcal{G}$ contains a spanning tree. Note that $\mathcal{C} = \operatorname{im}(\mathbf{1})$. Concluding, we have two statements which prove Theorem 2.1. First, the null space $\ker(L_{\mathcal{G}})$ is asymptotically stable with respect to (2.3) for arbitrary graphs $\mathcal{G}$. Second, the consensus subspace is identical to the null space of the Laplacian matrix, $\mathcal{C} = \ker(L_{\mathcal{G}})$, if and only if $\mathcal{G}$ contains a spanning tree.

When the multi-agent system is guaranteed to reach consensus, the next question is what the final consensus value will be. Of particular interest is the average consensus problem, where x^* is supposed to be the average of the initial states of all agents, i.e., $x^* = 1/N \sum_{k=1}^{N} x_k(0)$. It was shown by Olfati-Saber and Murray (2004) that the network (2.3) reaches average consensus,

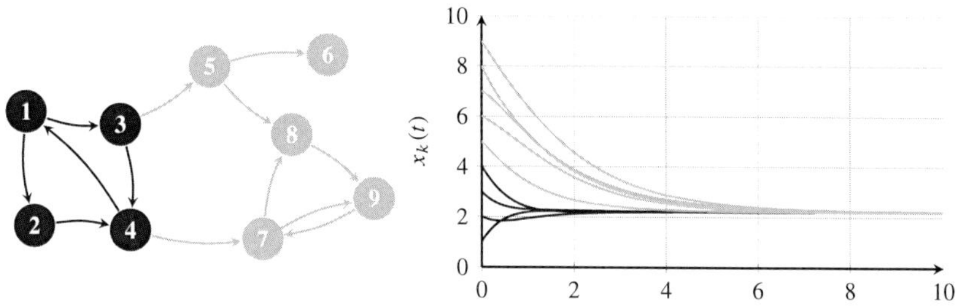

(a) Directed graph $\mathcal{G}$ (b) Agent states $x_k(t)$, $k \in \mathcal{N}$

Figure 2.1: Underlying graph and evolution of the agent states of the consensus network (2.3) in Example 2.1. All states within (—) and outside (—) the iSCC converge to the consensus value x^*.

if and only if the graph $\mathcal{G}$ is strongly connected and balanced, i.e., $\sum_{j=1}^{N} a_{kj} = \sum_{j=1}^{N} a_{jk}$ for all $k \in \mathcal{N}$. The final consensus value for arbitrary connected graphs was characterized by Ren et al. (2005b). It turns out that

$$x^* = p^\mathsf{T} x_0, \tag{2.5}$$

where x_0 is the stack vector of the initial states and p^T is the normalized left eigenvector of $L_\mathcal{G}$ corresponding to the zero eigenvalue, i.e., $p^\mathsf{T} L_\mathcal{G} = \mathbf{0}^\mathsf{T}$ and $p^\mathsf{T}\mathbf{1} = 1$. For connected graphs which are undirected or strongly connected and balanced, it holds that $p^\mathsf{T} = 1/N\mathbf{1}^\mathsf{T}$. Hence, (2.5) contains average consensus as a special case. Moreover, (2.5) reveals the influence of the graph topology on the consensus value. From Theorem A.27, we know that all elements p_k of p^T are nonnegative and p_k is positive if and only if the corresponding node v_k is contained in the independent strongly connected component (iSCC) of $\mathcal{G}$. Hence, the consensus value is a weighted sum of the initial conditions of the agents within the iSCC, those outside the iSCC do not contribute to the consensus value. If the iSCC consists of a single node, the corresponding agent is referred to as a leader and the states of all agents eventually converge to the state of the leader. The speed of convergence is determined by the smallest real part of all non-zero eigenvalues of $L_\mathcal{G}$, i.e., the decay rate of the disagreement defined as $\delta = x - \mathbf{1}x^*$ is given by $\gamma = \mathbf{Re}(\lambda_2(L_\mathcal{G})) > 0$. For undirected graphs, $\lambda_2(L_\mathcal{G})$ is real and called the Fiedler eigenvalue or algebraic connectivity of $L_\mathcal{G}$. Intuitively, the higher the algebraic connectivity of the graph, the higher is the speed of convergence of (2.3).

Example 2.1. We consider a multi-agent system consisting of $N = 9$ single-integrator agents (2.1) with the communication graph $\mathcal{G}$ shown in Fig. 2.1(a) (cf., Example A.1) and the consensus protocol (2.2). Then, the network (2.3) reaches consensus since $\mathcal{G}$ is connected. The speed of convergence is limited by $\mathbf{Re}(\lambda_2(L_\mathcal{G})) = 0.6753$. For the simulation in Fig. 2.1(b), the initial conditions are chosen as $x_k(0) = k$ for each $k \in \mathcal{N}$. Only the nodes $v_1, \ldots, v_4$ within the iSCC of $\mathcal{G}$ contribute to the resulting consensus value $x^* = p^\mathsf{T} x_0 = 2.2$. Fig. 2.1(b) shows the evolution of the states over time. As expected, a consensus is reached at x^*. △

2.2 Emergence of a new Research Field

The consensus problem had a tremendous impact on the field of systems theory and automatic control. The reason for the great interest in the consensus problem and its success is the fact that it represents the core of a new class of control problems which have emerged over the past decade. This novel research field can be summarized under the title *distributed and cooperative control* and is dedicated to analysis and design methods for so-called multi-agent systems, which consist of a group of autonomous dynamical agents, an information structure which defines communication capabilities among the agents, and a group objective which the agents shall achieve. Such systems have no central data collection or processing entity, which makes distributed control schemes relying on local interactions necessary. The tasks are typically cooperative in the sense that the agents in the group need to coordinate their behavior in order to achieve the group objective.

Although the consensus problem is the most basic problem of this class, it is already of great importance in practical applications. The most prominent application domain is coordination of groups of vehicles, ranging from unmanned underwater vehicles (UUV) over mobile robots, land vehicles, and unmanned aerial vehicles (UAV) to spacecrafts. Distributed control algorithms which realize a flocking behavior in groups of mobile agents are constructed by Tanner et al. (2003a); Tanner et al. (2003b); Olfati-Saber (2006) and vehicle formations are studied by Fax and Murray (2004); Lin et al. (2005); Lafferriere et al. (2005); Ren (2007). Attitude synchronization problems, e.g. for groups of satellites, are addressed by Sarlette et al. (2007); Sarlette et al. (2009). An extensive survey of multi-agent formation control is provided by Oh et al. (2015). Further applications include synchronization phenomena in coupled oscillator networks (Paley et al. (2007); Dörfler and Bullo (2014)), distributed filtering and sensor fusion algorithms for sensor networks (Xiao et al. (2005); Olfati-Saber (2005); Olfati-Saber and Shamma (2005); Carli et al. (2008); Garin and Schenato (2010)), as well as distributed optimization algorithms (Rabbat and Nowak (2004); Nedic and Ozdaglar (2009); Boyd et al. (2011); Bürger et al. (2012); Bürger et al. (2014)) which have a very wide application range on their own.

Besides its direct practical relevance, the consensus problem forms the starting point for research in various directions. The research it initiated was categorized into three dimensions of complexity, cf., Wieland (2010). The first dimension is the topological complexity, which increases toward switching, time and state-dependent communication topologies. The second dimension is the link complexity, aiming at communication imperfections such as loss, delay, and noise, and communication constraints such as limited bandwidth, sampling, and quantization. The third dimension is the system complexity, which increases from single-integrator agents to agents with higher-order linear and nonlinear dynamics, as well as to heterogeneous groups consisting of non-identical agents. Besides these research directions, the complexity of the group objectives has been pushed forward. Recent research activities focus on performance requirements and issues like external disturbances and model uncertainties and aim at robust coordination algorithms, as well as optimality.

Concluding, the consensus problem is seminal and has initiated a great amount of research in the field of systems theory and automatic control. This fact becomes evident from the number of papers with the keyword consensus in their title per year. We have analyzed all papers which have appeared in the two top conferences *IEEE Conference on Decision and Control (CDC)* and *American Control Conference (ACC)* (annual conference of the American Automatic Control Council (AACC)) and the two most renowned journals *Automatica* and *IEEE Transactions on Automatic Control* in the years from 2000 to 2014, and additionally the newly-established journal *IEEE Transactions on Control of Network Systems* in 2014. Fig. 2.2(a) and 2.2(b) show the number of papers with the keyword *"consensus"* in their title. The number has been increasing over the past decade from two conference papers and zero journal articles in 2003 to 67 conference papers and 50 journal articles in 2014.

Research Topic Overview

In the following, we give an overview on the research topic of the present thesis. The field of distributed and cooperative control has been growing tremendously over the past decade, reaching far beyond the consensus problem. If we expand the keyword list to *"distributed, cooperative, multi-agent, multi-vehicle, multi-robot, agreement, consensus, synchronization, formation control, coordination"*, the statistics in Fig. 2.2 reveal that a significant amount of research is dedicated to this topic. Fig. 2.2(c) and 2.2(d) show the respective absolute numbers, which have been increasing up to 281 conference papers and 136 journal articles in 2014. Even more impressively, Fig. 2.2(e) and 2.2(f) show that the relative amount of these papers with respect to the total number of papers has been increasing to 13.6% of the conference papers and 16.8% of the journal articles in 2014. A comparison to other topics is shown in Fig. 2.3. Hence, it is fair to say that we have witnessed the emergence of a new research field within the field of systems and control theory. It forms a part of the wider area named *control of network systems* (Paschalidis and Egerstedt (2014)). This area captures systems with interconnected components and includes, besides networks of autonomous agents, also networked control systems, communication networks, sensor networks, cyber-physical systems, electric power networks as well as biological or social networks.

As a consequence, a complete overview on the research activities in the field of distributed and cooperative control in networks of autonomous agents would go beyond the constraints of the present thesis. As an excellent introduction to the field, we recommend the monographs by Ren and Beard (2008); Bullo et al. (2009); Qu (2009); Mesbahi and Egerstedt (2010); Ren and Cao (2011); Lewis et al. (2014), which document major achievements. Some recent advances are also listed by Cao et al. (2013). In the following, we summarize selected milestones, which are relevant for the present thesis. Detailed discussions of related work are provided in the following respective chapters.

Early advancements from the consensus problem regarding the topological complexity include the study of switching graphs (Olfati-Saber and Murray (2004); Ren and Beard (2005)) and time-varying graphs (Moreau (2005)), which lead to minimalistic connectivity conditions, under which consensus can be guaranteed. Regarding the link complexity, the major advances are a characterization of admissible communication time-delays (Olfati-Saber and Murray (2004); Bliman and Ferrari-Trecate (2008); Seuret et al. (2008); Münz et al. (2010)), admissible sampling rates (Xie et al. (2009)), quantization techniques (Frasca et al. (2009); Cai and Ishii (2011)), randomized algorithms (Frasca et al. (2015)), and event-based implementations (Dimarogonas et al. (2011); Seyboth et al. (2013); De Persis and Frasca (2013)). A lot of effort has been put into the study of networks with increasing system complexity accompanied by increasingly complex group objectives beyond consensus. The advancements in this direction include the extension to agents with double-integrator dynamics (Ren and Atkins (2007); Zhu et al. (2009); Mei et al. (2014)), and general higher-order linear dynamics (Fax and Murray (2004); Tuna (2008); Scardovi and Sepulchre (2009); Li et al. (2010); Wieland et al. (2011a); Zhang et al. (2012)). In this context, the consensus objective is generalized to state and output synchronization, i.e., asymptotic agreement of the agents on a common trajectory. In order to include external reference signals and realize cooperative tracking tasks, a leader-follower architecture is proposed (Hong et al. (2006); Ni and Cheng (2010); Li et al. (2010); Zhang et al. (2011)). The idea is

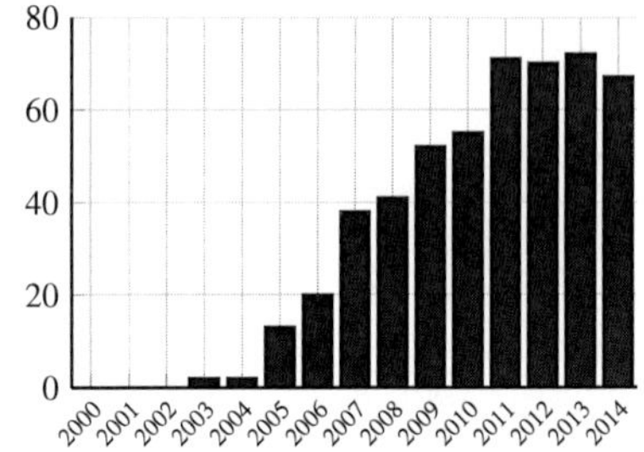

(a) Conference articles with *"consensus"* in the title.

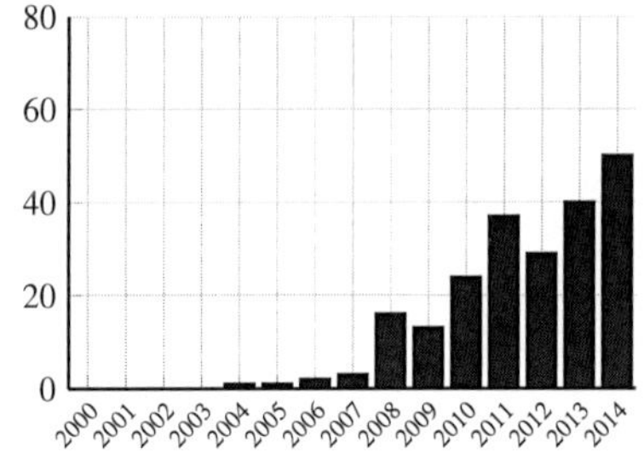

(b) Journal articles with *"consensus"* in the title.

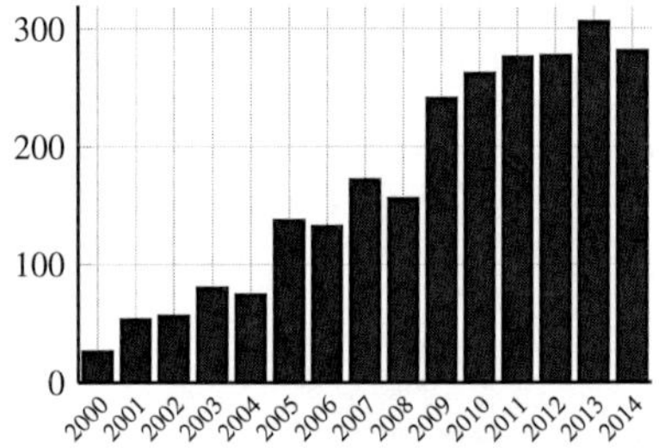

(c) Conference articles with any keywords related to *distributed and cooperative control* in the title.

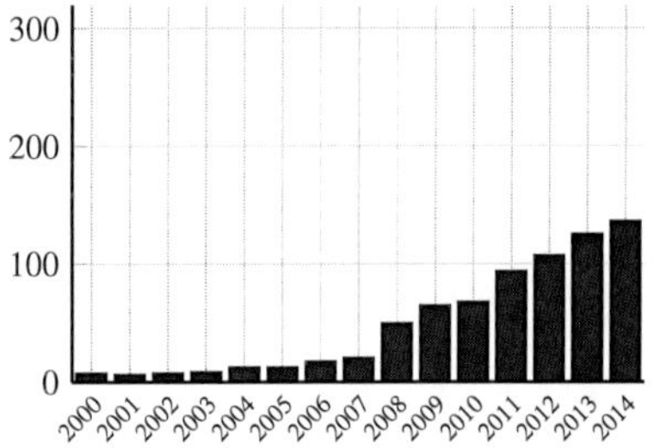

(d) Journal articles with any keywords related to *distributed and cooperative control* in the title.

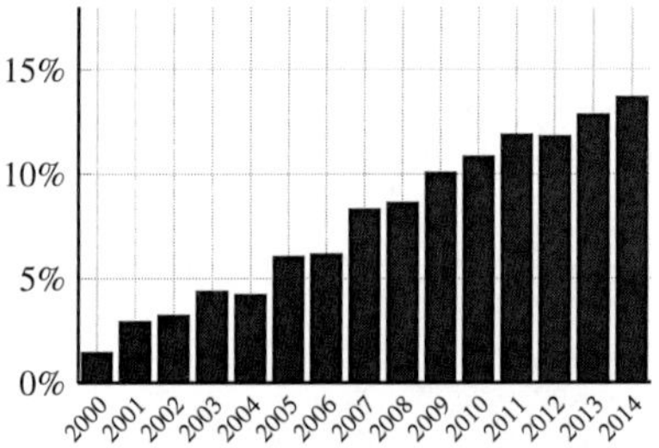

(e) Relative amount of conference articles with any keywords related to *distributed and cooperative control* in the title.

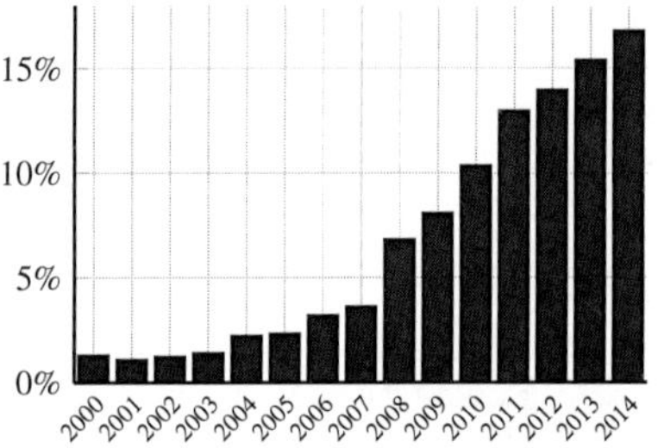

(f) Relative amount of journal articles with any keywords related to *distributed and cooperative control* in the title.

Figure 2.2: The amount of publications containing the keyword *"consensus"*, or any of the keywords {*distributed, cooperative, multi-agent, multi-vehicle, multi-robot, agreement, consensus, synchronization, formation control, coordination*} in their title, which appeared in the proceedings of the most renowned conferences and journals in the field of systems theory and automatic control since the year 2000. The considered journals are *Automatica*, *IEEE Transactions on Automatic Control*, as well as *IEEE Transactions on Control of Network Systems* (only 2014), and the conferences are the *IEEE Conference on Decision and Control (CDC)* and the *American Control Conference (ACC)*. The database was obtained from http://www.sciencedirect.com and http://ieeexplore.ieee.org.

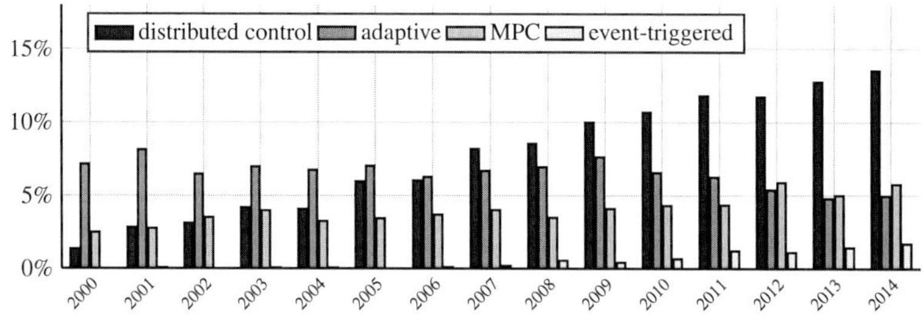

(a) Relative amount of conference articles with any keywords from the respective lists in the title.

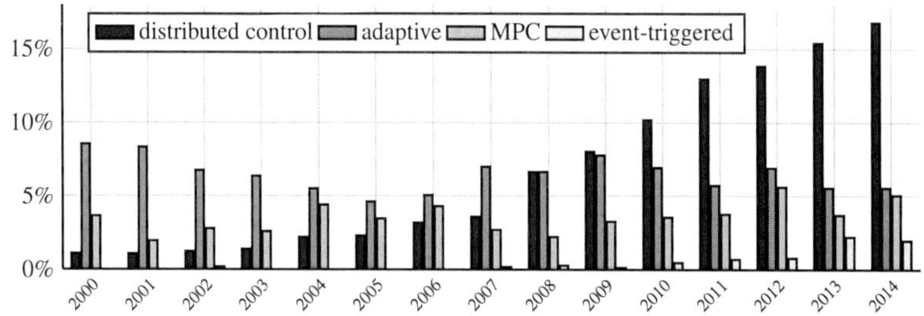

(b) Relative amount of journal articles with any keywords from the respective lists in the title.

Figure 2.3: Comparison of the amount of publications containing any keywords of the respective topical lists in their title. The previously defined list for *distributed control* is compared to the lists *adaptive* = {*adaptive, adapting, adaptation*}, *MPC* = {*MPC, predictive control, receding horizon, moving horizon*}, and *event-triggered* = {*event-triggered, event-based, self-triggered, event-scheduled*}. The same database as in Fig. 2.2 is evaluated. Adaptive control and model predictive control (MPC) have a slightly varying level with a mean of 6.5% and 4% (conferences) and 6.5% and 3.5% (journals), respectively. Event-triggered control began rising in 2007 and appears to be a growing topic since then. Today, distributed control outnumbers the other three topics by far.

to select a particular agent as leader for the group or introduce a virtual leader and design the distributed control law such that all agents synchronize to this leader. The motion of the group is then controlled through the motion of the active leader. Disturbance attenuation requirements in terms of and $\mathcal{H}_\infty$ and $\mathcal{H}_2$ performance criteria are addressed by Zelazo and Mesbahi (2011); Li et al. (2011); Zhao et al. (2012), Wang et al. (2013). Robust distributed controllers for networks with uncertain agent dynamics are constructed by Trentelman et al. (2013). Optimal distributed control laws with respect to quadratic cost functionals are discussed by Cao and Ren (2010); Hengster-Movric and Lewis (2014); Montenbruck et al. (2015b) for networks of linear systems. Distributed estimation problems with $\mathcal{H}_\infty$ performance requirements are tackled by Ugrinovskii and Langbort (2011); Ugrinovskii (2011b); Wu et al. (2014). Furthermore, networks of agents with nonlinear dynamics have been investigated. This includes motion coordination algorithms for groups of nonholonomic vehicles (Lin et al. (2005); Marshall et al. (2004, 2006); Dimarogonas and Kyriakopoulos (2007); Leonard et al. (2007)), attitude synchronization of rigid bodies (Sarlette et al. (2009)), cooperative control of Euler-Lagrange systems (Chung and Slotine (2009); Mei et al. (2011)), oscillator models (Paley et al. (2007); Dörfler and Bullo (2014)), and further classes of nonlinear systems (Arcak (2007); Wu (2007); Qu (2010); DeLellis et al. (2011); Bürger and De Persis (2015); Montenbruck et al. (2015a)). Another major step towards general networks of dynamical systems is the study of heterogeneous multi-agent systems consisting of agents with non-identical dynamics, in contrast to homogeneous networks consisting of identical agents. Output synchronization problems in heterogeneous networks of linear systems are formulated and solved by Wieland et al. (2011b); Kim et al. (2011); Lunze (2012); Grip et al. (2012); Wu and Allgöwer (2012). A major achievement in this context is the formulation of an internal model principle for synchronization, which constitutes a necessary structural condition on heterogeneous networks for output synchronization (Wieland and Allgöwer (2009); Wieland et al. (2011b); Lunze (2012); Wieland et al. (2013)). Based on this insight and the classical output regulation theory, various distributed control schemes are under development, which allow to solve a wide range of coordination problems for heterogeneous multi-agent systems. Xiang et al. (2009); Huang (2011); Su and Huang (2012a) suggest to formulate multi-agent coordination problems with external reference and disturbance signals as distributed output regulation problems and construct a distributed regulator based on distributed estimation and classical output regulation methods. Among the most prominent application domains of modern distributed and cooperative control methods are smart grids, cf., Bidram et al. (2014), and automated vehicle platoons based on Cooperative Adaptive Cruise Control (CACC) systems, cf., Ploeg et al. (2012, 2014).

Chapter 3

Output Synchronization

With the discussion in Chapter 2, the stage is set and we start out with the developments of the present thesis. We focus on output synchronization in networks of dynamical systems, which forms the core of the distributed and cooperative control theory. It generalizes the consensus problem to networks of more complex dynamical systems. The importance of this problem lies in the fact that it serves as building block and is instrumental for the development of control algorithms for more sophisticated coordination tasks. For this reason, it has been identified as a core problem in (displacement-based) formation control in the survey article by Oh et al. (2015).

In the present chapter, output synchronization problems are posed, analyzed, and solved for various types of multi-agent systems. The multi-agent systems are networks of dynamical systems, also referred to as dynamical networks in the following. Section 3.1 addresses *homogeneous* linear networks, i.e., networks of *identical* linear time-invariant dynamical agents. The focus is on synthesis methods for distributed control laws guaranteeing output synchronization. Section 3.2 addresses *heterogeneous* linear networks, i.e., networks of *non-identical* linear time-invariant dynamical agents. The focus is on structural requirements for output synchronization and robustness issues with respect to parameter perturbations in the agent dynamics. Section 3.3 addresses both homogeneous and heterogeneous networks consisting of linear parameter-varying dynamical agents. The focus is again on the development of synthesis methods for (gain-scheduled) distributed control laws guaranteeing output synchronization.

3.1 Homogeneous Linear Networks

This section is dedicated to the output synchronization problem for homogeneous linear networks, i.e., for multi-agent systems consisting of identical linear dynamical agents and an underlying graph. More accurately, the focus is on state synchronization which implies output synchronization. Section 3.1.1 presents the precise problem formulation including the agent models, information structure, and group objective. The existence of a solution to the state synchronization problem is characterized in Section 3.1.2. In Section 3.1.3, design methods for the distributed control laws are presented. After a brief review of state-of-the-art synthesis procedures, we present novel procedures taking performance requirements in terms of pole placement constraints into account. The corresponding results are partly based on Seyboth et al. (2014a).

3.1.1 Problem Formulation

The formulation of a cooperative control problem comprises the specification of the agent models, the information structure, and the group objective. In the following, the multi-agent system under consideration is introduced and each of these three elements is specified.

Agent Models We consider multi-agent systems consisting of N identical linear dynamical systems. The index set of the agents is defined as $\mathcal{N} = \{1, ..., N\}$, where $N > 1$ is the number of agents in the group. The dynamics of each agent are described by the linear state-space model

$$\dot{x}_k = Ax_k + Bu_k \tag{3.1a}$$

$$y_k = Cx_k \tag{3.1b}$$

with state $x_k(t) \in \mathbb{R}^{n_x}$, input $u_k(t) \in \mathbb{R}^{n_u}$, output $y_k(t) \in \mathbb{R}^{n_y}$, where $t \in \mathbb{R}_0^+$ represents time, and matrices A, B, C of appropriate dimensions, for all $k \in \mathcal{N}$. By means of the Kronecker product $\otimes$, the dynamics of all agents can be written in aggregate form as

$$\dot{x} = (I_N \otimes A)x + (I_N \otimes B)u \tag{3.2a}$$

$$y = (I_N \otimes C)x, \tag{3.2b}$$

where we used the stack vectors $x = [x_1^\mathsf{T} \ \cdots \ x_N^\mathsf{T}]^\mathsf{T}$, $u = [u_1^\mathsf{T} \ \cdots \ u_N^\mathsf{T}]^\mathsf{T}$, and $y = [y_1^\mathsf{T} \ \cdots \ y_N^\mathsf{T}]^\mathsf{T}$.

Information Structure The agents have communication capabilities. The communication topology is described by a directed graph $\mathcal{G} = (\mathcal{V}, \mathcal{E}, A_\mathcal{G})$. An introduction to graph theory including all the relevant definitions and results is provided in Section A.2. Every node $v_k \in \mathcal{V}$ of the graph is associated with one agent $k \in \mathcal{N}$ in the group. A directed edge $(v_j, v_k) \in \mathcal{E}$ corresponds to possible information flow from agent $j \in \mathcal{N}$ to agent $k \in \mathcal{N}$. There are different forms of information flow, which we do not distinguish yet. Typical forms of information flow are sensing or communication. Sensing means that agent k has a measurement device and is able to sense agent j, e.g., a mobile robot can obtain the relative distance and orientation to some other robots in the group via a vision-based sensing device. Communication means that agent j can transmit information to agent k through a communication medium, e.g., a packet-based wireless network. The crucial point is that information flow happens only in accordance with the directed edges $\mathcal{E}$. Each agent k may obtain information only from its direct neighbors $j \in \mathcal{N}_k = \{j \in \mathcal{N} : (v_k, v_j) \in \mathcal{E}\}$ but not from its two-hop or multi-hop neighbors. The particular form of information flow will be specified wherever it is necessary. A control law respecting this information structure is called a distributed control law.

Definition 3.1 (Distributed Control Law)**.** A *distributed control law* is an algorithm that computes the control input $u_k(t)$ locally at agent k, solely based on the information available at this node.

In contrast to a centralized control law, distributed control laws take only local information about the respective agent and its direct neighbors into account.

Group Objective In the present chapter, we focus on the most fundamental group objective. This objective is the synchronization problem and is defined precisely in the following.

Definition 3.2 (State Synchronization)**.** The agents (3.1) are said to achieve *state synchronization*, if for all initial conditions $x_k(0)$, $k \in \mathcal{N}$, it holds that for all pairs $j, k \in \mathcal{N}$,

$$\lim_{t \to \infty} (x_k(t) - x_j(t)) = \mathbf{0}. \tag{3.3}$$

Moreover, the agents (3.1) are said to achieve *nontrivial* state synchronization if $x = \mathbf{0}$ is not asymptotically stable. Every trajectory $s : \mathbb{R}_0^+ \to \mathbb{R}^{n_x}$ satisfying $\lim_{t\to\infty}(x_k(t) - s(t)) = \mathbf{0}$ for all $k \in \mathcal{N}$ is called a *synchronous state trajectory*.

State Synchronization is a generalization of the consensus problem in Section 2.1 in respect of the general linear agent dynamics (3.1) and the time-dependent synchronous state trajectory.

Definition 3.3 (Output Synchronization)**.** The agents (3.1) are said to achieve *output synchronization*, if for all initial conditions $x_k(0)$, $k \in \mathcal{N}$, it holds that for all pairs $j, k \in \mathcal{N}$,

$$\lim_{t\to\infty}(y_k(t) - y_j(t)) = \mathbf{0}. \tag{3.4}$$

Moreover, the agents (3.1) are said to achieve *nontrivial* output synchronization, if the subspace $\{x \in \mathbb{R}^{Nn_x} : y_1 = y_2 = \cdots = y_N = \mathbf{0}\} \subset \mathbb{R}^{Nn_x}$ is not asymptotically stable. Every trajectory $s : \mathbb{R}_0^+ \to \mathbb{R}^{n_y}$ satisfying $\lim_{t\to\infty}(y_k(t) - s(t)) = \mathbf{0}$ for all $k \in \mathcal{N}$ is called a *synchronous output trajectory*.

Note that the multi-agent system achieves state/output synchronization if and only if there exists a synchronous state/output trajectory for all initial conditions. The notion of nontrivial state/output synchronization allows to exclude the non-cooperative case, in which each individual agent is asymptotically stable or stabilized by local feedback separately and state/output synchronization would trivially be achieved with $s(t) = \mathbf{0}$ for all $t \geq 0$. Obviously, state synchronization implies output synchronization. The converse is in general not true. The distinction will be of particular importance for heterogeneous multi-agent systems (Section 3.2). In the present section, we study homogeneous linear multi-agent systems and focus on state synchronization.

Problem 3.1 (State Synchronization Problem)**.** Consider the multi-agent system consisting of $N > 1$ agents (3.1) and an underlying graph $\mathcal{G}$. Find a distributed control law, such that the group achieves nontrivial state synchronization.

3.1.2 Existence of a Solution

Analogously to the consensus subspace (2.4), we define the synchronous subspace $\mathcal{S}_x \subset \mathbb{R}^{Nn_x}$ as

$$\mathcal{S}_x = \{x \in \mathbb{R}^{Nn_x} : x_1 = x_2 = \cdots = x_N\}. \tag{3.5}$$

Moreover, we define the asynchronous subspace $\mathcal{A}_x$ as the orthogonal complement of $\mathcal{S}_x$, i.e., $\mathcal{A}_x = \mathcal{S}_x^\perp = \{x \in \mathbb{R}^{Nn_x} : x^\mathsf{T} z = 0 \text{ for all } z \in \mathcal{S}_x\}$. The subspace $\mathcal{S}_x$ can as well be written as

$$\mathcal{S}_x = \operatorname{im}(\mathbf{1} \otimes I_{n_x}). \tag{3.6}$$

Consequently, by Theorem A.14, it holds that $\mathcal{A}_x = \ker(\mathbf{1}^\mathsf{T} \otimes I_{n_x})$. The State Synchronization Problem 3.1 amounts to stabilization of the synchronous subspace $\mathcal{S}_x$ by means of a distributed control law. We focus on static diffusive couplings of the form

$$u_k = K \sum_{j=1}^{N} a_{kj}(x_j - x_k) \tag{3.7}$$

as distributed control laws. The matrix $K \in \mathbb{R}^{n_u \times n_x}$ is referred to as the coupling gain matrix and a_{kj} are the elements of the adjacency matrix $A_\mathcal{G}$ corresponding to the underlying graph $\mathcal{G}$. Since $a_{kj} = 0$ if $(v_j, v_k) \notin \mathcal{E}$, (3.7) constitutes a distributed control law. The static diffusive couplings are a direct generalization of the consensus protocol (2.2) for higher-dimensional state vectors.

The following celebrated result by Fax and Murray (2004) provides a necessary and sufficient condition for state synchronization involving the agent dynamics (A, B), the coupling gain matrix K, and the graph $\mathcal{G}$. Formulations of this result can also be found in (Wieland et al., 2008, Thm. 3); (Wieland et al., 2011a, Lem. 3.1), and partly in (Tuna, 2008, Thm. 1).

Theorem 3.1 (Necessary and Sufficient Condition for State Synchronization)**.** *The agents* (3.1) *achieve state synchronization under the distributed control law* (3.7)*, if and only if the matrices*

$$A - \lambda_k BK \tag{3.8}$$

are Hurwitz, i.e., $\sigma(A - \lambda_k BK) \subset \mathbb{C}^-$*, for all* $k \in \mathcal{N}\backslash\{1\}$*, where* λ_k*,* $k \in \mathcal{N}\backslash\{1\}$*, are the non-zero eigenvalues of the Laplacian* $L_\mathcal{G}$ *of* $\mathcal{G}$*. Moreover, in this case a synchronous state trajectory* $s(t)$ *is the solution of* $\dot{s} = As$ *with initial condition* $s(0) = (p^\mathsf{T} \otimes I_{n_x})x(0)$*, where* $p^\mathsf{T} L_\mathcal{G} = \mathbf{0}^\mathsf{T}$*,* $p^\mathsf{T}\mathbf{1} = 1$*.*

We provide a proof for Theorem 3.1 since its key idea will be used repeatedly in the remainder of this thesis and since it adds to a self-contained presentation.

Proof. Let us recall the compact description (3.2) of the multi-agent system and note that the stack vector u of the control laws (3.7) is given by

$$u = -(L_\mathcal{G} \otimes K)x. \tag{3.9}$$

Then, the closed-loop system consisting of (3.2a), (3.9) is given by

$$\dot{x} = (I_N \otimes A - L_\mathcal{G} \otimes (BK))\, x. \tag{3.10}$$

The key idea is to perform a coordinate transformation for the system (3.10) such that the transformed system has a suitable block triangular structure. For that purpose, we construct a transformation matrix $T \in \mathbb{R}^{N\times N}$ such that the first column of T is $\mathbf{1}$, the first row of T^{-1} is $p^\mathsf{T} \in \mathbb{R}^N$ satisfying $p^\mathsf{T} L_\mathcal{G} = \mathbf{0}^\mathsf{T}$, $p^\mathsf{T}\mathbf{1} = 1$, and $p \geq \mathbf{0}$ element-wise, and

$$T^{-1}L_\mathcal{G}T = \Lambda = \left[\begin{array}{c|c} 0 & 0 \\ \hline 0 & \Delta_\mathrm{r} \end{array}\right] \in \mathbb{R}^{N\times N} \quad \text{with} \quad \Delta_\mathrm{r} = \begin{bmatrix} \Lambda_2 & \cdots & \star \\ & \ddots & \vdots \\ 0 & & \Lambda_m \end{bmatrix} \in \mathbb{R}^{(N-1)\times(N-1)}, \tag{3.11}$$

where Λ_i, $i = 2, \ldots, m$, is a real scalar, or a real 2×2 matrix with a non-real pair of complex conjugate eigenvalues. Such a T exists always and can be constructed as shown in Appendix B.1. Application of the state transformation $z = (T^{-1} \otimes I_{n_x})x$ to the system (3.10) yields the system $\dot{z} = (I_N \otimes A - \Lambda \otimes (BK))z$. Let z be partitioned into $z = [z'^\mathsf{T}\ z''^\mathsf{T}]^\mathsf{T}$ with $z' \in \mathbb{R}^{n_x}$ and $z'' \in \mathbb{C}^{(N-1)n_x}$. Then, the transformed system decomposes into

$$\dot{z}' = Az' \tag{3.12}$$

$$\dot{z}'' = (I_{N-1} \otimes A - \Delta_\mathrm{r} \otimes (BK))\, z''. \tag{3.13}$$

The synchronous subspace $\mathcal{S}_x$ defined in (3.5) is invariant with respect to (3.10) and spanned by the first n_x columns of $T \otimes I_{n_x}$, cf., (3.6). Hence, in view of Theorem A.16, $\mathcal{S}_x$ is asymptotically stable if and only if (3.13) is asymptotically stable. Moreover, (3.12) describes the dynamics restricted to $\mathcal{S}_x$. Due to Schur's triangularization theorem, there exists a unitary matrix $H_3 \in \mathbb{C}^{N-1\times N-1}$ such that $\Delta_\mathrm{c} = H_3^\mathsf{H}\Delta_\mathrm{r}H_3 \in \mathbb{C}^{(N-1)\times(N-1)}$ is an upper triangular matrix with $\lambda_2, \ldots, \lambda_N$ on the diagonal, where $\cdot^\mathsf{H}$ denotes the Hermitian adjoint. With the similarity matrix $H_3 \otimes I_{N-1}$, the system matrix of (3.13) is similar to

$$\begin{bmatrix} A - \lambda_2 BK & \cdots & \star \\ & \ddots & \vdots \\ 0 & & A - \lambda_N BK \end{bmatrix}.$$

Due to the block triangular structure, it follows that (3.13) is asymptotically stable if and only if $\sigma(A - \lambda_k BK) \subset \mathbb{C}^-$ for all $k \in \mathcal{N}\backslash\{1\}$, which proves the first statement. In this case, $z''(t) \to \mathbf{0}$ as $t \to \infty$ for all initial conditions. Moreover, $z'(0) = (p^\top \otimes I_{n_x})x(0)$. It holds that $x = (T \otimes I_{n_x})z = (\mathbf{1} \otimes I_{n_x})z' + ((UH_2) \otimes I_{n_x})z''$, and therefore $x(t) - (\mathbf{1} \otimes z'(t)) \to \mathbf{0}$ as $t \to \infty$, which proves the second statement. □

As a consequence of Theorem 3.1, nontrivial state synchronization can only happen if the agents (3.1) are not asymptotically stable. The following theorem shows that the State Synchronization Problem 3.1 has a solution if and only if the underlying graph $\mathcal{G}$ contains a spanning tree. This is the main result of Tuna (2008) and has also been reported by Ma and Zhang (2010); Xi et al. (2010); (Wieland et al., 2011a, Thm. 3.3); (Wieland, 2010, Thm. 3.15); Zhang et al. (2011).

Theorem 3.2 (State Synchronization: Existence of K). *Consider the agents* (3.1) *with underlying graph $\mathcal{G}$. Suppose that A is not Hurwitz and (A, B) is stabilizable. Then, there exists a coupling gain matrix $K \in \mathbb{R}^{n_u \times n_x}$ such that the network* (3.1), (3.7) *achieves state synchronization, if and only if $\mathcal{G}$ is connected.*

Note that by Definition A.10, $\mathcal{G}$ being connected means that $\mathcal{G}$ contains a spanning tree. We provide a proof for Theorem 3.2 since it is constructive and presents a design procedure for the coupling gain K and it adds to a self-contained presentation.

Proof of Theorem 3.2. Necessity follows by the following argument. If $\mathcal{G}$ is not connected, $L_\mathcal{G}$ has a second zero eigenvalue $\lambda_2 = 0$. The matrix $A - \lambda_2 BK = A$ cannot be stabilized by K and hence, by Theorem 3.1, state synchronization cannot be achieved. The sufficiency proof is constructive and uses the gain margins of a Linear Quadratic Regulator (LQR) in order to find K which simultaneously stabilizes $A - \lambda_k BK$, $k \in \mathcal{N}\backslash\{1\}$. We choose any $R > 0$, $Q > 0$ and compute P as the unique positive definite solution of the Algebraic Riccati Equation (ARE) (A.9). Then, we define the coupling gain matrix

$$K = cR^{-1}B^\top P, \tag{3.14}$$

with some constant parameter

$$c \geq \frac{1}{2\,\mathbf{Re}(\lambda_2)}. \tag{3.15}$$

Due to the gain margins specified in Corollary A.19, it holds that $A - \rho BK$ is Hurwitz for all $\rho \in \mathbb{C}$ with $\mathbf{Re}(\rho) \geq \mathbf{Re}(\lambda_2)$, which includes all λ_k, $k \in \mathcal{N}\backslash\{1\}$. □

Remark 3.1 (Static Diffusive Output Couplings). The static diffusive couplings (3.7) are based on the agent states. In case the relative state information is not available, one might employ

$$u_k = K \sum_{j=1}^{N} a_{kj}(y_j - y_k) \tag{3.16}$$

instead. Analogously to Theorem 3.1, state synchronization is achieved if and only if

$$A - \lambda_k BKC$$

is Hurwitz for all $k \in \mathcal{N}\backslash\{1\}$. In contrast to Theorem 3.2, stabilizability of (A, B) and connectedness of $\mathcal{G}$ are not sufficient for the existence of such K. In fact, we are facing a static output feedback problem, for which no general necessary and sufficient conditions are known (Syrmos et al. (1997)). Moreover, the problem of simultaneous stabilization by output feedback is known to be NP-hard (Blondel and Tsitsiklis (1997)). Nevertheless, there exist algorithmic solution approaches, e.g., via iterative LMIs (Cao et al. (1998)). Regarding (3.16), it was shown by Ma and Zhang (2010) that detectability of (A, C) is necessary as well and the following condition is sufficient for synchronizability: Suppose that A is not Hurwitz and

$$\operatorname{rank}(C) = \operatorname{rank}\left(\begin{bmatrix} C \\ B^\top P \end{bmatrix}\right),$$

where P is the positive semi-definite solution of the ARE $A^\top P + PA - PBB^\top P + I_{n_x} = 0$. Then, there exists K such that (3.1a), (3.16) achieve state synchronization if and only if (A, B) is stabilizable and $\mathcal{G}$ contains a spanning tree. In this case, a suitable coupling gain matrix is $K = cK_0$, where K_0 solves $K_0 C = B^\top P$ and $c = \max\{1, (\mathbf{Re}(\lambda_2))^{-1}\}$. It should be noted that the rank condition above is rather restrictive. ○

3.1.3 Distributed Control Design Methods

The proof of Theorem 3.2 provides a design method for the coupling gain matrix K in (3.7) based on an ARE. The purpose of this section is to present further design methods based on Linear Matrix Inequality (LMI) conditions and Semi-Definite Programming (SDP). Firstly, we review two LMI based design methods proposed by Li et al. (2010) and Wieland et al. (2011a). Secondly, we present a novel design method which allows to include performance specifications in terms of pole placement constraints into the design. Note that with distributed control design, we refer to the design of distributed control laws according to Definition 3.1 while the design procedure itself needs not to be distributed. These results are partly based on Seyboth et al. (2014a).

If (A, B) is stabilizable, then there exists a matrix K such that $A - BK$ is Hurwitz. The following result by Boyd et al. (1994) shows how such K can be found via LMI conditions and SDP.

Lemma 3.3 (Static State Feedback Synthesis via LMIs (Boyd et al. (1994))). *Suppose that (A, B) is stabilizable. Then, there exists a matrix $X = X^\top > 0$ and a real scalar $\tau > 0$ such that*

$$XA^\top + AX - \tau BB^\top < 0, \tag{3.17}$$

and the state feedback gain $K = \frac{\tau}{2} B^\top X^{-1}$ stabilizes (A, B), i.e., $A - BK$ is Hurwitz. Moreover, $A - \rho BK$ is Hurwitz for all $\rho \in \mathbb{C}$ with $\mathbf{Re}(\rho) \geq 1$. Without loss of generality, we can take $\tau = 2$.

Proof. See Appendix B.2. □

A proof is provided for completeness and in order to show the gain margin property, which is not reported in Boyd et al. (1994). Lemma 3.3 is well suited for the design of the coupling gain matrix K in (3.7). Analogously to the LQR design in Theorem 3.2, the gain margins guarantee simultaneous stabilization of the matrices (3.8) for all $k \in \mathcal{N}\backslash\{1\}$.

Theorem 3.4 (Static Diffusive Coupling Synthesis via LMIs I (Li et al. (2010))). *Consider the multi-agent system* (3.1) *with underlying graph $\mathcal{G}$. Suppose that (A, B) is stabilizable and $\mathcal{G}$ is connected. Then, there exists $X \in \mathbb{R}^{n_x \times n_x}$ such that $X = X^\top > 0$ and X satisfies the LMI* (3.17) *for $\tau = 2$. Moreover, the coupling gain matrix*

$$K = cB^\top X^{-1}, \tag{3.18}$$

with any constant parameter

$$c \geq \frac{1}{\mathbf{Re}(\lambda_2)} \tag{3.19}$$

solves the State Synchronization Problem 3.1.

Note that the threshold for the coupling strength (3.19) is twice as large as (3.15), which is due to the larger gain margins of the LQR. An alternative synthesis procedure for the coupling gain matrix based on LMI conditions was proposed by Wieland et al. (2011a); Wieland (2010).

Theorem 3.5 (Static Diffusive Coupling Synthesis via LMIs II (Wieland, 2010, Thm. 3.17)). *Consider the multi-agent system* (3.1) *with underlying graph* $\mathcal{G}$. *If there exist matrices* $X \in \mathbb{R}^{n_x \times n_x}$, $Y \in \mathbb{R}^{n_u \times n_x}$ *and a real scalar* $\gamma > 0$ *such that* $X = X^\mathsf{T} > 0$ *and*

$$XA^\mathsf{T} + AX + 2\gamma X - \lambda_k BY - \overline{\lambda_k} Y^\mathsf{T} B^\mathsf{T} \leq 0 \tag{3.20}$$

for all $k \in \mathcal{N}\backslash\{1\}$, *then the coupling gain matrix* $K = YX^{-1}$ *solves the State Synchronization Problem 3.1. Moreover, the decay rate of the synchronization error is at least* γ.

The idea behind Theorem 3.5 is to construct a common Lyapunov function $V(z) = z^\mathsf{H} P z$ for the systems $\dot{z} = (A - \lambda_k BK)z$, $k \in \mathcal{N}\backslash\{1\}$. In order to achieve a desired decay rate $\gamma > 0$, the matrix A is replaced by $A + \gamma I_{n_x}$ in the design. We will elaborate on this point later. The matrix P has to satisfy $P > 0$ and $P(A - \lambda_k BK) + (A - \lambda_k BK)^\mathsf{H} P < 0$ for all $k \in \mathcal{N}\backslash\{1\}$. The standard change of variables $X = P^{-1}$ and $Y = KX$ leads to (3.20), which is linear in the decision variables X and Y. This synthesis procedure for K entails two sources of conservatism: the fact that P is restricted to be real symmetric instead of complex Hermitian, and the fact P is required to be identical for all $k \in \mathcal{N}\backslash\{1\}$. Consequently, it is not guaranteed that there is a feasible solution for the set of LMIs (3.20) in all cases. Moreover, a coupling gain K found in this way does not have guaranteed gain margins in contrast to (3.14) and (3.18). However, as we will see next, an approach as in Theorem 3.5 allows to incorporate performance specifications in terms of pole placement constraints on the synchronization error.

Static Diffusive Couplings with Pole Placement Constraints

We have seen that (3.12) describes the synchronous motion of the system while the synchronization error is governed by the dynamics (3.13). State synchronization is achieved exponentially with a convergence rate $\gamma > 0$ if the equilibrium $z'' = \mathbf{0}$ of (3.13) is exponentially stable with decay rate $\gamma > 0$. Under the assumption that $(A + \gamma I_{n_x}, B)$ is stabilizable, this performance requirement can be incorporated in the synthesis procedures in Theorem 3.2 and Theorem 3.4 by replacing A by $A + \gamma I_{n_x}$ in (A.9) and (3.17), respectively. Such a design guarantees that all eigenvalues of $A - \lambda_k BK$ are contained in the complex half plane $H^-(\gamma) = \{z \in \mathbb{C} : \mathbf{Re}(z) < -\gamma\}$ shown in Fig. 3.1(a). As a generalization of this idea, we present a novel design procedure based on LMIs which takes into account performance specifications in terms of pole placement constraints. For this purpose, we make use of the LMI region introduced by Chilali and Gahinet (1996), which is defined as

$$\mathcal{R} = \{z \in \mathbb{C} : L + zM + \bar{z}M^\mathsf{T} < 0\}, \tag{3.21}$$

for some fixed matrices $L = L^\mathsf{T} \in \mathbb{R}^{m \times m}$ and $M \in \mathbb{R}^{m \times m}$. The region $\mathcal{R}$ is a convex subset of $\mathbb{C}$. A variety of convex sets such as vertical strips, horizontal strips, disks, conic sectors as well as intersections thereof can be expressed as LMI region. Pole placement constraints in terms of such regions allow to enforce not only a desired decay rate of the closed loop but also a desired damping ratio, a maximum undamped frequency, or a bound on the control gain. The following result is the key to incorporate pole placement constraints into LMI based controller synthesis procedures.

Lemma 3.6 (Pole Placement in an LMI Region (Chilali and Gahinet, 1996, Thm. 2.2))**.** *The matrix $A \in \mathbb{C}^{n_x \times n_x}$ satisfies $\sigma(A) \subset \mathcal{R}$ if and only if there exists a matrix $X \in \mathbb{C}^{n_x \times n_x}$ such that $X = X^{\mathsf{H}} \succ 0$ and*

$$L \otimes X + M \otimes (AX) + M^{\mathsf{T}} \otimes (XA^{\mathsf{H}}) \prec 0. \tag{3.22}$$

If A is real, then X can be chosen to be real symmetric.

This result is originally formulated for real matrices A, but its proof reveals that the more general complex version holds. It allows us to derive the following result.

Theorem 3.7 (Static Diffusive Coupling Synthesis with Pole Placement Constraints I)**.** *Consider the agents* (3.1) *with underlying graph $\mathcal{G}$. Let $\mathcal{R} \subset \mathbb{C}^-$ be an LMI region defined by some matrices $L = L^{\mathsf{T}} \in \mathbb{R}^{m \times m}$ and $M \in \mathbb{R}^{m \times m}$. If there exist matrices $X \in \mathbb{R}^{n_x \times n_x}$ and $Y \in \mathbb{R}^{n_u \times n_x}$ such that $X = X^{\mathsf{T}} \succ 0$ and*

$$L \otimes X + M \otimes (AX - \lambda_k BY) + M^{\mathsf{T}} \otimes (AX - \overline{\lambda_k} BY)^{\mathsf{T}} \prec 0 \tag{3.23}$$

for all $k \in \mathcal{N} \backslash \{1\}$, then the coupling gain matrix $K = YX^{-1}$ solves the State Synchronization Problem 3.1. Moreover, all poles corresponding to the synchronization error are contained in $\mathcal{R}$.

Proof. In view of Theorem 3.1, it remains to show that the constructed K guarantees that $\sigma(A - \lambda_k BK) \subset \mathcal{R}$ for all $k \in \mathcal{N} \backslash \{1\}$. According to Lemma 3.6, a sufficient condition is that there exists a matrix $X \in \mathbb{R}^{n_x \times n_x}$ such that $X = X^{\mathsf{T}} \succ 0$ and

$$L \otimes X + M \otimes ((A - \lambda_k BK)X) + M^{\mathsf{T}} \otimes (X(A - \overline{\lambda_k} BK)^{\mathsf{T}}) \prec 0$$

for all $k \in \mathcal{N} \backslash \{1\}$. Note that X is restricted to be real and identical for all $k \in \mathcal{N} \backslash \{1\}$, which introduces some conservatism. The change of variable $Y = KX$ leads to the LMI (3.23), which concludes the proof. □

Theorem 3.7 is an important generalization of Theorem 3.5 with practical relevance since it allows to include performance specifications such as a desired decay rate and damping ratio for the synchronization error into the design of the coupling gain. Two exemplary LMI regions of practical interest are illustrated in Fig. 3.1. The half plane in Fig. 3.1(a) is expressed by the scalars $L = 2\gamma$ and $M = 1$. For this special case, the synthesis LMI (3.23) reduces to (3.20) in Theorem 3.5. A description of the region $S(\gamma, r, \theta)$ is obtained as an intersection of a half plane, a disk, and a conical sector. These basic regions are listed in Table 3.1. The matrices L and M describing $S(\gamma, r, \theta)$ are block diagonal combinations of those in Table 3.1. Pole placement in the region $S(\gamma, r, \theta)$ guarantees a minimum decay rate γ, minimum damping ratio $\zeta = \cos(\theta)$, and a maximum undamped frequency $\omega_d = r \sin \theta$.

The synthesis procedure for the coupling gain matrix K presented in Theorem 3.7 has the limitation that the LMIs (3.23) depend on the eigenvalues of the Laplacian matrix $L_{\mathcal{G}}$. The synthesis procedures in Theorem 3.2 and 3.4 do not have this limitation since they exploit the respective gain margins, but these procedures cannot take pole placement constraints into account. An application of Finsler's Lemma A.2 analogously to Theorem 3.4 does not lead to a feasible

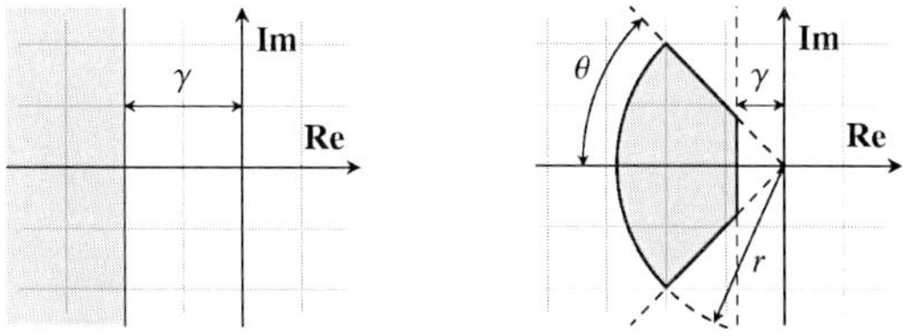

(a) Half plane $H^-(\gamma)$. (b) Region $S(\gamma, r, \theta)$.

Figure 3.1: Two exemplary LMI regions.

synthesis procedure in presence of pole placement constraints due to the Kronecker products in (3.22). It is worth noting that the number of LMI conditions in Theorem 3.23 may be reduced since some LMIs may be redundant. In particular, it suffices to consider one LMI for the smallest real non-zero eigenvalue, one LMI for the largest real non-zero eigenvalue, and one LMI for each complex conjugate pair of eigenvalues of $L_{\mathcal{G}}$. This idea can be generalized as will become clear in the sequel.

We present two further LMI based synthesis procedures which yield a coupling gain matrix K solving the State Synchronization Problem 3.1 under pole placement constraints for a whole class of connected graphs $\mathcal{G}$. These two classes are graphs for which all the non-zero eigenvalues of $L_{\mathcal{G}}$ are contained (i) in the convex hull of a set of given vertices in the complex plane, or (ii) in a circle of given radius centered at $+1$ in the complex plane.

The synthesis procedure corresponding to the first class has the advantage that no exact knowledge of the eigenvalues of $L_{\mathcal{G}}$ is required. A suitable convex hull $\mathrm{conv}(\{\mu_1, ..., \mu_p, \overline{\mu_1}, ..., \overline{\mu_p}\})$ defined by some vertices $\mu_j \in \mathbb{C}$, $j = 1, ..., p$ can be constructed, e.g., based only on a lower bound on $\mathbf{Re}(\lambda_2)$ and on the number N of agents. The following result is a corollary to Theorem 3.7.

Corollary 3.8 (Static Diffusive Coupling Synthesis with Pole Placement Constraints II). *Consider the agents* (3.1). *Let $\mathcal{R} \subset \mathbb{C}^-$ be an LMI region defined by the matrices $L = L^\top \in \mathbb{R}^{m\times m}$ and $M \in \mathbb{R}^{m\times m}$. If there exist matrices $X \in \mathbb{R}^{n_x\times n_x}$ and $Y \in \mathbb{R}^{n_u\times n_x}$ such that $X = X^\top > 0$ and*

$$L \otimes X + M \otimes (AX - \mu_j BY) + M^\top \otimes (AX - \overline{\mu_j} BY)^\top < 0 \tag{3.24}$$

for all $j = 1, ..., p$, then the coupling gain matrix $K = YX^{-1}$ solves the State Synchronization Problem 3.1 for all connected graphs $\mathcal{G}$ which satisfy $\lambda_k \in \mathrm{conv}(\{\mu_1, ..., \mu_p, \overline{\mu_1}, ..., \overline{\mu_p}\})$ for all $k \in \mathcal{N}\backslash\{1\}$. Moreover, all poles corresponding to the synchronization error are contained in $\mathcal{R}$ for all such graphs.

Table 3.1: Examples for basic LMI regions

	L	M
half plane $\{z \in \mathbb{C} : \mathbf{Re}(z) < \gamma\}$	2γ	1
disk $\{z \in \mathbb{C} : \lvert z\rvert < r\}$	$\begin{bmatrix} -r & 0 \\ 0 & -r \end{bmatrix}$	$\begin{bmatrix} 0 & 1 \\ 0 & 0 \end{bmatrix}$
conical sector with angle θ	0	$\begin{bmatrix} \sin\theta & \cos\theta \\ -\cos\theta & \sin\theta \end{bmatrix}$

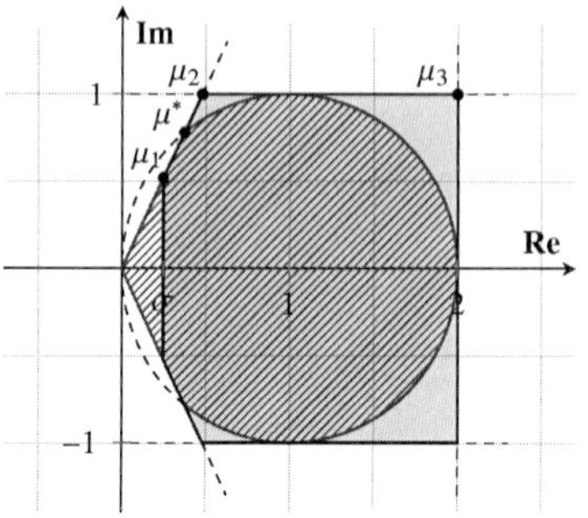

Figure 3.2: For any directed graph $\mathcal{G}$ with N nodes, the spectrum of the normalized Laplacian $\tilde{L}_{\mathcal{G}}$ is contained in the hatched region (▨) with $\mu^* = 1 - \exp(-2\pi\mathbf{i}/N)$, which is visualized for $N = 7$. The shaded polygon (□) is the convex hull $\mathrm{conv}(\{\mu_1, \mu_2, \mu_3, \overline{\mu_1}, \overline{\mu_2}, \overline{\mu_3}\})$ with $\mu_1 = \sigma + \sigma(\mathbf{Im}(\mu^*)/\mathbf{Re}(\mu^*))\mathbf{i}$, $\mu_2 = \mathbf{Re}(\mu^*)/\mathbf{Im}(\mu^*) + 1\mathbf{i}$, $\mu_3 = 2 + 1\mathbf{i}$.

Proof. Following (Wieland et al., 2011a, Lem. 4.2), we define the map $C(\mu) : \mathbb{C} \to \mathbb{C}^{mn_x \times mn_x}$ as $C(\mu) = L \otimes X + M \otimes (AX - \mu BY) + M^\mathsf{T} \otimes (AX - \overline{\mu} BY)^\mathsf{T}$. Then, $C(\overline{\mu}) \prec 0 \Leftrightarrow C(\mu) \prec 0$ since $C^\mathsf{T}(\mu) = C(\overline{\mu})$. Moreover, the set $\{\mu \in \mathbb{C} : C(\mu) \prec 0\}$ is convex. Therefore (3.24) implies that $C(\mu) \prec 0$ for all $\mu \in \mathrm{conv}(\{\mu_1, \dots, \mu_p, \overline{\mu_1}, \dots, \overline{\mu_p}\})$. Consequently, $\sigma(A - \lambda_k BK) \subset \mathcal{R} \subset \mathbb{C}^-$ for all $k \in \mathcal{N}\backslash\{1\}$ if the spectrum of the Laplacian matrix $L_{\mathcal{G}}$ of the graph $\mathcal{G}$ satisfies $\lambda_k \in \mathrm{conv}(\{\mu_1, \dots, \mu_p, \overline{\mu_1}, \dots, \overline{\mu_p}\})$ for all $k \in \mathcal{N}\backslash\{1\}$. With Theorem 3.1, this completes the proof. □

For every directed graph $\mathcal{G}$ with N nodes, the spectrum of its normalized Laplacian matrix $\tilde{L}_{\mathcal{G}} = D_{\mathcal{G}}^{-1} L_{\mathcal{G}}$ is contained in the hatched region in Fig. 3.2, cf., (Bauer, 2012, Prop. 3.3). We consider the class of directed graphs $\mathcal{G}$ with N nodes and with $\mathbf{Re}(\lambda_2(\tilde{L}_{\mathcal{G}})) \geq \sigma$ for some fixed $\sigma > 0$. For all such graphs, all non-zero eigenvalues of the normalized Laplacian $\tilde{L}_{\mathcal{G}}$ are contained in the convex set defined by μ_1, μ_2, μ_3 (or $\sigma + 1\mathbf{i}, \mu_3$ in case $\sigma \geq \mathbf{Re}(\mu_2)$), cf., Fig. 3.2. This region is less conservative than the region constructed by Wieland et al. (2011a). Hence, these vertices are a suitable choice for Corollary 3.8. Note that the Laplacian matrix can be normalized by rescaling the control law (3.7) with $1/d_k$ for each agent.

A suitable approach for the second class of connected graphs $\mathcal{G}$ with all the non-zero eigenvalues of $L_{\mathcal{G}}$ contained in a circle of given radius r centered at +1 in the complex plane is robust pole placement in LMI regions. The following lemma is a direct consequence of the more general result on robust pole placement in LMI regions (Chilali et al., 1999, Thm. 3.3).

Lemma 3.9 (Robust Pole Placement in an LMI Region). *Consider the matrices $A \in \mathbb{R}^{n_x \times n_x}$, $B \in \mathbb{R}^{n_x \times n_u}$, $K \in \mathbb{R}^{n_u \times n_x}$ and $\mathcal{R}$ as defined in* (3.21). *Let $q = \mathrm{rank}(M)$ and factorize M as $M = M_1^\mathsf{T} M_2$ with $M_1, M_2 \in \mathbb{R}^{q \times m}$. Suppose that there exist matrices $P \in \mathbb{R}^{n_x \times n_x}$ and $R \in \mathbb{R}^{q \times q}$ and a real scalar $r > 0$ such that $P = P^\mathsf{T} \succ 0$, $R = R^\mathsf{T} \succ 0$, and*

$$\begin{bmatrix} L \otimes P + M \otimes (P(A - BK)) + M^\mathsf{T} \otimes ((A - BK)^\mathsf{T} P) & M_1^\mathsf{T} \otimes (PB) & (M_2^\mathsf{T} R) \otimes K^\mathsf{T} \\ M_1 \otimes (B^\mathsf{T} P) & -\frac{1}{r} R \otimes I_{n_u} & 0 \\ (RM_2) \otimes K & 0 & -\frac{1}{r} R \otimes I_{n_u} \end{bmatrix} \prec 0. \tag{3.25}$$

Then, $\sigma(A - \rho BK) \subset \mathcal{R}$ for all $\rho \in \mathbb{C}$ contained in a disk of radius r centered at +1, i.e., $\rho \in \{z \in \mathbb{C} : |z - 1| \leq r\}$.

Proof. The statement follows by identification of $A(\Delta) = A + B(I - \Delta D)^{-1}\Delta C$ in (Chilali et al., 1999, Thm. 3.3) with $A - (1 - \epsilon)BK$, where $\rho = 1 - \epsilon$ and the scalar ϵ represents the uncertainty Δ. □

With the help of Lemma 3.9, we obtain the following synthesis result.

Theorem 3.10 (Static Diffusive Coupling Synthesis with Pole Placement Constraints III). *Consider the agents* (3.1). *Let $\mathcal{R} \subset \mathbb{C}^-$ be an LMI region defined by the real matrices $L = L^\mathsf{T} \in \mathbb{R}^{m\times m}$ and $M \in \mathbb{R}^{m\times m}$. Let $q = \mathrm{rank}(M)$ and factorize M as $M = M_1^\mathsf{T} M_2$ with $M_1, M_2 \in \mathbb{R}^{q\times m}$. Suppose that there exist matrices $X \in \mathbb{R}^{n_x\times n_x}$, $Y \in \mathbb{R}^{n_u\times n_x}$, $Z \in \mathbb{R}^{q\times q}$ and a real scalar $r \in]0,1[$ such that $X = X^\mathsf{T} > 0$, $Z = Z^\mathsf{T} > 0$, and*

$$\begin{bmatrix} L\otimes X + M\otimes(AX - BY) + M^\mathsf{T}\otimes(XA^\mathsf{T} - Y^\mathsf{T}B^\mathsf{T}) & (M_1^\mathsf{T}Z)\otimes B & M_2^\mathsf{T}\otimes Y^\mathsf{T} \\ (ZM_1)\otimes B^\mathsf{T} & -\frac{1}{r}Z\otimes I_{n_u} & 0 \\ M_2\otimes Y & 0 & -\frac{1}{r}Z\otimes I_{n_u} \end{bmatrix} \prec 0. \quad (3.26)$$

Then, the coupling gain matrix $K = YX^{-1}$ solves the State Synchronization Problem 3.1 for all connected graphs $\mathcal{G}$ which satisfy $\lambda_k \in \{z \in \mathbb{C} : |z - 1| \leq r\}$ for all $k \in \mathcal{N}\backslash\{1\}$. Moreover, all poles corresponding to the synchronization error are contained in $\mathcal{R}$ for all such graphs.

Proof. The LMI (3.26) is obtained from (3.25) by the change of variables $X = P^{-1}$, $Y = KX$, $Z = R^{-1}$ and left and right multiplication with $\mathrm{diag}(I_m \otimes X, Z \otimes I_{n_u}, Z \otimes I_{n_u})$. Hence, by Lemma 3.9, the coupling gain matrix $K = YX^{-1}$ guarantees that $\sigma(A - \rho BK) \subset \mathcal{R}$ for all $\rho \in \{z \in \mathbb{C} : |z - 1| \leq r\}$. Consequently, $\sigma(A - \lambda_k BK) \subset \mathcal{R} \subset \mathbb{C}^-$ for all $k \in \mathcal{N}\backslash\{1\}$ if the spectrum of the Laplacian matrix $L_\mathcal{G}$ of the graph $\mathcal{G}$ satisfies $\lambda_k \in \{z \in \mathbb{C} : |z - 1| \leq r\}$ for all $k \in \mathcal{N}\backslash\{1\}$. In view of Theorem 3.1, this completes the proof. □

Remark 3.2 (Synchronization Region). As shown in Theorem 3.1, state synchronization amounts to simultaneous stabilization of the matrices $A - \lambda_k BK$ for $k \in \mathcal{N}\backslash\{1\}$. Hence, it is natural to ask for the set $\mathcal{L} = \{\rho \in \mathbb{C} : \sigma(A - \rho BK) \subset \mathbb{C}^-\}$ which is denoted as consensus region (Li et al. (2010)) or synchronization region (Zhang et al. (2011)). In fact, state synchronization is achieved for all directed graphs $\mathcal{G}$ satisfying $\lambda_k \in \mathcal{L}$ for all $k \in \mathcal{N}\backslash\{1\}$. An advantage of the coupling gain matrices obtained in Theorem 3.2 and 3.4 is the fact that the corresponding consensus regions are unbounded. Analogously, we can define the consensus region $\mathcal{L}_\mathcal{R} = \{\rho \in \mathbb{C} : \sigma(A - \rho BK) \subset \mathcal{R}\}$ respecting the pole placement constraint in terms of the LMI region $\mathcal{R} \subset \mathbb{C}^-$. In this light, Corollary 3.8 and Theorem 3.10 yield $\mathcal{L}_\mathcal{R} = \mathrm{conv}(\{\mu_1, \ldots, \mu_p, \overline{\mu_1}, \ldots, \overline{\mu_p}\})$ and $\mathcal{L}_\mathcal{R} = \{\rho \in \mathbb{C} : |\rho - 1| \leq r\}$, respectively. ○

Example 3.1. We consider a multi-agent system consisting of $N = 9$ agents each described by

$$\dot{x}_k = \begin{bmatrix} 0 & 1 & 0 \\ -1 & 0 & 0.2 \\ 0 & 0 & -0.9 \end{bmatrix} x_k + \begin{bmatrix} 0 \\ 0 \\ 1 \end{bmatrix} u_k,$$

$k = 1, \ldots, 9$, and with the communication graph $\mathcal{G}$ defined in Example A.1. We synthesize coupling gain matrices K in (3.7) according to Theorem 3.2, 3.4, and 3.7. For the LQR based synthesis described in the proof of Theorem 3.2, we choose $Q = I_3$, $R = 1$ and $c = 1/(2\,\mathbf{Re}(\lambda_2)) + \epsilon$ with $\epsilon = 0.001$. For the synthesis via Theorem 3.4, we choose $c = 1/(\mathbf{Re}(\lambda_2)) + \epsilon$. In order to achieve a desired decay rate of $\gamma = 1.01$, the matrix A is replaced by $A + \gamma I_3$ in both cases. For the synthesis via Theorem 3.7, we choose $\mathcal{R} = S(1.01, 10, \pi/4)$, cf., Fig. 3.1(b). The resulting coupling

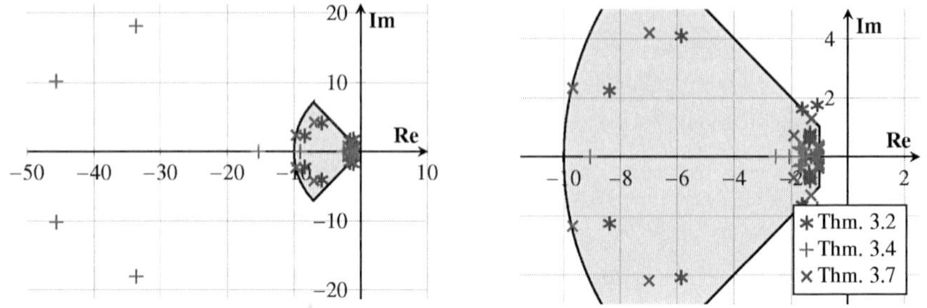

Figure 3.3: Location of the poles corresponding to the synchronization error for the different coupling gain matrices and the region $S(1.01, 10, \pi/4)$ (▭) from Example 3.1.

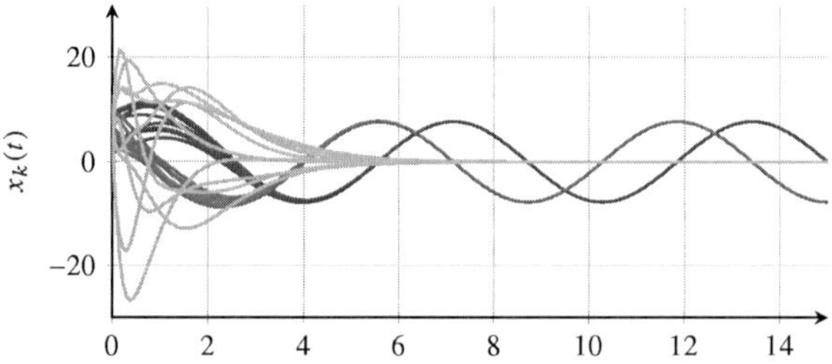

Figure 3.4: Synchronization of the 1st (—), 2nd (—), 3rd (—) states of all agents from Example 3.1.

gain matrices are $K_{\text{Thm. 3.2}} = [\,15.6812\ 45.4663\ 3.8286\,]$, $K_{\text{Thm. 3.4}} = [\,154.0019\ 289.6904\ 18.0335\,]$, and $K_{\text{Thm. 3.7}} = [\,-1.0549\ 37.6783\ 4.0681\,]$. The location of the poles corresponding to the synchronization error for these coupling gain matrices is shown in Fig. 3.3. The poles resulting from $K_{\text{Thm. 3.7}}$ indeed satisfy the pole placement constraint and are contained in $S(1.01, 10, \pi/4)$, whereas $K_{\text{Thm. 3.2}}$ results in some lightly damped poles and $K_{\text{Thm. 3.4}}$ is a high control gain. A closed loop simulation with random initial conditions and coupling gain matrix $K_{\text{Thm. 3.7}}$ is shown in Fig. 3.4. As expected, the agents achieve state synchronization. △

Remark 3.3 ($\mathcal{H}_\infty$ and $\mathcal{H}_2$ Performance Specifications)**.** It is possible to include further performance specifications beyond a desired decay rate or damping ratio in the design of the distributed control laws. Most notably, disturbance attenuation requirements in terms of $\mathcal{H}_\infty$ and $\mathcal{H}_2$ norms have recently been addressed in this context. Ugrinovskii (2011b) studies a distributed filtering problem and shows how to construct a network of filters with a guaranteed $\mathcal{H}_\infty$ level of disagreement of estimates at the nodes. Li et al. (2011) study the problem of stabilizing a network of identical linear systems via distributed control with a guaranteed $\mathcal{H}_\infty$ or $\mathcal{H}_2$ performance level. For this purpose, the concept of consensus regions is extended to $\mathcal{H}_\infty$ and $\mathcal{H}_2$ performance regions. Zhao et al. (2012) address a state synchronization problem with a guaranteed bound on the $\mathcal{H}_\infty$ norm of the transfer function from local exogenous inputs to the output disagreement and Wang et al. (2013) additionally include transient performance specifications in the control design procedure. A generalization from undirected to directed graphs is presented by Wang et al. (2014). ○

Remark 3.4 (Optimality)**.** The relation of static diffusive couplings (3.7) and optimality with respect to a quadratic cost functional was explored by Cao and Ren (2010), Hengster-Movric and Lewis (2014) and Montenbruck et al. (2015b). Cao and Ren (2010) focus on single-integrator networks and derive the optimal Laplacian matrix solving the consensus problem while minimizing a quadratic cost functional. Hengster-Movric and Lewis (2014) use partial stability theory and an inverse optimality condition in order to construct optimal coupling gains for homogeneous linear multi-agent systems solving the state synchronization problem while minimizing a quadratic cost functional. Moreover, Montenbruck et al. (2015b) investigate the structure of the coupling that is optimal with respect to cost functionals integrating quadratic synchronization error and quadratic input signals and show that such a coupling is in fact necessarily diffusive. A detailed discussion and further references regarding optimal cooperative control are provided in the book by Lewis et al. (2014). ○

Remark 3.5 (Observer-based Implementation)**.** The static diffusive couplings (3.7) are based on the relative states of neighboring agents. An important generalization are distributed control laws, which require only absolute or relative output information instead of state information. There is a vast body of literature on this topic, including the work of Scardovi and Sepulchre (2009); Li et al. (2010); Zhang et al. (2011); Ugrinovskii (2011b); Xi et al. (2012), where various observer-based implementations are presented and discussed, and the book by Lewis et al. (2014). ○

3.2 Heterogeneous Linear Networks

A major challenge in the field of distributed and cooperative control is the analysis and controller design for heterogeneous groups of agents and agents with complex dynamics. A multi-agent system is called heterogeneous if the individual agents in the group have non-identical dynamics. The practical relevance of heterogeneous multi-agent systems is apparent: In reality, different agents in a group, such as mobile robots, will never be perfectly identical due to different parameters such as non-identical masses, damping coefficients, operating conditions and other reasons. Hence it is an important task to study the effects of heterogeneity in the agent dynamics on the behavior of the multi-agent system. An understanding of such effects is essential for the development of suitable distributed control techniques guaranteeing a desired cooperative behavior and performance of the network. The importance of the study of heterogeneous multi-agent systems was already pointed out by Ren et al. (2005a). There have been significant advances in this direction since then. Nevertheless, there are still many questions to be answered. As an example, Oh et al. (2015) still identify heterogeneous agent dynamics as an important future research direction in the area of formation control.

In this section, we focus on output synchronization problems in heterogeneous networks of non-identical linear agents in order to give insight into the effects of heterogeneous agent dynamics. Section 3.2.1 presents the problem formulation. In Section 3.2.2, the internal model principle for synchronization representing a necessary structural condition for synchronizability is reviewed. Then, in Section 3.2.3, the robustness of synchronization in heterogeneous networks subject to parameter perturbations is studied. Finally, Section 3.2.4 gives a brief overview on distributed dynamic control laws which were developed for heterogeneous linear multi-agent systems.

In parts, this section is based on the publications Seyboth et al. (2012a, 2015).

3.2.1 Problem Formulation

In the following, the agent models, the information structure, and the group objective of the multi-agent systems under consideration are introduced.

Agent Models We consider multi-agent systems consisting of non-identical linear dynamical systems. The index set of the agents is $\mathcal{N} = \{1, ..., N\}$, where $N > 1$ is the number of agents in the group. The dynamics of the agents are described by the linear state-space models

$$\dot{x}_k = A_k x_k + B_k u_k \tag{3.27a}$$

$$y_k = C_k x_k \tag{3.27b}$$

with states $x_k(t) \in \mathbb{R}^{n_k^x}$, inputs $u_k(t) \in \mathbb{R}^{n_k^u}$, outputs $y_k(t) \in \mathbb{R}^{n_y}$, and matrices A_k, B_k, C_k of appropriate dimensions, for all $k \in \mathcal{N}$. The sum of all agents' state dimensions is denoted by $\hat{n}_x = \sum_{k=1}^{N} n_k^x$.

Information Structure The same information structure as previously in Section 3.1 is assumed, i.e., the communication topology is described by a directed graph $\mathcal{G} = (\mathcal{V}, \mathcal{E}, A_{\mathcal{G}})$.

Group Objective The group objective is output synchronization, formally defined as follows.

Definition 3.4 (Output Synchronization)**.** The agents (3.27) are said to achieve *output synchronization*, if for all initial conditions $x_k(0)$, $k \in \mathcal{N}$, it holds that for all pairs $j, k \in \mathcal{N}$,

$$\lim_{t\to\infty} (y_k(t) - y_j(t)) = \mathbf{0}. \tag{3.28}$$

Moreover, the agents (3.27) are said to achieve *nontrivial* output synchronization, if the subspace $\{x \in \mathbb{R}^{\hat{n}_x} : y_1 = y_2 = \cdots = y_N = \mathbf{0}\} \subset \mathbb{R}^{\hat{n}_x}$ is not asymptotically stable. Every trajectory $s : \mathbb{R}_0^+ \to \mathbb{R}^{n_y}$ satisfying $\lim_{t\to\infty}(y_k(t) - s(t)) = \mathbf{0}$ for all $k \in \mathcal{N}$ is called a *synchronous output trajectory*.

Problem 3.2 (Output Synchronization Problem)**.** Consider the multi-agent system consisting of $N > 1$ agents (3.27) and an underlying graph $\mathcal{G}$. Find a distributed control law, such that the group achieves nontrivial output synchronization.

Analogously to the consensus subspace $\mathcal{C}$ in (2.4) and the synchronous subspace $\mathcal{S}_x$ in (3.5), we define the output synchronous subspace $\mathcal{S}_y \subset \mathbb{R}^{\hat{n}_x}$ as

$$\mathcal{S}_y = \{x \in \mathbb{R}^{\hat{n}_x} : y_1 = y_2 = \cdots = y_N\} \tag{3.29}$$

The output synchronization problem amounts to asymptotic stabilization of $\mathcal{S}_y$, or more precisely, an invariant subspace contained in $\mathcal{S}_y$, by a distributed control law.

Remark 3.6 (Homogenization of Heterogeneous Multi-Agent Systems)**.** For special classes of dynamical agents, the heterogeneous output synchronization problem may be cast as a homogeneous output synchronization problem by homogenization of the agent dynamics. Several homogenization techniques were reported for this purpose, including local pre-compensation of right-invertible agents (Yang et al. (2014)), decoupling and pole assignment of multi-variable systems (Khodaverdian and Adamy (2014)), homogenization of second-order nonlinear systems (Zhu and Chen (2014)), and homogenization via feedback linearization and asymptotic model matching (Seyboth and Allgöwer (2014)). ○

3.2.2 Internal Model Principle for Synchronization

For homogeneous linear multi-agent systems, a necessary and sufficient condition for state synchronization is presented in Section 3.1. It turned out that for groups of identical stabilizable agents, there exist static diffusive couplings guaranteeing state synchronization if and only if the underlying graph $\mathcal{G}$ contains a spanning tree. For heterogeneous linear multi-agent systems, the situation is more intricate. It was revealed by Wieland and Allgöwer (2009) that the dynamics of the agents in the group must satisfy an internal model requirement in order to be synchronizable. This result became well-known as the internal model principle for synchronization.

Theorem 3.11 (Internal Model Principle for Synchronization (Wieland et al., 2011b, Thm. 3)). *Consider $N > 1$ linear state-space models* (3.27) *coupled through general dynamic controllers*

$$\dot{z}_k = D_k z_k + E_k y_k + F_k v_k \tag{3.30a}$$
$$u_k = G_k z_k + M_k y_k + O_k v_k \tag{3.30b}$$
$$\zeta_k = P_k z_k + Q y_k \tag{3.30c}$$
$$v_k = \textstyle\sum_{j=1}^{N} a_{kj}(\zeta_j - \zeta_k) \tag{3.30d}$$

with controller state $z_k(t) \in \mathbb{R}^{n_k^z}$, virtual output $\zeta_k(t) \in \mathbb{R}^{n_\zeta}$, relative virtual output $v_k(t) \in \mathbb{R}^{n_\zeta}$, and system input $u_k(t) \in \mathbb{R}^{n_k^u}$ for all $k \in \mathcal{N}$. Assume that (A_k, C_k) is detectable for all $k \in \mathcal{N}$ and the closed loop has no asymptotically stable equilibrium set on which $y_k(t) = \mathbf{0}$ for all $k \in \mathcal{N}$.

If $(y_k(t) - y_j(t)) \to \mathbf{0}$ and $(\zeta_k(t) - \zeta_j(t)) \to \mathbf{0}$ for all pairs $j, k \in \mathcal{N}$ exponentially as $t \to \infty$, then there exist an integer $m > 0$, matrices $S \in \mathbb{R}^{m \times m}$ and $R \in \mathbb{R}^{n_y \times m}$, where $\sigma(S) \subset \mathbb{C}^0 \cup \mathbb{C}^+$ and (S, R) is observable, and matrices $\Pi_k \in \mathbb{R}^{n_k^x \times m}$, $\Gamma_k \in \mathbb{R}^{n_k^u \times m}$ for all $k \in \mathcal{N}$ such that

$$A_k \Pi_k + B_k \Gamma_k = \Pi_k S \tag{3.31a}$$
$$C_k \Pi_k = R \tag{3.31b}$$

for all $k \in \mathcal{N}$. Furthermore, there exists $w_0 \in \mathbb{R}^m$ such that

$$\lim_{t \to \infty} \| y_k(t) - R e^{St} w_0 \| = \mathbf{0}, \tag{3.32}$$

for all $k \in \mathcal{N}$.

In order to get more insight into this result, we focus on static diffusive couplings instead of dynamic diffusive couplings in the following. This situation occurs when we do not distinguish between agent and controller states and consider general networks of non-identical diffusively coupled subsystems. It also occurs when we restrict ourselves to static distributed control laws for the agents (3.27). The remainder of this section is based on Seyboth et al. (2012a, 2015). When considering static diffusive couplings instead of (3.30), then the internal model principle for synchronization takes the following form.

Theorem 3.12. *Consider $N > 1$ linear state-space models* (3.27) *coupled through static controllers*

$$u_k = K_k \sum_{j=1}^{N} a_{kj}(y_j - y_k) \tag{3.33}$$

with coupling gain matrices $K_k \in \mathbb{R}^{n_k^u \times n_y}$, $k \in \mathcal{N}$. Assume that (A_k, C_k) is detectable for all $k \in \mathcal{N}$ and the closed loop has no asymptotically stable equilibrium set on which $y_k(t) = \mathbf{0}$ for all $k \in \mathcal{N}$.

If $(y_k(t) - y_j(t)) \to \mathbf{0}$ *for all pairs* $j, k \in \mathcal{N}$ *exponentially as* $t \to \infty$*, then there exist an integer* $m > 0$*, matrices* $S \in \mathbb{R}^{m \times m}$ *and* $R \in \mathbb{R}^{n_y \times m}$*, where* $\sigma(S) \subset \mathbb{C}^0 \cup \mathbb{C}^+$ *and* (S, R) *is observable, and matrices* $\Pi_k \in \mathbb{R}^{n_k^x \times m}$ *with full column rank such that*

$$A_k \Pi_k = \Pi_k S \tag{3.34a}$$

$$C_k \Pi_k = R \tag{3.34b}$$

for all $k \in \mathcal{N}$*. Furthermore, there exists* $w_0 \in \mathbb{R}^m$ *such that* (3.32) *holds for all* $k \in \mathcal{N}$*.*

Proof. See Appendix B.3. □

Corollary 3.13. *Under the conditions of Theorem 3.12, every eigenvalue of* S *is an eigenvalue of* A_k*, i.e.,*

$$\sigma(S) \subseteq \sigma(A_k), \quad \textit{for all } k \in \mathcal{N}.$$

In other words, the eigenvalues of S *are a subset of the largest common subset of all agents' spectra,* $\sigma(S) \subseteq \bigcap_{k=1}^{N} \sigma(A_k)$*.*

Proof. The statement follows from (3.34a), the fact that matrices Π_k have full column rank, and Theorem A.12. □

It is possible to check in a systematic way whether the necessary condition in Theorem 3.12 is fulfilled. The possible spectra of S can be listed according to the conditions in Corollary 3.13 since the candidates for matrix S can be chosen in Jordan canonical form.

Lemma 3.14. *Under the conditions of Theorem 3.12, the matrix* S *can be found in Jordan canonical form without loss of generality.*

Proof. Suppose that $\tilde{\Pi}_k$, $\tilde{S}$, $\tilde{R}$ satisfy (3.34) for all $k \in \mathcal{N}$ and $\tilde{S}$ does not have Jordan canonical form. Then, there exists a similarity matrix $T \in \mathbb{R}^{m \times m}$ such that $S = T^{-1}\tilde{S}T$ has Jordan canonical form, and we obtain $A_k \tilde{\Pi}_k T = \tilde{\Pi}_k T S$, $C_k \tilde{\Pi}_k T = \tilde{R} T$. This is equivalent to (3.34) with new variables $\Pi_k = \tilde{\Pi}_k T$, $S = T^{-1}\tilde{S}T$, and $R = \tilde{R}T$. □

The result in Theorem 3.12 is called internal model principle for synchronization for the following reason.

Lemma 3.15. *Under the conditions of Theorem 3.12, there exist matrices* $T_k \in \mathbb{R}^{n_k^x \times n_k^x}$ *for all* $k \in \mathcal{N}$ *such that transformation* $x_k = T_k \tilde{x}_k$ *for* (3.27) *yields* $\dot{\tilde{x}}_k = \tilde{A}_k \tilde{x}_k + \tilde{B}_k u_k$*,* $y_k = \tilde{C}_k \tilde{x}_k$ *with*

$$\tilde{A}_k = T_k^{-1} A_k T_k = \begin{bmatrix} S & \star \\ 0 & \star \end{bmatrix} \qquad \tilde{B}_k = T_k^{-1} B_k, \qquad \tilde{C}_k = C_k T_k = \begin{bmatrix} R & \star \end{bmatrix}.$$

Proof. The statement follows from Theorem A.7 and A_k-invariance of $\mathcal{S}_{y,k}^* = \operatorname{im}(\Pi_k)$. □

Concluding, the internal model principle for synchronization in heterogeneous networks of linear systems constitutes a necessary structural condition for synchronizability. Heterogeneous agents can only achieve nontrivial output synchronization if their dynamics embed a common internal model (S, R) as shown in Lemma 3.15. The synchronous output trajectory is generated by this internal model, cf., (3.32). A different formulation of the internal model principle for synchronization is provided by Lunze (2012). Moreover, the importance of the detectability assumption is discussed by Grip et al. (2013).

Remark 3.7 (Necessary and Sufficient Conditions for Synchronizability via Static Couplings). Necessary conditions for the existence of coupling gain matrices K_k, $k \in \mathcal{N}$, such that the network (3.27), (3.33) achieves nontrivial output synchronization are stabilizability of (A_k, B_k), detectability of (A_k, C_k), and connectedness of $\mathcal{G}$. Moreover, as stated in Theorem 3.12, a necessary condition is that there exist Π_k, S, R solving (3.34). All these conditions are still not sufficient for the existence of suitable K_k, $k \in \mathcal{N}$ since we are faced with a static output feedback problem (cf., Remark 3.1) for the heterogeneous network (3.27), (3.33) can be written as

$$\dot{x} = \left(\mathcal{A} - \mathcal{B}\mathcal{K}(L_{\mathcal{G}} \otimes I_{n_y})\mathcal{C}\right) x,$$
$$y = \mathcal{C}x$$

with block diagonal matrices $\mathcal{A} = \text{diag}(A_1, ..., A_N)$, $\mathcal{B} = \text{diag}(B_1, ..., B_N)$, $\mathcal{C} = \text{diag}(C_1, ..., C_N)$, $\mathcal{K} = \text{diag}(K_1, ..., K_N)$. Unfortunately, a state transformation as in the proof of Theorem 3.1 which translates this structured output feedback problem into a simultaneous stabilization problem for $N - 1$ subsystems is not applicable to heterogeneous networks. ○

Remark 3.8 (Necessary and Sufficient Conditions for Synchronizability via Dynamic Couplings). The necessary conditions stated in Theorem 3.11, along with the basic assumptions that all pairs (A_k, B_k) are stabilizable and $\mathcal{G}$ is connected, are also sufficient for the existence of *dynamic* diffusive couplings of the form (3.30) such that output synchronization is achieved by (3.27), (3.30). This constitutes the main result of Wieland et al. (2011b). The proof is constructive and relies on the output regulation theory. Each agent is equipped with a copy of the internal model (S, R). These virtual models are coupled over the network and achieve state synchronization. Local Luenberger observers and static output regulation controllers guarantee that the output of each agent tracks the output of its virtual model. Such regulators are guaranteed to exist under the given assumptions, cf., Theorem A.20. The proposed dynamic diffusive couplings are

$$\begin{aligned} \dot{\zeta}_k &= S\zeta_k + K \textstyle\sum_{j=1}^{N} a_{kj}(\zeta_j - \zeta_k) \\ \dot{\hat{x}}_k &= A_k\hat{x}_k + Bu_k + L_k(y_k - \hat{y}_k) \\ u_k &= -F_k(\hat{x}_k - \Pi_k\zeta_k) + \Gamma_k\zeta_k, \end{aligned}$$

where K, F_k, H_k are chosen such that $S - \lambda_k K$, $k \in \mathcal{N}\backslash\{1\}$, $A_k - B_kF_k$, $A_k - L_kC_k$, $k \in \mathcal{N}$ are Hurwitz and S, R, Π_k, Γ_k solve the regulator equations (3.31). This distributed control law solves the Output Synchronization Problem 3.2. ○

Remark 3.9 (Connection to the Output Regulation Problem). The results in Theorem 3.11 and Remark 3.8 reveal the strong connection between output synchronization in heterogeneous networks and the classical output regulation problem, which is highlighted by Wieland (2010). In view of Theorem A.20, solvability of the output synchronization problem translates into the existence of an exosystem described by (S, R) such that a corresponding output regulation problem is solvable for each agent in the group. This insight is exploited in the construction of the dynamic diffusive couplings in Remark 3.8. ○

Remark 3.10 (Nonlinear Dynamical Agents). Formulations of the internal model principle for synchronization for heterogeneous networks of nonlinear dynamical networks are provided by Wieland et al. (2013); Isidori et al. (2014). ○

3.2.3 Robustness of Synchronization

This section is dedicated to the study of robustness of synchronization in heterogeneous linear networks, i.e., in diffusively coupled multi-agent systems with general high-order linear dynamics subject to parameter perturbations. These perturbations result in non-identical agent dynamics of the form (3.27) coupled through (3.33). This section is based on Seyboth et al. (2015).

The goal is to develop a deeper understanding of the effects of heterogeneity in the agent dynamics on the dynamic behavior of the diffusively coupled multi-agent system and its implications for distributed control design. We analyze the dynamic behavior of selected heterogeneous linear multi-agent systems. For each network, we discuss the implications of the internal model principle for synchronization, highlight the importance of the network topology, and assess the robustness of synchronization with respect to parameter uncertainties in the agent dynamics. Firstly, we consider a network of non-identical double-integrators, which achieves output synchronization if the output is position only. Afterwards, we study state synchronization in the same network. The structural requirements for synchronization are not met in this case, but it turns out that the synchronization error remains small, depending on the graph topology and the heterogeneity in the network. Secondly, we consider a network of harmonic oscillators with perturbed frequencies. We show that the internal model condition is not satisfied and that static diffusive couplings have a stabilizing effect in such networks. In particular, the network is rendered asymptotically stable if and only if there are oscillators with different frequencies in a certain region of the network. Moreover, we present two application examples: a clock synchronization problem and a motion coordination problem for mobile robots. The latter shows that heterogeneity may significantly impair the performance of cooperative control strategies designed for identical agents.

Double-Integrators with Partial Output

We focus on a network of non-identical double-integrator agents described by

$$\dot{x}_k = \begin{bmatrix} 0 & 1+\delta_k \\ 0 & 0 \end{bmatrix} x_k + \begin{bmatrix} 1 & 0 \\ 0 & 1 \end{bmatrix} u_k, \tag{3.35a}$$

$$y_k = \begin{bmatrix} 1 & 0 \end{bmatrix} x_k, \tag{3.35b}$$

with parameter $\delta_k \in \mathbb{R}$, $k \in \mathcal{N}$. The couplings are given by

$$u_k = \begin{bmatrix} 1 \\ \alpha \end{bmatrix} \sum_{j=1}^{N} a_{kj}(y_j - y_k) \tag{3.36}$$

for some $\alpha > 0$. The objective is synchronization of the outputs $y_k(t)$ to a common ramp function. Such networks appear, for instance, in distributed clock synchronization as discussed in Example 3.2. It is easy to check that the internal model principle for synchronization (Theorem 3.12) is satisfied with

$$\Pi_k = \begin{bmatrix} 1 & 0 \\ 0 & (1+\delta_k)^{-1} \end{bmatrix}, \qquad S = \begin{bmatrix} 1 & 0 \\ 0 & 0 \end{bmatrix}, \qquad R = \begin{bmatrix} 1 & 0 \end{bmatrix}, \tag{3.37}$$

where it is assumed that $\delta_k \neq -1$, for all $k \in \mathcal{N}$. As stated next, the network (3.35), (3.36) indeed reaches output synchronization to a ramp function generated by (S, R).

Theorem 3.16. *Consider a network of $N > 1$ double-integrator agents* (3.35) *interconnected by static diffusive couplings* (3.36)*. Suppose that the graph $\mathcal{G}$ is undirected and connected and let $\alpha > 0$. Then, for all parameters $\delta_k > -1$, $k \in \mathcal{N}$, the network reaches output synchronization and the synchronous output trajectory is a ramp function, i.e., for all $j,k \in \mathcal{N}$, it holds that $y_k(t) - y_j(t) \to \mathbf{0}$ and $\frac{\mathrm{d}}{\mathrm{d}t} y_k(t) \to c$ as $t \to \infty$ for some constant $c \in \mathbb{R}$.*

Proof. See Appendix B.4. □

The network (3.35), (3.36) reaches non-trivial output synchronization robustly with respect to the parameter perturbations. Exact synchronization is achieved despite the heterogeneous agent dynamics. Fig. 3.5 shows simulation results for an exemplary network with five nodes.

Example 3.2 (Distributed Clock Synchronization)**.** It has been shown by Carli et al. (2011); Carli and Lovisari (2012) that a distributed synchronization protocol for a network of non-identical clocks can be derived based on a discrete-time model similar to (3.35), (3.36). Each node k in the network has a register $\tau_k(t)$ which periodically increments its value by one with period Δ_k. The period Δ_k is an unknown and perturbed value of some nominal Δ. Hence the time estimate of node k is given by $y_k(t) = \Delta\tau_k(t)$. In order to synchronize clocks, every node is equipped with a processor that communicates with neighboring nodes in the network and post-processes the register value $\tau_k(t)$. The evolution of the time estimate is described by the linear discrete-time system

$$x_k((h+1)T) = \begin{bmatrix} 1 & d_k T \\ 0 & 1 \end{bmatrix} x_k(hT) + u_k(hT)$$

and $y_k(hT) = [1 \;\; 0] x_k(hT)$, where $d_k = \Delta/\Delta_k$ and T is the communication period. The synchronization protocol proposed by Carli et al. (2011) is given by $u_k(hT) = [1 \;\; \alpha]^{\mathsf{T}} \sum_{j=1}^{N} a_{kj}(y_j(hT) - y_k(hT))$. For undirected connected graphs $\mathcal{G}$, necessary and sufficient conditions on α and the spectrum of the Laplacian $L_{\mathcal{G}}$ for synchronization of the nominal network are given by Carli et al. (2011). In the proof therein, it is assumed that $D = \mathrm{diag}(d_1, ..., d_N) = I_N$, i.e., the clocks are identical. Then, by continuity arguments, it is concluded that the network also synchronizes for sufficiently small perturbations $D \neq I_N$. Explicit bounds on the tolerable parameter variation are presented by Carli and Lovisari (2012). Suppose that the register $\tau_k(t)$ is a continuous ramp function $\tau_k(t) = d_k t$. Then, the limit $T \to 0$ corresponds to continuous communication between neighboring clocks and the dynamics of the network are described by (3.35), where $\delta_k = d_k - 1$, and the couplings are given by (3.36). A similar model was studied by Lovisari and Jönsson (2011). In this context, Theorem 3.16 shows that, in case of continuous communication, the control law (3.36) achieves synchronization for arbitrary $d_k > 0$, $k \in \mathcal{N}$. In discrete-time, with $T > 0$, synchronization is not reached for all $\alpha > 0$, $d_k > 0$. However, Theorem 3.16 suggests that the discrete-time heterogeneous network can always be synchronized by sufficiently fast sampling, i.e., sufficiently small T. This is confirmed by the results of Carli and Lovisari (2012). △

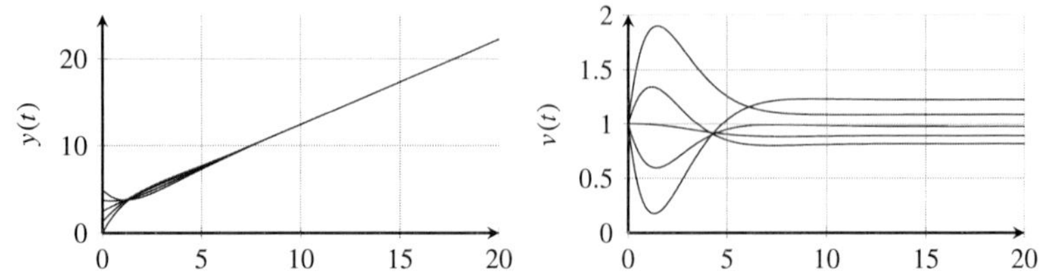

Figure 3.5: This simulation of a network (3.35), (3.36) with 5 nodes, $\alpha = 0.5$, and $d_1 = 0.9$, $d_2 = 1.2$, $d_3 = 1.1$, $d_4 = 1.0$, $d_5 = 0.8$ and undirected cycle $\mathcal{G}$ shows that the outputs $y_k(t)$ synchronize. The right plot shows the second components $v_k(t) \in \mathbb{R}$ of the state vectors $x_k = [s_k\ v_k]^\mathsf{T}$.

Double-Integrators with Full State Output

We focus again on synchronization in a network of non-identical double-integrator agents. In contrast to the previous section, we consider the full state output $y_k = x_k$, i.e.,

$$\dot{x}_k = \begin{bmatrix} 0 & 1+\delta_k \\ 0 & 0 \end{bmatrix} x_k + \begin{bmatrix} 1 & 0 \\ 0 & 1 \end{bmatrix} u_k, \tag{3.38a}$$

$$y_k = \begin{bmatrix} 1 & 0 \\ 0 & 1 \end{bmatrix} x_k, \tag{3.38b}$$

where $\delta_k \in \mathbb{R}$, $k \in \mathcal{N}$, and the state couplings are given by

$$u_k = \sum_{j=1}^{N} a_{kj}(x_j - x_k). \tag{3.39}$$

According to Corollary 3.13, a candidate matrix S (3.34a) has to fulfill $\sigma(S) \subseteq \sigma(A_k) = \{0,0\}$ for all $k \in \mathcal{N}$. Thus, there are three candidates in Jordan canonical form

$$S_1 = \begin{bmatrix} 0 & 0 \\ 0 & 0 \end{bmatrix}, \quad S_2 = \begin{bmatrix} 0 & 1 \\ 0 & 0 \end{bmatrix}, \quad S_3 = 0.$$

Since $C_k = I_2$, condition (3.34b) yields $\Pi_k = \Pi_j$ for all $j,k \in \mathcal{N}$. Therefore, a necessary condition for non-trivial state synchronization in the double-integrator network is that there exists a matrix Π with full column rank such that $A_k\Pi = \Pi S_i$ for some $i \in \{1,2,3\}$ and all $k \in \mathcal{N}$. In general, $\delta_k \neq -1$ and thus there is no solution for S_1. There is also no solution for S_2. The necessary condition in Theorem 3.12 is not fulfilled for any $S \in \mathbb{R}^{2\times 2}$. Note that there exist A_k-invariant subspaces on which the dynamics of all agents are identical, given by $\mathrm{im}(\Pi_k)$ with Π_k as in (3.37). However, (3.34a) and (3.34b) can not be satisfied at the same time for $C_k = I_2$ and S_1 or S_2. In other words, in the present network, the dynamics of the agents are compatible but the outputs do not match, which is a structural difference to the network of non-identical harmonic oscillators studied later.

Note that for S_3, the necessary condition is fulfilled for $\Pi = [1\ 0]^\mathsf{T}$. This is not surprising since S_3 is contained in A_k as the lower right entry. Hence, exact synchronization to a trajectory generated by a single-integrator may be possible. However, exact synchronization to a trajectory generated by a double-integrator model is impossible. The following theorem characterizes the dynamic behavior of the network (3.38), (3.39).

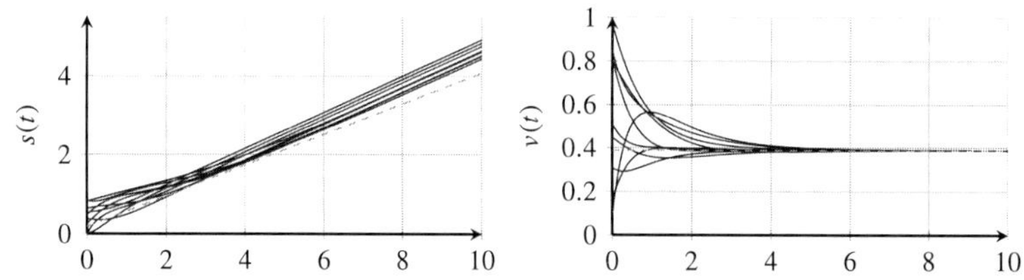

Figure 3.6: Simulation result for a network of 9 agents (3.38) with (3.39), $\mathcal{G}$ as in Fig. A.2, random non-identical parameters $\delta_k \in [-0.5, 0.5]$ with $\mathbf{1}^\mathsf{T}\delta = 0$, and random initial conditions. The dashed lines (– – –) indicate the asymptotic solution in the nominal case ($\delta = \mathbf{0}$). The second states $v_k(t)$ of all agents reach consensus (right). The first states $s_k(t)$ grow with constant and identical slope but with constant offsets $s_\perp$ according to (3.40) (left).

Theorem 3.17. *Consider a network of $N > 1$ double-integrator agents* (3.38) *interconnected by static diffusive couplings* (3.39). *Suppose that the directed graph $\mathcal{G}$ is connected. Furthermore, suppose that there exist $j, k \in \mathcal{N}$ such that $\delta_k \neq \delta_j$. Let $x_k = [s_k\ v_k]^\mathsf{T}$, $p^\mathsf{T} L_\mathcal{G} = \mathbf{0}^\mathsf{T}$, and $p^\mathsf{T}\mathbf{1} = 1$. Then, $v(t) \to \mathbf{1}p^\mathsf{T} v_0$ as $t \to \infty$ and the states s_k do not synchronize but asymptotically grow with constant and identical speed. In particular, $(s(t) - s_\perp) \to \operatorname{im}(\mathbf{1})$ and $\frac{\mathrm{d}}{\mathrm{d}t}s(t) \to \mathbf{1}(p^\mathsf{T} v_0 + c)$ as $t \to \infty$, where $c \in \mathbb{R}$ and the asymptotic disagreement $s_\perp \in \mathbb{R}^N$ with $\mathbf{1}^\mathsf{T} s_\perp = 0$ are given by*

$$\begin{bmatrix} s_\perp \\ c \end{bmatrix} = \begin{bmatrix} L_\mathcal{G} & \mathbf{1} \\ \mathbf{1}^\mathsf{T} & 0 \end{bmatrix}^{-1} \begin{bmatrix} \delta p^\mathsf{T} v_0 \\ 0 \end{bmatrix}. \tag{3.40}$$

Proof. See Appendix B.5. □

Theorem 3.17 shows that networks of double-integrators (3.38) with static diffusive couplings (3.39) have a certain robustness with respect to heterogeneity in the dynamics, in the sense that they synchronize approximately for small perturbations δ_k, $k \in \mathcal{N}$. The quantity $\|s_\perp\|$ can be seen as an asymptotic synchronization error since $\lim_{t\to\infty} \operatorname{dist}(s(t), \operatorname{im}(\mathbf{1})) = \|s_\perp\|$. According to (B.4), $s_\perp$ scales inversely with $L_\mathcal{G}$, i.e., for scaled $\tilde{L}_\mathcal{G} = \gamma L_\mathcal{G}$, $\gamma > 0$, we obtain $\tilde{s}_\perp = \gamma^{-1} s_\perp$. This shows that the underlying graph plays an important role: stronger couplings decrease the error proportionally. The velocities of the agents synchronize for arbitrary parameters δ_k. Both the final velocity and the asymptotic offsets between the agents can be computed explicitly according to (3.40), depending on the graph topology, parameters δ, and the initial states. A numerical example is shown in Fig. 3.6. In the context of coupled Kuramoto models, such a behavior (motion with common frequency and constant phase offsets) is also called phase locking, cf., Dörfler and Bullo (2012). The analysis above demonstrates that a heterogeneous network may fail to reach exact synchronization but can still reach synchronization approximately, even if the internal model principle for synchronization is not fulfilled.

Harmonic Oscillators

In the sequel, networks of non-identical harmonic oscillators are analyzed. In these networks, exact non-trivial synchronization is impossible. The structural difference to the networks considered so far is that there exists no solution to (3.34a), even if (3.34b) is disregarded. As an application example, it is shown that a certain multi-agent control algorithm is not robust with respect to parameter uncertainty. The dynamical agents are described by

$$\dot{x}_k = \begin{bmatrix} 0 & 1 \\ -(\omega + \delta_k)^2 & 0 \end{bmatrix} x_k + \begin{bmatrix} 0 \\ 1 \end{bmatrix} u_k, \tag{3.41a}$$

$$y_k = \begin{bmatrix} 0 & 1 \end{bmatrix} x_k \tag{3.41b}$$

and coupled through

$$u_k = \sum_{j=1}^{N} a_{kj}(y_j - y_k), \tag{3.42}$$

for $k \in \mathcal{N}$. The frequencies of the individual oscillators are perturbed by the parameters $\delta_k \in \mathbb{R}$ and deviate from the nominal frequency $\omega \in \mathbb{R}$. It is shown by Ren (2008) that the oscillator network reaches state synchronization when $\delta_k = 0$ for all $k \in \mathcal{N}$ and $\mathcal{G}$ is connected. Suppose instead there exist two agents $j, k \in \mathcal{N}$ such that $\delta_k \neq \delta_j$, i.e., not all oscillators have identical frequencies. Then the intersection of the agents' spectra $\bigcap_{k=1}^{N} \sigma(A_k)$ is empty and exact non-trivial synchronization is impossible since the necessary condition in Theorem 3.12 is not fulfilled, cf., Corollary 3.13. In geometric terms, there exist no A_k-invariant subspaces on which the dynamics of all agents are identical. Equation (3.34a) can not be satisfied, even if (3.34b) is disregarded. The following result characterizes the dynamic behavior of the network.

Theorem 3.18. *Consider a network of $N > 1$ harmonic oscillators* (3.41) *interconnected by static diffusive couplings* (3.42). *Suppose that the directed graph $\mathcal{G}$ is connected. Then, the network is asymptotically stable if and only if there exist $j, k \in \mathcal{N}$ within the iSCC of $\mathcal{G}$ such that $\delta_k \neq \delta_j$. Otherwise, the oscillators within the iSCC reach non-trivial output synchronization.*

Proof. See Appendix B.6. □

Theorem 3.18 shows that the network topology again plays a crucial role for the dynamic behavior of the oscillator network. The network is asymptotically stable, meaning that the origin $x_k = \mathbf{0}$ for all $k \in \mathcal{N}$ is asymptotically stable, if and only if there is a pair of oscillators inside the iSCC of the underlying graph, which do not have identical frequencies. If the graph is strongly connected, then all nodes belong to the iSCC and the network is stabilized whenever there exist any two oscillators with non-identical frequencies. Furthermore, Theorem 3.18 shows that (non-trivial) synchronization of harmonic oscillators via static diffusive couplings is not at all robust with respect to parametric uncertainties causing variations of the frequencies. It suffices to change the frequency of one single oscillator in the iSCC by an arbitrarily small δ_k in order to render the entire network asymptotically stable. Fig. 3.7 shows two numerical examples.

Example 3.3 (Distributed Motion Coordination)**.** Ren (2008) presents a motion coordination problem for a group of mobile robots as an application of distributed oscillator synchronization. The mobile robots are modeled as point-mass agents in the plane with force inputs and are equipped with distributed controllers, which coordinate their motion such that all agents move synchronously on identical elliptic paths. Here, we assume that the agents have non-identical and unknown masses $m_k > 0$, i.e., each agent is modeled as

$$\dot{s}_k = v_k, \qquad m_k \dot{v}_k = u_k, \tag{3.43}$$

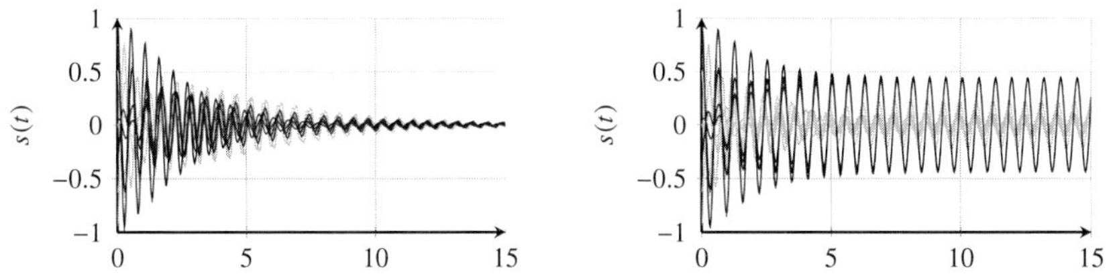

(a) Non-identical oscillators in the iSCC stabilize the network.

(b) Identical oscillators in the iSCC synchronize and excite the perturbed oscillators outside the iSCC.

Figure 3.7: Simulation results for a network of 9 oscillators with $\mathcal{G}$ as in Fig. A.2, nominal frequency $\omega = 10$, and random offsets $\delta_k \in [0,2]$ (a) for all $k = 1,...,9$ and (b) identical frequencies within the iSCC. The plots show the first states of the oscillators within (——) and outside (——) the iSCC of $\mathcal{G}$.

where $s_k, v_k \in \mathbb{R}^2$ are the position and velocity of agent k in the plane, $k \in \mathcal{N}$. The control law proposed by Ren (2008) is

$$u_k = -\alpha(s_k - c_{s,k}) - \sum_{j=1}^{N} a_{kj}(v_k - v_j), \tag{3.44}$$

where $c_{s,k} \in \mathbb{R}^2$ is a constant offset which defines the relative position of agent k in the formation and $\alpha > 0$. Note that Ren (2008) assumes that the network has a dedicated leader node which is the only root of a spanning tree of the graph $\mathcal{G}$. Here we consider general connected directed graphs, and leader-follower topologies are included as a special case. The constant offsets $c_{s,k}$, $k \in \mathcal{N}$, can be set to zero in the stability analysis since they represent a constant shift of the motion in the plane. The dynamics of agent k with its controller are

$$\dot{s}_k = v_k, \qquad \dot{v}_k = -\frac{\alpha}{m_k} s_k - \sum_{j=1}^{N} \frac{a_{kj}}{m_k}(v_k - v_j)$$

Let $L_{\mathcal{G}}$ be the Laplacian associated to the graph $\hat{\mathcal{G}}$ with weights $\hat{a}_{kj} = a_{kj}/m_k$, and $D = \operatorname{diag}(\alpha_1, ..., \alpha_N)$, $\alpha_k = \alpha/m_k$. Then, the dynamics of the network (3.43), (3.44) match (3.41), (3.42). Consequently, the network of N point-mass agents (3.43), (3.44) is asymptotically stable if and only if there exists a pair $j, k \in \mathcal{N}$ in the iSCC of $\mathcal{G}$ such that $m_k \neq m_j$. This fact shows that the motion coordination algorithm proposed by Ren (2008) is not robust with respect to non-identical and possibly uncertain masses of the robots. In this example, the stabilizing effect due to the parameter uncertainties in the network is an issue. Numerical simulations are shown in Fig. 3.8. △

3.2.4 Dynamic Diffusive Couplings

In general, it is a hard task to find static diffusive couplings based on output differences (3.33), which ensure output synchronization. Already for homogeneous multi-agent systems, this results in a static output feedback problem (cf., Remark 3.1). For heterogeneous multi-agent systems, we

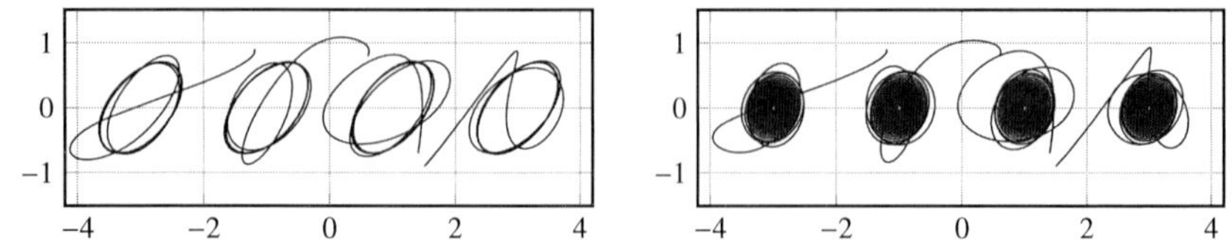

(a) Control law (3.44), identical $m_k = 1$.

(b) Control law (3.44), non-identical $m_k \in [0.75\ 1.25]$.

Figure 3.8: Motion coordination of $N = 4$ mobile robots from Example 3.3. The simulation with $\alpha = 1.8$, masses $m_1 = 0.75$, $m_2 = 0.92$, $m_3 = 1.08$, $m_4 = 1.25$, and directed cycle graph $\mathcal{G}$ demonstrates that the agents positions s_k indeed asymptotically converge to fixed points in the plane, see (b). For comparison, the nominal case with $m_k = 1$ for all $k \in \mathcal{N}$ is shown in (a).

are faced with a structured static output feedback problem for the entire network (cf., Remark 3.8). This motivates dynamic instead of static distributed control laws, which provide much more flexibility and lend themselves to systematic synthesis procedures for large classes of homogeneous and heterogeneous multi-agent systems. As an example, we mentioned observer-based distributed control laws for homogeneous networks (cf., Remark 3.5). In fact, it is a key message of Wieland (2010) that dynamic couplings are favorable and allow for consensus and synchronization in networks with increased system and topological complexity. In the following, we provide a brief overview on selected distributed dynamic control laws developed for synchronization problems in heterogeneous linear multi-agent systems.

For heterogeneous multi-agent systems, hierarchical control architectures have proved useful. A typical approach is to decompose the heterogeneous output synchronization problem into two simpler subproblems: a virtual homogeneous synchronization problem on the network level and local reference tracking problems on the agent level. The dynamic diffusive couplings proposed by Wieland et al. (2011b) (cf., Remark 3.8) realize such a hierarchical distributed control architecture. Identical virtual models are synchronized over the network and provide coordinated local reference signals to the individual agents. Local output regulation controllers with Luenberger observers achieve asymptotic tracking of these reference signals, which ultimately solves the output synchronization problem. Kim et al. (2011) consider minimum-phase SISO agents subject to parametric uncertainties. Motivated by the internal model principle for synchronization, the proposed distributed dynamic control law consists of type generators, dirty derivative observers, and robust regulators. Absolute and relative output measurements are required, but not the exchange of controller states. Listmann (2012) splits the problem into three subproblems: homogeneous synchronization of reference generators, decentralized estimation of the agent states, and output regulation. The control law relies on relative state information and requires an exchange of controller states among neighbors. It should be noted that distributed estimation and observer design is an active area of research in its own right, cf., Carli et al. (2008); Zelazo and Mesbahi (2011); Ugrinovskii (2011b); Wu et al. (2014); and references therein. Such distributed estimation algorithms may form the basis for distributed cooperative control laws. As an example, it is a challenging task to construct distributed control laws which are solely based on relative output measurements with respect to neighboring agents. The output synchronization problem under such a limitation is tackled, e.g., by Grip et al. (2012) and Wu and Allgöwer (2012). Furthermore, distributed dynamic control laws guaranteeing robust synchronization in groups of identical linear agents with non-identical additive perturbations are constructed by Trentelman et al. (2013).

A powerful framework for distributed cooperative control design for heterogeneous networks is the so-called distributed output regulation framework, which is a major subject of Chapter 4 in the present thesis. It has recently been developed based on the output regulation theory and allows to formulate and solve a variety of cooperative control tasks including reference tracking and disturbance rejection for groups of non-identical linear dynamical agents in a systematic fashion.

3.3 Networks of Linear Parameter-Varying Systems

Linear parameter-varying (LPV) systems are a system class which appears in many engineering applications. They have been identified as a suitable model class for gain-scheduling controller design by Shamma and Athans (1991). Based on the notion of quadratic stability presented by Becker et al. (1993), a design procedure for self-scheduled $\mathcal{H}_\infty$ controllers was proposed by Apkarian et al. (1995). Such methods have been applied successfully in a wide range of applications, for example for the control of airplanes (Ganguli et al. (2002); Ameho and Prempain (2011)), wind turbines (Lescher et al. (2005); Heß and Seyboth (2010)), autonomous underwater vehicles (AUVs) (Roche et al. (2009)), and spacecrafts (Nagashio et al. (2010)). The book by Briat (2015) provides a good introduction to analysis and control of LPV systems.

Multi-agent systems consisting of LPV agents are appealing since this model class is well suited for the description of dynamical agents such as airplanes, AUVs, or satellites. Moreover, LPV systems provide a nice possibility to model networks of non-identical subsystems by allowing the parameters of all agents to be different. In general, such networks can represent groups of identical physical agents in a heterogeneous environment described by the parameters, e.g., varying wind speeds in airplane formations, varying masses in groups of AUVs, or parametric uncertainties in the actuators of mobile robots.

LPV systems have rarely been studied in the context of cooperative control. A related synchronization problem was investigated by Ugrinovskii (2011a), where a group of identical LPV agents is synchronized to the trajectory of a common external reference system under noisy measurements via dissipativity techniques. These results build on the $\mathcal{H}_\infty$ consensus approach of Ugrinovskii (2011b) and are extended and applied to the problem of synchronization for networks of uncertain single-input bilinear systems in Ugrinovskii (2014). Moreover, Eichler et al. (2014) and Hoffmann et al. (2014) develop robust control methods for decomposable LPV systems suitable for the design of distributed controllers. The developed framework relies on a combination of the notion of decomposable systems (Massioni and Verhaegen (2009)) and the full block S-procedure for LPV controller design (Scherer (2001)).

In the present section, we formulate and solve output synchronization problems for networks of LPV systems. The parameters are assumed to be time-varying and available as real-time measurements. The LPV systems may involve parameters which influence all agents (global parameters) and parameters which influence only individual agents (local parameters). The local parameters can be different for each agent, which causes heterogeneity in the multi-agent system. As an example, in a group of airplanes a common altitude and individual masses can be captured as a global and a local parameter, respectively. After introducing the state and output synchronization problems for networks of LPV systems, we analyze the synchronous subspace of the state-space and its invariance properties, both for homogeneous and heterogeneous groups. As main contribution, we provide LMI-based synthesis methods for suitable distributed gain-scheduled control laws guaranteeing synchronization for polytopic LPV systems. The results in this section are based on Seyboth et al. (2012b,c).

3.3.1 Problem Formulation

In the following, the agent models, the information structure, and the group objective of the multi-agent systems under consideration are introduced.

Agent Models We consider multi-agent systems consisting of agents with LPV dynamics. A brief introduction to LPV systems is provided in Section A.1.6. All agents have an identical structure but may depend on time-varying non-identical local parameters as well as on time-varying global parameters. Given a compact subset $\mathcal{P} \subset \mathbb{R}^{n_\rho}$, the parameter variation set $\mathcal{F}_\mathcal{P}$ is the set of all piece-wise continuous functions $\rho(\cdot)$ mapping $\mathbb{R}_0^+$ (time) into $\mathcal{P}$ with a finite number of discontinuities in any interval, following the notation of Becker et al. (1993). Here, we distinguish between global parameters $\rho^g(t)$ and local parameters $\rho_k^\ell(t)$ with

$$\begin{aligned}\rho^g &\in \mathcal{F}_{\mathcal{P}^g} = \{\rho : \mathbb{R}_0^+ \to \mathcal{P}^g, \rho \text{ piece-wise continuous}\},\\ \rho_k^\ell &\in \mathcal{F}_{\mathcal{P}^\ell} = \{\rho : \mathbb{R}_0^+ \to \mathcal{P}^\ell, \rho \text{ piece-wise continuous}\},\end{aligned}$$

for all $k \in \mathcal{N}$, where $\mathcal{P}^g \subset \mathbb{R}^{n_{\rho^g}}$ and $\mathcal{P}^\ell \subset \mathbb{R}^{n_{\rho^\ell}}$ are compact sets. Note that the superscript g and ℓ stand for *global* and *local*, respectively. For notational convenience, let

$$\rho_k(t) = \begin{bmatrix}\rho^g(t)\\ \rho_k^\ell(t)\end{bmatrix}$$

be the stack vector of all parameters affecting agent k. Note that $\rho_k \in \mathcal{F}_\mathcal{P}$ with $\mathcal{P} = \mathcal{P}^g \times \mathcal{P}^\ell \subset \mathbb{R}^{n_\rho}$, where $n_\rho = n_{\rho^g} + n_{\rho^\ell}$. Each agent $k \in \mathcal{N}$ is described by

$$\begin{aligned}\dot{x}_k &= A(\rho_k(t))x_k + B(\rho_k(t))u_k &\text{(3.45a)}\\ y_k &= C(\rho_k(t))x_k, &\text{(3.45b)}\end{aligned}$$

where $x_k(t) \in \mathbb{R}^{n_x}$ is the state, $y_k(t) \in \mathbb{R}^{n_y}$ is the output and $u_k(t) \in \mathbb{R}^{n_u}$ the input of agent k. The system matrices $A(\cdot)$, $B(\cdot)$, $C(\cdot)$ depend continuously on the parameters. In particular, $A : \mathbb{R}^{n_\rho} \to \mathbb{R}^{n_x \times n_x}$, $B : \mathbb{R}^{n_\rho} \to \mathbb{R}^{n_x \times n_u}$, $C : \mathbb{R}^{n_\rho} \to \mathbb{R}^{n_y \times n_x}$ are continuous, bounded functions. Note that $A(\cdot)$, $B(\cdot)$, $C(\cdot)$ are identical for all agents but the time-varying parameters $\rho_k(t)$ are in general not identical due to the different trajectories of the local parameters $\rho_k^\ell(t)$ for each agent. In the sequel, it is assumed that the matrices B and C in (3.45) are parameter-independent, i.e.,

$$\begin{aligned}\dot{x}_k &= A(\rho_k(t))x_k + Bu_k &\text{(3.46a)}\\ y_k &= Cx_k. &\text{(3.46b)}\end{aligned}$$

This can be achieved by pre- and/or post-filtering of the input and/or output as noted by Apkarian et al. (1995). In the following, we focus on polytopic LPV systems according to Definition A.7. In this case, $A(\cdot)$ depends affinely on $\rho_k(t)$, i.e., $A(\rho_k(t)) = A_0 + \rho_k^{(1)}(t)A_1 + \cdots + \rho_k^{(n_\rho)}(t)A_{n_\rho}$. Moreover, the sets $\mathcal{P}^g$ and $\mathcal{P}^\ell$ are convex polytopes defined by

$$\mathcal{P}^g = \operatorname{conv}\{\omega_1^g, \ldots, \omega_{p^g}^g\} \quad \text{and} \quad \mathcal{P}^\ell = \operatorname{conv}\{\omega_1^\ell, \ldots, \omega_{p^\ell}^\ell\},$$

respectively. The parameter vector $\rho_k(t)$ is hence contained in $\mathcal{P} = \mathcal{P}^g \times \mathcal{P}^\ell$, which is itself a convex polytope of the form $\mathcal{P} = \operatorname{conv}\{\omega_1, \ldots, \omega_p\}$ with $p = p^g p^\ell$ vertices. For a polytopic LPV system, the system matrix can be written as convex combination of corresponding vertex matrices,

$$A(\rho_k(t)) = A\left(\textstyle\sum_{i=1}^p \alpha_i(t)\omega_i\right) = \textstyle\sum_{i=1}^p \alpha_i(t)A(\omega_i) \in \operatorname{conv}\{A(\omega_1), \ldots, A(\omega_p)\}, \qquad (3.47)$$

where $\alpha_i(t) \geq 0$, $\sum_{i=1}^{p} \alpha_i(t) = 1$ for all t. Polytopic LPV systems are a subclass of LPV systems which is suitable for a wide range of practical applications, cf., Apkarian et al. (1995). We impose the following additional assumption (without loss of generality in view of Lemma A.24).

Assumption 3.1. *The matrices $A_1, A_2, \ldots, A_{n_\rho}$ are linearly independent.*

Information Structure The same information structure as in Section 3.1.1 is assumed, i.e., the communication topology is described by a directed graph $\mathcal{G} = (\mathcal{V}, \mathcal{E}, A_{\mathcal{G}})$. Additionally, the time-varying parameters are assumed to be available as real-time measurements which means that they can be incorporated in the control law. Each agent k has access to the global parameter $\rho^g(t)$ as well as its own local parameter $\rho_k^\ell(t)$ at time t.

Group Objective The group objectives under consideration are state synchronization as well as output synchronization for groups of N dynamical agents (3.46), analogously to Definition 3.2 and Definition 3.3, respectively. For that purpose, both robust and gain-scheduled static and dynamic distributed control laws are constructed.

In Section 3.3.3, we focus on the state synchronization problem and resort to static diffusive couplings as distributed control laws. We distinguish between gain-scheduled and robust static diffusive couplings given by

$$\text{gain-scheduled:} \quad u_k = K(\rho_k(t)) \textstyle\sum_{k=1}^{N} a_{kj}(x_j - x_k) \tag{3.48}$$

$$\text{robust:} \quad u_k = K \textstyle\sum_{k=1}^{N} a_{kj}(x_j - x_k). \tag{3.49}$$

In (3.48), $K(\cdot)$ is a continuous bounded function $K : \mathbb{R}^{n_\rho} \to \mathbb{R}^{n_u \times n_x}$. In case of polytopic agents (3.46), K is assumed to be affine as well. In (3.49), $K \in \mathbb{R}^{n_u \times n_x}$ is a constant matrix.

In Section 3.3.4, we focus on the output synchronization problem and resort to gain-scheduled dynamic diffusive couplings of the general form

$$\dot{z}_k = D(\rho_k(t))z_k + E(\rho_k(t))y_k + F(\rho_k(t))v_k \tag{3.50a}$$

$$u_k = G(\rho_k(t))z_k + M(\rho_k(t))y_k + O(\rho_k(t))v_k \tag{3.50b}$$

$$\zeta_k = P(\rho_k(t))z_k + Q(\rho_k(t))y_k \tag{3.50c}$$

$$v_k = \textstyle\sum_{j=1}^{N} a_{kj}(\zeta_j - \zeta_k) \tag{3.50d}$$

with $z_k(t) \in \mathbb{R}^{n_z}$, $\zeta_k(t) \in \mathbb{R}^{n_\zeta}$, $v_k(t) \in \mathbb{R}^{n_\zeta}$, $u_k(t) \in \mathbb{R}^{n_u}$, and continuously parameter-dependent bounded matrices of appropriate dimensions, for all $k \in \mathcal{N}$. The controller (3.50) of agent k has access to the output y_k and communicates over the network as sketched in Fig. 3.9.

For the developments in the sequel, it is convenient to define the projection matrices $P_{\mathcal{S}} = \frac{1}{N}\mathbf{1}\mathbf{1}^\top$ and $P_{\mathcal{A}} = I_N - \frac{1}{N}\mathbf{1}\mathbf{1}^\top$. The matrix $P_{\mathcal{S}} \otimes I_{n_x}$ defines a projection onto the synchronous subspace $\mathcal{S}_x$ and $P_{\mathcal{A}} \otimes I_{n_x}$ defines a projection onto the asynchronous subspace $\mathcal{A}_x$. Using these projections, the state $x \in \mathbb{R}^{Nn_x}$ can be decomposed into $x_{\mathcal{S}} = (P_{\mathcal{S}} \otimes I_{n_x})x \in \mathcal{S}_x$, $x_{\mathcal{A}} = (P_{\mathcal{A}} \otimes I_{n_x})x \in \mathcal{A}_x$. Note that $P_{\mathcal{S}} + P_{\mathcal{A}} = I_N$ and therefore $x = x_{\mathcal{S}} + x_{\mathcal{A}}$. Consequently, the subspace $\mathcal{S}_x$ defined in (3.5) can also be written as

$$\mathcal{S}_x = \ker(P_{\mathcal{A}} \otimes I_{n_x}). \tag{3.51}$$

Definition 3.5 (State Synchronization of LPV Agents). The LPV agents (3.46) are said to achieve *state synchronization* under (3.48) or (3.49), if there exist constants $k, \gamma > 0$ such that

$$\|(P_{\mathcal{A}} \otimes I_{n_x})x(t)\| \leq k\|x(t_0)\|e^{-\gamma(t-t_0)} \tag{3.52}$$

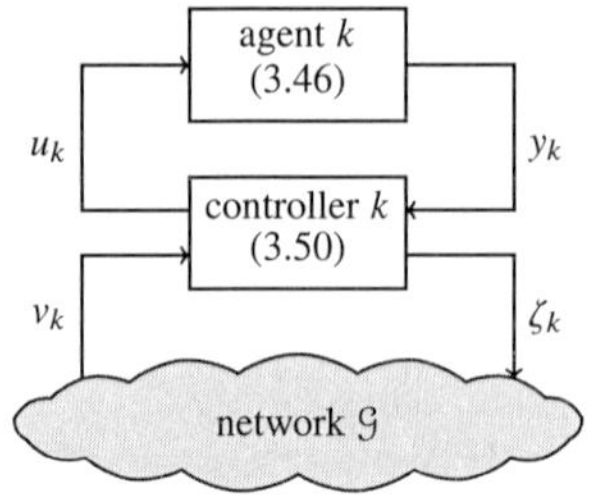

Figure 3.9: Dynamically coupled LPV systems (3.46), (3.50).

for all $t \geq t_0$ and all parameter trajectories $\rho^g \in \mathcal{F}_{\mathcal{P}^g}$, $\rho^\ell_k \in \mathcal{F}_{\mathcal{P}^\ell}$, $k \in \mathcal{N}$. Moreover, the agents (3.46) are said to achieve *nontrivial* state synchronization if $x = \mathbf{0}$ is not asymptotically stable. Every trajectory $s : \mathbb{R}^+_0 \to \mathbb{R}^{n_x}$ satisfying $\lim_{t\to\infty}((P_\mathcal{S} \otimes I_{n_x})x(t) - \mathbf{1} \otimes s(t)) = \mathbf{0}$ is called a *synchronous state trajectory*.

The condition (3.52) implies state synchronization according to Definition 3.2, i.e., for all initial conditions and all pairs $j, k \in \mathcal{N}$, it holds that $\lim_{t\to\infty}(x_k(t) - x_j(t)) = \mathbf{0}$. Since the constants k, γ are independent of t_0 and the parameter trajectory, Definition 3.5 requires global uniform exponential stability of $\mathcal{S}_x$, which is a stronger requirement than Definition 3.2. The motivation for Definition 3.5 is twofold. First, we require the same convergence properties for networks of LPV agents as for networks of LTI agents. Note that the two definitions are equivalent for LTI agents. Second, the notion of quadratic stability is commonly used in the context of LPV systems. The advantages of this notion are that it implies uniform exponential stability and hence a strong form of robust stability (Becker et al. (1993)) and that the common Lyapunov function is well suited for stability analysis and controller synthesis purposes.

Output synchronization of LPV agents is defined analogously.

Definition 3.6 (Output Synchronization of LPV Agents). The LPV agents (3.46) are said to achieve *output synchronization* under (3.50), if there exist constants $k, \gamma > 0$ such that

$$\left\| \begin{bmatrix} P_\mathcal{A} \otimes I_{n_y} & 0 \\ 0 & P_\mathcal{A} \otimes I_{n_\zeta} \end{bmatrix} \begin{bmatrix} y(t) \\ \zeta(t) \end{bmatrix} \right\| \leq k \left\| \begin{bmatrix} x(t_0) \\ z(t_0) \end{bmatrix} \right\| e^{-\gamma(t-t_0)} \tag{3.53}$$

for all $t \geq t_0$ and all parameter trajectories $\rho^g \in \mathcal{F}_{\mathcal{P}^g}$, $\rho^\ell_k \in \mathcal{F}_{\mathcal{P}^\ell}$, $k \in \mathcal{N}$. Moreover, the agents are said to achieve *nontrivial* output synchronization if the subspace $\{x \in \mathbb{R}^{Nn_x} : y_1 = y_2 = \cdots = y_N = \mathbf{0}\}$ is not asymptotically stable. Every trajectory $s : \mathbb{R}^+_0 \to \mathbb{R}^{n_y}$ satisfying $\lim_{t\to\infty}((P_\mathcal{S} \otimes I_{n_y})y(t) - \mathbf{1} \otimes s(t)) = \mathbf{0}$ is called a *synchronous output trajectory*.

3.3.2 Parameter-Varying Invariance and the Synchronous Subspace

The study of networks of LTI systems in Sections 3.1 and 3.2 has made apparent that nontrivial synchronization corresponds to the existence of an asymptotically stable invariant subspace contained in the synchronous subspace. The following consideration illustrates this fact.

An abstract synchronization problem for LTI systems may be formulated as follows. Consider the LTI system $\dot{x} = Ax$, $x \in \mathbb{R}^{n_x}$ (representing the closed-loop network) and the subspace $\mathcal{S} \subset \mathbb{R}^{n_x}$ (representing the synchronous subspace) described by $\mathcal{S} = \ker(D)$ for some matrix D with full row rank. Define the (artificial) output $y = Dx$. The objective is that for all initial conditions $x(0)$, it holds that $\lim_{t\to\infty} y(t) = \mathbf{0}$ (asymptotic synchronization). In this formulation, the problem at hand is a classical output stabilization problem and was solved by Wonham (1985). The requirement $\lim_{t\to\infty} De^{At}x(0) = \mathbf{0}$ for all $x(0)$ is equivalent to $\mathcal{S}^* \subseteq \ker(D)$, where $\mathcal{S}^*$ is the largest among all subspaces $\mathcal{U}$ that satisfy $A\mathcal{U} \subseteq \mathcal{U}$ and $\sigma(A|_{\mathcal{U}}) \subset \mathbb{C}^0 \cup \mathbb{C}^+$. In particular, $\mathcal{S}^* = \mathcal{E}_{0+}(A)$, i.e., the complement of the stable subspace of A. Hence, if $x(t) \to \mathcal{S}$ as $t \to \infty$ for all initial conditions, then there exists an invariant subspace $\mathcal{S}^* \subseteq \mathcal{S}$ which is asymptotically stable.

For networks of LPV systems, the situation is different. In fact, the existence of an invariant subspace contained in the synchronous subspace may in general not be guaranteed. In this section, we impose the existence of such an invariant subspace as an assumption in order to recover results in the spirit of the internal model principle for synchronization discussed in Section 3.2.2. This section is based on Seyboth et al. (2012c).

Definition 3.7 (Parameter-Varying Invariant Subspace (Balas et al. (2003)))**.** A subspace $\mathcal{U}$ is a *parameter-varying invariant subspace* with respect to an LPV system $\dot{x} = A(\rho(t))x$ with $A : \mathbb{R}^{n_\rho} \to \mathbb{R}^{n_x \times n_x}$ continuous and bounded, $\rho \in \mathcal{F}_{\mathcal{P}}$, if

$$A(\rho)\mathcal{U} \subseteq \mathcal{U} \quad \text{for all } \rho \in \mathcal{P}. \tag{3.54}$$

Theorem 3.19. *Consider the agents* (3.46) *with underlying graph $\mathcal{G}$ and couplings* (3.50)*. Suppose that the group achieves nontrivial output synchronization (according to Definition 3.6).*

Moreover, suppose that there exists a nontrivial parameter-varying invariant subspace $\mathcal{S}^* \subset \mathbb{R}^{N(n_x+n_z)}$ with $m = \dim(\mathcal{S}^*)$ for the network (3.46), (3.50)*, which is globally uniformly exponentially stable and on which $y_k = y_j$, $\zeta_k = \zeta_j$ for all pairs $j,k \in \mathcal{N}$ and all parameters.*

Then, there exist a constant matrix $\Pi \in \mathbb{R}^{n_x \times m}$ and continuously parameter-dependent bounded matrices $\Gamma(\rho_k)$ and $S(\rho^g)$, $\Gamma : \mathcal{P} \to \mathbb{R}^{n_u \times n_x}$ and $S : \mathcal{P}^g \to \mathbb{R}^{m\times m}$, such that

$$A(\rho_k)\Pi + B\Gamma(\rho_k) = \Pi S(\rho^g) \tag{3.55}$$

for all $\rho^g \in \mathcal{P}^g$, $\rho_k^\ell \in \mathcal{P}^\ell$. If the system and controller matrices depend affinely on the parameters, then the parameter-dependence of S and Γ is affine as well.

Proof. See Appendix B.7. □

Equation (3.55) is a parameter-dependent regulator equation. As we will show later, a solution of this equation allows to construct a distributed control law solving the synchronization problem, given that the underlying graph is connected and the agents are stabilizable and detectable. In contrast to LTI networks, however, the existence of a solution to (3.55) appears to be not necessary. Solvability of the synchronization problem is closely related to solvability of an output regulation problem. It is known that the solvability conditions of output regulation problems for linear time-varying (LTV) and LPV systems involve matrix differential equations instead of algebraic equations, cf., Shim et al. (2006), Köroğlu and Scherer (2011) and references therein. This complicates the constructive solution of such output regulation problems.

The following corollary is obtained for networks with static distributed control laws (3.48).

Corollary 3.20. *Consider the agents* (3.46) *with underlying graph $\mathcal{G}$ and couplings* (3.48)*. Suppose that the group achieves nontrivial state synchronization. Moreover, suppose that there exists a nontrivial parameter-varying invariant subspace $\mathcal{S}^* \subset \mathbb{R}^{Nn_x}$ with $m = \dim(\mathcal{S}^*)$ for the network* (3.46), (3.48)*, which is globally uniformly exponentially stable and on which $x_k = x_j$ for all pairs $j,k \in \mathcal{N}$.*

Then, there exist a matrix $\Pi \in \mathbb{R}^{n_x \times m}$ *with full column rank and a continuously parameter-dependent bounded matrix* $S(\rho^g)$, $S : \mathcal{P}^g \to \mathbb{R}^{m\times m}$, *such that for all* $\rho^g \in \mathcal{P}^g$, $\rho_k^\ell \in \mathcal{P}^\ell$, $k \in \mathcal{N}$,

$$A(\rho_k)\Pi = \Pi S(\rho^g).$$

Proof. The static control law (3.48) is contained in (3.50) by setting $O(\rho_k) = K(\rho_k)$, $M(\rho_k) = 0$, $Q(\rho_k) = I_{n_x}$ and omitting the controller states z. State synchronization corresponds to setting $C = I_{n_x}$ in the agent model (3.46). □

The following result shows that the subspace $\mathcal{S}_x$ is invariant for homogeneous groups of agents.

Lemma 3.21. *Consider the agents* (3.46) *with underlying graph* $\mathcal{G}$ *and couplings* (3.48). *Suppose that Assumption 3.1 is satisfied. Then, the synchronous subspace* $\mathcal{S}_x$ *is invariant with respect to the closed loop, if and only if* $A(\cdot)$ *in* (3.46) *depends solely on the global parameters* $\rho^g(t)$.

Proof. If $x(t) \in \mathcal{S}_x$, then $u_k(t) = \mathbf{0}$ for all $k \in \mathcal{N}$, i.e., the diffusive couplings (3.48) vanish on $\mathcal{S}_x$. Hence, the dynamics on $\mathcal{S}_x$ are $\dot{x} = \operatorname{diag}(A(\rho_1(t)), ..., A(\rho_N(t)))x$. The invariance condition (3.54) for $\mathcal{S}_x$ is $\operatorname{diag}(A(\rho_1), ..., A(\rho_N))\mathcal{S}_x \subseteq \mathcal{S}_x$ for all $\rho_k \in \mathcal{P}$, $k \in \mathcal{N}$. Equivalently, for all $w \in \mathbb{R}^{n_x}$ and $\rho_k \in \mathcal{P}$, $k \in \mathcal{N}$, it holds that

$$A(\rho_1)w = \cdots = A(\rho_N)w. \tag{3.56}$$

If $\rho_1(t) = \cdots = \rho_N(t) = \rho^g(t)$ for all $t \geq 0$, then (3.56) is clearly satisfied and $\mathcal{S}_x$ is invariant.

If $\mathcal{S}_x$ is invariant, then (3.56) holds for all $w \in \mathbb{R}^{n_x}$. If we choose w as the canonical basis vectors $\mathbf{e}_1, ..., \mathbf{e}_n$, we see that $A(\rho_k) = A(\rho_j)$, for all $j, k \in \mathcal{N}$. Due to injectivity of $A(\cdot)$, it follows that $\rho_1 = \cdots = \rho_N$. Since the local parameters are in general non-identical, it follows that $\rho_k = \rho^g$ for all $k \in \mathcal{N}$. □

Lemma 3.21 and Corollary 3.20 show that the dynamics of all agents must be homogeneous on an invariant subspace contained in $\mathcal{S}_x$. If the synchronous subspace is invariant itself, i.e., $\mathcal{S}_x^* = \mathcal{S}_x$, the agents must not depend on non-identical local parameters ρ_k^ℓ. If $\mathcal{S}_x^* \subset \mathcal{S}_x$, then the agents may depend on local parameters but the dynamics restricted to $\mathcal{S}_x^*$ must be homogeneous. These observations are well aligned with the results for homogeneous and heterogeneous groups of LTI agents discussed in the previous sections (cf., Theorem 3.12). In particular, under the conditions of Corollary 3.20, there exists a matrix $T \in \mathbb{R}^{n_x \times n_x}$ such that the transformation $x_k = T\tilde{x}_k$ for (3.46) yields $\dot{\tilde{x}}_k = \tilde{A}(\rho_k(t))\tilde{x}_k + \tilde{B}u_k$, $y_k = \tilde{C}\tilde{x}_k$ with $\tilde{B} = T^{-1}B$, $\tilde{C} = CT$, and

$$\tilde{A}(\rho_k) = T^{-1}A(\rho_k)T = \begin{bmatrix} S(\rho^g) & \tilde{A}_{12}(\rho_k) \\ 0 & \tilde{A}_{22}(\rho_k) \end{bmatrix}.$$

The system matrix embeds a block which depends on global parameters only, in accordance with Lemma 3.15.

Next, the resulting synchronous trajectories of such networks of LPV systems are derived. Recall that the synchronous state and output trajectories for networks of LTI systems are described in Theorem 3.1 and Theorem 3.11. The statement in Theorem 3.11 follows from the observation that each solution of an exponentially unforced LTI systems converges exponentially to a solution of the unforced system (Lemma A.10). A system is called asymptotically unforced if its input asymptotically converges to zero. A similar result holds for a certain class of LPV systems.

Lemma 3.22. *Consider an LPV system of the form*

$$\dot{x} = A(\rho(t))x + \delta(t) \tag{3.57}$$

with $x(0) = x_0$, $\rho \in \mathcal{F}_{\mathcal{P}}$, $A : \mathbb{R}^{n_\rho} \to \mathbb{R}^{n_x \times n_x}$ *continuous and bounded. Suppose that* $A(\cdot)$ *is a skew-symmetric map, i.e.,* $A(\rho) = -A(\rho)^{\mathsf{T}}$ *for all* $\rho \in \mathcal{P}$. *Moreover, suppose that* $\|\delta(t)\| \leq c_1 e^{-\mu t}$ *for some* $c_1, \mu > 0$. *Then, there exists* $c_2 > 0$ *and for all* x_0 *there exists an* $\tilde{x}_0$ *such that*

$$\|x(t) - \Phi(t,0)\tilde{x}_0\| \leq c_2 e^{-\mu t},$$

where $\Phi(t,0)$ *is the state transition matrix of the system* $\dot{x} = A(\rho(t))x$.

Proof. See Appendix B.8. □

In words, Lemma 3.22 states that each solution $x(t)$ converges exponentially to a solution of the unforced system $\dot{x} = A(\rho(t))x$. Note that a variety of practically relevant LPV systems have the property that the system matrix is skew-symmetric for all parameter values. Most importantly, with $A = 0$ and

$$A(\rho(t)) = \begin{bmatrix} 0 & -\rho(t) \\ \rho(t) & 0 \end{bmatrix},$$

this class contains systems generating constant and non-stationary sinusoidal signals. These are the most relevant signal types for consensus and synchronization problems. Lemma 3.22 leads to the following result.

Theorem 3.23. *Suppose that all conditions of Theorem 3.19 hold and, additionally, that* $S : \mathcal{P}^g \to \mathbb{R}^{m \times m}$ *is a skew-symmetric map, i.e.,* $S(\rho^g) = -S(\rho^g)^{\mathsf{T}}$ *for all* $\rho^g \in \mathcal{P}^g$. *Then, there exists* $w_0 \in \mathbb{R}^m$ *such that* $s(t) = C\Pi\Phi_S(t,0)w_0$ *is a synchronous output trajectory of* (3.46), (3.50), *i.e.,*

$$\lim_{t\to\infty} \|y_k(t) - C\Pi\Phi_S(t,0)w_0\| = \mathbf{0} \tag{3.58}$$

for all $k \in \mathcal{N}$, *where* $\Phi_S(t,0)$ *is the state transition matrix of the system* $\dot{w} = S(\rho^g(t))w$.

Proof. The statement is a direct consequence of Theorem 3.19 and Lemma 3.22. □

3.3.3 Homogeneous LPV Networks

This section addresses the state synchronization problem in homogeneous networks consisting of identical LPV systems. The results are partly based on Seyboth et al. (2012b). We consider homogeneous groups of $N > 1$ polytopic LPV agents (3.46) depending on global parameters only and coupled by (3.48), i.e.,

$$\dot{x}_k = A(\rho^g(t))x_k + Bu_k, \tag{3.59}$$

$$u_k = K(\rho^g(t)) \textstyle\sum_{k=1}^N a_{kj}(x_j - x_k) \tag{3.60}$$

and $\rho^g \in \mathcal{F}_{\mathcal{P}^g}$, $k \in \mathcal{N}$. According to Lemma 3.21, the synchronous subspace $\mathcal{S}_x$ is invariant with respect to (3.59), (3.60). In the following, we present both analysis and synthesis results for such groups of LPV agents.

Problem 3.3 (State Synchronization Problem)**.** Consider the multi-agent system consisting of $N > 1$ LPV agents (3.59) and an underlying graph $\mathcal{G}$. Find a distributed control law of the form (3.60), such that the group achieves state synchronization according to Definition 3.5.

Theorem 3.24 (Necessary and Sufficient Condition for State Synchronization). *The LPV agents* (3.59) *achieve state synchronization under* (3.60), *if and only if the systems*

$$\dot{z} = (A(\rho^g(t)) - \lambda_k BK(\rho^g(t)))\, z \tag{3.61}$$

are globally uniformly exponentially stable for all $k \in \mathcal{N}\backslash\{1\}$, *where* λ_k, $k \in \mathcal{N}\backslash\{1\}$, *are eigenvalues of* $L_\mathcal{G}$. *Moreover, in this case a synchronous state trajectory* $s(t)$ *is the solution of* $\dot{s} = A(\rho^g(t))s$ *with initial condition* $s(0) = (p^\mathsf{T} \otimes I_{n_x})x(0)$, *where* $p^\mathsf{T} L_\mathcal{G} = \mathbf{0}^\mathsf{T}$, $p^\mathsf{T}\mathbf{1} = 1$.

Proof. In compact form, the closed loop system (3.59) and (3.60) is given by

$$\dot{x} = (I_N \otimes A(\rho^g(t)) - L_\mathcal{G} \otimes (BK(\rho^g(t))))\, x. \tag{3.62}$$

Analogously to the proof of Theorem 3.1, we apply a state transformation $x = (T \otimes I_n)z$ to (3.62) such that $\dot{z} = (I_N \otimes A(\rho^g(t)) - \Lambda \otimes (BK(\rho^g(t))))\, z$. This system has block triangular structure with blocks $A(\rho^g(t))$ and $A(\rho^g(t)) - \lambda_k BK(\rho^g(t))$, $k \in \mathcal{N}\backslash\{1\}$, on the diagonal. The first block corresponds to the dynamics restricted to $\mathcal{S}_x$ and characterizes the synchronous state trajectory. State synchronization is achieved if and only if the (decoupled) dynamics restricted to $\mathcal{A}_x$ are globally uniformly exponentially stable which is equivalent to global uniform exponential stability of the systems (3.61) on the diagonal. Exponential stability implies input-to-state stability (cf. Khalil (2002)) of each block with respect to the states of the lower blocks, thus guaranteeing exponential stability of the cascade, i.e., the state component in $\mathcal{A}_x$. □

Remark 3.11. Note that the eigenvalues λ_k of $L_\mathcal{G}$ may take complex values since we consider directed graphs $\mathcal{G}$. In this case, stability of the complex-valued system (3.61) with $z \in \mathbb{C}^{n_x}$ is assessed by embedding it into $\mathbb{R}^{2n_x}$ according to

$$\dot{\hat{z}} = \begin{bmatrix} A(\rho^g(t)) + \sigma_k BK(\rho^g(t)) & -\omega_k BK(\rho^g(t)) \\ \omega_k BK(\rho^g(t)) & A(\rho^g(t)) + \sigma_k BK(\rho^g(t)) \end{bmatrix} \hat{z},$$

where $\lambda_k = \sigma_k + \mathbf{i}\omega_k$ and $\hat{z} = [\mathbf{Re}(z)^\mathsf{T}\ \mathbf{Im}(z)^\mathsf{T}]^\mathsf{T}$, cf., Lemma A.1. ○

Theorem 3.24 extends the well-known Theorem 3.1 for LTI systems to LPV systems. For a given group of polytopic LPV agents (3.59) with $A(\rho^g) \in \mathrm{conv}\{A(\omega_1^g), ..., A(\omega_{p^g}^g)\}$, suitable parameter-dependent gain matrices $K(\rho^g) \in \mathrm{conv}\{K(\omega_1^g), ..., K(\omega_{p^g}^g)\}$ can be found with the following synthesis procedure.

Theorem 3.25 (Gain-Scheduled Diffusive Coupling Synthesis via LMIs). *Consider the multi-agent system* (3.59), (3.60) *with underlying graph* $\mathcal{G}$. *Suppose that* $(A(\rho^g), B)$ *is quadratically stabilizable and* $\mathcal{G}$ *is connected. Then, there exist* $X \in \mathbb{R}^{n_x \times n_x}$ *and* $\tau_i \in \mathbb{R}$, $i = 1, ..., p^g$, *such that* $X = X^\mathsf{T} > 0$, $\tau_i > 0$, *and*

$$XA(\omega_i^g)^\mathsf{T} + A(\omega_i^g)X - \tau_i BB^\mathsf{T} < 0 \tag{3.63}$$

are satisfied. The parameter-dependent coupling gain matrix $K(\rho^g)$ *defined by*

$$K(\omega_i^g) = c\frac{\tau_i}{2}B^\mathsf{T}X^{-1} \tag{3.64}$$

with any constant parameter $c \geq \mathbf{Re}(\lambda_2)^{-1}$ *solves the State Synchronization Problem 3.3. Furthermore, the constant coupling gain matrix* K *defined by*

$$K = c\frac{\tau}{2}B^\mathsf{T}X^{-1} \tag{3.65}$$

with $\tau = max\{\tau_1, ..., \tau_{p^g}\}$ *and* c *as in* (3.64) *solves the State Synchronization Problem 3.3 as well.*

Proof. According to Lemma A.22 and Lemma A.23, the control laws defined by (3.64) and (3.65) ensure that $\dot{z} = (A(\rho^g(t)) - \mu BK(\rho^g(t)))\, z$ is quadratically stable for all $\mu \in \mathbb{C}$ with $\mathbf{Re}(\mu) \geq \mathbf{Re}(\lambda_2)$. Consequently, the systems (3.61) are quadratically stable for all λ_k, $k \in \mathcal{N}\backslash\{1\}$, and state synchronization follows from Theorem 3.24. □

Theorem 3.25 represents a counterpart to Theorem 3.4 addressing LPV agents instead of LTI agents. The couplings (3.60) designed in this way yield an unbounded synchronization region (see Remark 3.2) due to the guaranteed gain margins obtained in Lemma A.22 and Lemma A.23. Only the constant c in (3.64) or (3.65) has to be adjusted in order to implement the couplings for a specific graph $\mathcal{G}$. The implementation can be carried out as follows. The parameter vector $\rho^g(t)$ is available in real-time by assumption. Convex coefficients $\alpha_i(t)$, $i = 1, ..., p^g$, for the polytopic representation of the parameter vector are obtained online by solving the linear system of equations $\rho^g(t) = \sum_{i=1}^{p^g} \alpha_i(t)\omega_i^g$, $\sum_{i=1}^{p^g} \alpha_i(t) = 1$, with the constraints $\alpha_i(t) \geq 0$ for $i = 1, ..., p^g$. This can be done numerically with very efficient algorithms. With these coefficients, $K(\rho_k(t)) = \sum_{i=1}^{p^g} \alpha_i(t)K(\omega_i^g)$ can be computed and the control law (3.60) can be evaluated locally at each agent.

A desired decay rate $\gamma > 0$ on the synchronization error of the network can be imposed by adding the term $2\gamma X$ to the left-hand side of the LMI (3.63). In this case, a feasible solution can be found whenever $(A(\rho^g) + \gamma I_{n_x}, B)$ is quadratically stabilizable.

It is possible to construct a variety of specialized and more sophisticated synthesis procedures for distributed control laws for homogeneous LPV networks consisting of (3.59). For example, the conservatism may be reduced significantly by taking bounds on the parameter variation rates into account. Based on the notion of quadratic detectability, observer-based distributed control laws can be developed similarly to those for LTI networks mentioned in Remark 3.5. Moreover, $\mathcal{L}_2$-gain requirements can be included in the design. In this case, it can be expected that parameter-dependent control-laws outperform parameter-independent robust control-laws. A good overview on LPV controller synthesis techniques is given in the book by Briat (2015).

3.3.4 Heterogeneous LPV Networks

This section addresses the output synchronization problem in heterogeneous networks consisting of non-identical LPV systems. The results are based on Seyboth et al. (2012c).

Problem 3.4 (Output Synchronization Problem)**.** Consider the multi-agent system consisting of $N > 1$ LPV agents (3.46) and an underlying graph $\mathcal{G}$. Find a distributed control law of the form (3.50), such that the group achieves output synchronization according to Definition 3.6.

In the sequel, a constructive solution to Problem 3.4 is presented. The problem is solved by a decomposition into two subproblems. On the network level, a synchronization problem for a homogeneous network of LPV systems is solved, and on the agent level, local tracking problems are solved by output regulation techniques.

Theorem 3.26. *Consider a heterogeneous group of $N > 1$ LPV agents* (3.46)*. Suppose that $(A(\rho_k), B)$ is quadratically stabilizable, $(A(\rho_k), C)$ is quadratically detectable, and the underlying graph $\mathcal{G}$ is connected. Moreover, suppose that there exist an integer $m > 0$, a constant matrix $\Pi \in \mathbb{R}^{n_x \times m}$ and continuously parameter-dependent bounded matrices $\Gamma(\rho_k)$ and $S(\rho^g)$, $\Gamma : \mathcal{P} \to \mathbb{R}^{n_u \times n_x}$ and $S : \mathcal{P}^g \to \mathbb{R}^{m \times m}$, such that the parameter-dependent regulator equation* (3.55) *holds for all $\rho^g \in \mathcal{P}^g$, $\rho_k^\ell \in \mathcal{P}^\ell$, $k \in \mathcal{N}$.*

Let $F(\cdot)$, $L(\cdot)$, $K(\cdot)$ be such that $A(\rho_k)-BF(\rho_k)$, $A(\rho_k)-L(\rho_k)C$, $S(\rho_k)-\lambda_i K(\rho_k)$, $i \in \mathcal{N}\backslash\{1\}$, are quadratically stable. Then, the distributed dynamic control law

$$\dot{\zeta}_k = S(\rho^g(t))\zeta_k + K(\rho^g(t)) \textstyle\sum_{j=1}^N a_{kj}(\zeta_j - \zeta_k), \tag{3.66a}$$

$$\dot{\hat{x}}_k = A(\rho_k(t))\hat{x}_k + Bu_k + L(\rho_k(t))(y_k - \hat{y}_k), \tag{3.66b}$$

$$u_k = -F(\rho_k(t))(\hat{x}_k - \Pi\zeta_k) + \Gamma(\rho_k(t))\zeta_k, \tag{3.66c}$$

solves the Output Synchronization Problem 3.4. Moreover, the solution $w(t)$ of $\dot{w} = S(\rho^g(t))w$ with $w(0) = (p^\mathsf{T} \otimes I_{n_x})\zeta(0)$ describes the synchronous state trajectory $\Pi w(t)$ and the synchronous output trajectory $s(t) = C\Pi w(t)$ of the network (3.46), (3.66).

Proof. See Appendix B.9. □

The proposed controllers (3.66) have the form (3.50), where $z_k = [\hat{x}_k^\mathsf{T}\ \zeta_k^\mathsf{T}]^\mathsf{T}$ is the controller state. Each controller contains a dynamical subsystem in (3.66a) which is identical for all controllers in the network. These homogeneous subsystems with states ζ_k are synchronized over the network via static diffusive couplings. The other parts of the dynamic controller are a Luenberger-type observer (3.66b) and an output regulation controller (3.66c) regulating the agent output y_k to the reference ζ_k.

This controller structure is adapted from Wieland et al. (2011b), cf., Remark 3.8. The main differences are the following. First, the design of controllers (3.66) is scalable. Due to the identical LPV structure of all agents, the controller matrices $F(\rho_k)$, $L(\rho_k)$, $K(\rho^g)$, $\Gamma(\rho_k)$, $S(\rho^g)$, and Π have to be designed only once and can then be applied for each agent. The synthesis problem is thus independent of the number of agents N in the network, whereas in Wieland et al. (2011b) the controllers have to be designed for each agent separately. The design can be performed with standard LPV control methods, cf. Apkarian et al. (1995), Wu et al. (1996), Briat (2015). Second, while a necessary and sufficient condition was obtained for LTI systems, only a sufficient condition can be obtained for the existence of a solution to Output Synchronization Problem 3.4 for LPV systems in Theorem 3.26. Third, LPV systems provide more flexibility in modeling heterogeneous multi-agent systems than LTI systems. In particular, the multi-agent system (3.46) contains the case of heterogeneous LTI agents by setting the parameter ρ_k^ℓ to a different constant value for each $k \in \mathcal{N}$. The system $\dot{w} = S(\rho^g(t))w$, embedded in the dynamics of each system in the network, is itself an LPV system depending on the global parameters $\rho^g(t)$.

The main results of the present section are illustrated in the following numerical example.

Example 3.4. We consider a network of $N = 5$ agents with LPV dynamics

$$\dot{x}_k = \begin{bmatrix} 0 & 1 & 0 & 0 & 0 \\ 0 & 0 & 1 & 0 & 0 \\ 0 & -\kappa(t) & -\varsigma(t) & 1 & 0 \\ 0 & 0 & 0 & 1 & 1 \\ 3-0.2\rho_k^\ell(t) & 1 & 0 & -\rho_k^\ell(t) & 3\rho_k^\ell(t) \end{bmatrix} x_k + \begin{bmatrix} 0 \\ 0 \\ 0 \\ 0 \\ 1 \end{bmatrix} u_k, \quad y_k = \begin{bmatrix} 1 & 0 & 0 & 0 \\ 0 & 1 & 0 & 0 \end{bmatrix} x_k,$$

$k \in \mathcal{N}$. The global time-varying parameters affecting all agents are $\rho^g(t) = [\kappa(t)\ \varsigma(t)]^\mathsf{T}$ and $\rho_k^\ell(t)$ is a scalar local time-varying parameter. The convex polytopes $\mathcal{P}^g$ and $\mathcal{P}^\ell$ containing all parameter values are defined by $1 \le \kappa(t) \le 10$, $0 \le \varsigma(t) \le 5$, and $-2 \le \rho_k^\ell(t) \le 2$. This is a polytopic LPV system since the system matrix is affine in the parameters and $\mathcal{P} = \mathcal{P}^g \times \mathcal{P}^\ell$ is

$$\mathcal{P} = \operatorname{conv}\left\{ \begin{bmatrix} 1 \\ 5 \\ 2 \end{bmatrix}, \begin{bmatrix} 1 \\ 5 \\ -2 \end{bmatrix}, \begin{bmatrix} 1 \\ 0 \\ 2 \end{bmatrix}, \begin{bmatrix} 1 \\ 0 \\ -2 \end{bmatrix}, \begin{bmatrix} 10 \\ 0 \\ 2 \end{bmatrix}, \begin{bmatrix} 10 \\ 0 \\ -2 \end{bmatrix}, \begin{bmatrix} 10 \\ 5 \\ 2 \end{bmatrix}, \begin{bmatrix} 10 \\ 5 \\ -2 \end{bmatrix} \right\}.$$

With $\rho_k(t) = [\rho^g(t)^\mathsf{T} \ \rho_k^\ell(t)]^\mathsf{T}$, it can be written as $A(\rho_k(t)) = A_0 + \kappa(t)A_1 + \varsigma(t)A_2 + \rho_k^\ell(t)A_3$ for some constant matrices A_0, A_1, A_2, A_3. For the construction of distributed dynamic control laws according to Theorem 3.26, the first step is to check whether the parameter-dependent regulator equation (3.55) admits a (nontrivial) solution. With an affine ansatz $\Gamma(\rho_k(t)) = \Gamma_0 + \kappa(t)\Gamma_1 + \varsigma(t)\Gamma_2 + \rho_k^\ell(t)\Gamma_3$ and $S(\rho^g(t)) = S_0 + \kappa(t)S_1 + \varsigma(t)S_2$, equating coefficients yields

$$A_0\Pi + B\Gamma_0 = \Pi S_0, \quad A_1\Pi + B\Gamma_1 = \Pi S_1, \quad A_2\Pi + B\Gamma_2 = \Pi S_2, \quad A_3\Pi + B\Gamma_3 = 0. \tag{3.67}$$

Thus, the problem of finding a solution of (3.55) in the present example amounts to solving the system of linear equations (3.67). A feasible choice for $S(\rho^g(t))$ is

$$S(\rho^g(t)) = \begin{bmatrix} 0 & 1 & 0 \\ 0 & 0 & 1 \\ 0 & -\kappa(t) & -\varsigma(t) \end{bmatrix}.$$

The lower-right 2×2-block describes the dynamics of a mass-spring-damper system with time-varying stiffness $\kappa(t)$ and damping $\varsigma(t)$. The mass-spring-damper system is augmented by an integrator state. For $\varsigma(t) = 0$, the system is undamped and generates non-stationary sinusoidal trajectories which may have a constant offset due to the additional integrator state. The matrices

$$\Pi = \begin{bmatrix} 1 & 0 & 0 & 0 & 0 \\ 0 & 1 & 0 & 0 & 0 \\ 0 & 0 & 1 & 0 & 0 \end{bmatrix}^\mathsf{T}, \qquad \Gamma(\rho_k(t)) = \begin{bmatrix} -3 + 0.2\rho_k^\ell(t) & -1 & 0 \end{bmatrix}.$$

solve (3.55). Since the agents are polytopic LPV systems and quadratically stabilizable as well as quadratically detectable, suitable parameter-dependent gain matrices $F(\rho_k(t))$ and $L(\rho_k(t))$ can be synthesized via Lemma A.22 (exploiting duality). The underlying graph $\mathcal{G}$ is a directed cycle graph and the coupling gain matrix $K(\rho_k(t))$ is synthesized via Theorem 3.25. A desired minimum decay rate of $\gamma = 0.05$ is imposed in the design of all gain matrices.

In order to illustrate the functionality of the distributed control law (3.66), numerical simulations were performed. The non-identical local parameters vary periodically between their maximal and minimal values according to $\rho_k^\ell(t) = 2 \cdot 0.9^k \sin(0.9^k t + \pi/k)$. The global parameters $\kappa(t)$ and $\varsigma(t)$ are chosen as step functions as shown in Fig. 3.10. The initial conditions for $x_k(0)$, $\zeta_k(0)$ are chosen randomly and $\hat{x}_k(0) = \mathbf{0}$ for all $k \in \mathcal{N}$. The simulation result in Fig. 3.10 shows a typical evolution of the agent outputs $y_k(t)$, $k \in \mathcal{N}$. Synchronization is achieved, as expected from Theorem 3.26. The synchronous output trajectory is indeed described by $s(t) = C\Pi w(t)$, where $w(t)$ is the solution of $\dot{w} = S(\rho^g(t))w$, $w(0) = 1/5 \sum_{k=1}^{5} \zeta_k(0)$. △

3.4 Summary and Discussion

Chapter 3 is dedicated to the output synchronization problem for networks of linear dynamical systems via distributed control, more precisely via static and dynamic diffusive couplings.

Section 3.1 is focused on homogeneous networks of LTI systems. Existence conditions and design methods for appropriate distributed control laws are reviewed. As a novel contribution, we propose a design method which allows to include performance specifications in terms of pole placement constraints. This method is particularly advantageous for practical applications when requirements such as a minimum decay rate, minimum damping ratio, and a maximum undamped

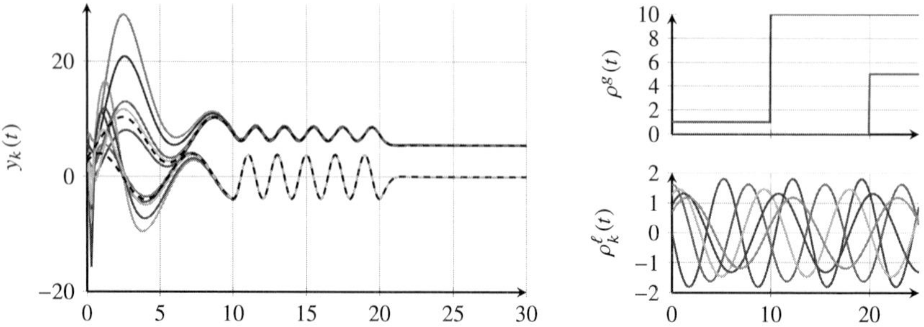

Figure 3.10: Output synchronization of the LPV network. The left plot shows the evolution of the agent outputs (solid) and the synchronous output trajectory $s(t)$ (dashed) from Example 3.4. The plots on the right show the global parameters $\kappa(t)$ (—) and $\varsigma(t)$ (—) (top) and the local parameters (bottom).

frequency need to be imposed on the synchronization error. This design method constitutes a building block to which we resort repeatedly in the sequel.

Section 3.2 is focused on heterogeneous networks of LTI systems which are significantly more complex to analyze and to control. Structural requirements for output synchronization are discussed in detail. In order to illustrate these requirements and to give insight into the effect of non-identical agent dynamics on the behavior of the dynamical networks, we analyze important network types, including coupled double-integrator agents and coupled harmonic oscillators. For each network, the role of the network topology is analyzed and the robustness of synchronization with respect to parameter uncertainties in the agent dynamics is assessed. Interestingly, we discovered that diffusively coupled networks of harmonic oscillators are rendered asymptotically stable by an arbitrarily small frequency mismatch within the iSCC, which may be harmful in practical applications. On the contrary, the double-integrator networks show a certain robustness of synchronization regarding the parameter perturbations, which facilitates the solution of a clock synchronization problem. Our results provide a deeper insight into the structural properties and dynamic behavior of heterogeneous linear networks.

In Section 3.3, the essential foregoing results are generalized to multi-agent systems consisting of agents with LPV dynamics. In particular, state and output synchronization problems are posed and solved for homogeneous and heterogeneous networks of LPV systems. Both static and dynamic gain-scheduled distributed control laws and corresponding design methods are developed. Our extension of the class of agent dynamics from LTI systems to LPV systems is of practical importance as this class is well suited to model a variety of systems such as airplanes, AUVs, or satellites, to name a few. These results push the boundaries of current knowledge towards multi-agent systems with higher agent complexity and also pave the way for the introduction of more advanced control design methods for LPV systems to the field of distributed control.

The theoretical results throughout the chapter are supported by numerical examples that illustrate the analysis results and show the feasibility of the control design methods.

Chapter 4

Cooperative Disturbance Rejection and Reference Tracking

As the most fundamental cooperative control problem, consensus and synchronization in groups of linear dynamical agents were studied extensively in Chapter 3 for different types of networks. Under the constructed distributed control laws, the network as a whole is an autonomous dynamical system without external input signals.

From a practical point of view, it is of great importance to take external signals into account in the controller design. External reference signals or unknown disturbances must be handled in almost any practical application of control theory. In typical single-loop control problems, it has to be ensured that the plant is not only stabilized but also remains stable and behaves in a desirable way despite inevitable disturbances acting on the plant. Moreover, a wide range of applications demand the ability to follow external commands and track a desired reference trajectory. The present chapter is dedicated to such control tasks for networks of dynamical systems.

The behavior of a multi-agent system can be dictated via external reference signals in the leader-follower setup (Hong et al. (2006); Li et al. (2010); Ni and Cheng (2010); Zhang et al. (2011)). The idea of this approach is to select a particular agent as leader for the group or introduce a virtual leader and design the distributed control law such that all agents synchronize to this leader. The motion of the group is then controlled through the motion of the active leader. The frameworks presented in the present chapter encompass a wide range of cooperative reference tracking scenarios which go beyond the leader-follower setup. Moreover, it is important to consider external disturbances acting on the multi-agent system, to analyze the performance of the closed-loop system, and to incorporate disturbance rejection or attenuation requirements in the design procedure. Rejection of constant disturbances is addressed by Yucelen and Egerstedt (2012); Andreasson et al. (2012); Seyboth and Allgöwer (2014). Disturbance attenuation with $\mathcal{H}_\infty$ performance criteria is addressed by Li et al. (2011); Zhao et al. (2012); Wang et al. (2013). The frameworks presented in this chapter provide a generalization of the class of agents as well as the class of external signals and cover a variety of cooperative disturbance rejection tasks.

In Section 4.1, distributed control laws with integral action are constructed. These control laws allow to solve both disturbance rejection and cooperative reference tracking problems in networks of identical linear systems subject to constant exogenous input signals and guarantee a zero steady-state error. Section 4.2 is dedicated to the distributed output regulation framework which brings together the classical output regulation theory and networks of dynamical systems. The distributed output regulation framework allows to systematically formulate and solve a wide range of cooperative reference tracking and disturbance rejection tasks for both homogeneous and heterogeneous multi-agent systems.

4.1 Distributed Integral Action

In practical applications, the individuals in a multi-agent system are typically subject to external disturbances. Therefore, it is important to include disturbances in the agent models and to develop distributed control laws which achieve a desired behavior of the group despite such disturbances. In the present section, we consider non-vanishing external disturbances, in particular constant disturbances, acting on the agents. As in classical control problems, integral action is employed in order to reject these disturbances without asymptotic offset. This section is based on Seyboth and Allgöwer (2015).

Related work includes the consensus protocols with integral action for single and double-integrator agents offered by Freeman et al. (2006); Carli et al. (2011); Yucelen and Egerstedt (2012); Yucelen and Johnson (2013); Andreasson et al. (2014a). Freeman et al. (2006) study a distributed estimation problem. The authors propose a "dynamic average consensus estimator" which allows to estimate the average of external input signals in a distributed way. Yucelen and Egerstedt (2012) and Yucelen and Johnson (2013) study the consensus problem for single-integrator agents under persistent disturbances. Andreasson et al. (2014a) present consensus protocols for both single-integrator and double-integrator networks, which guarantee convergence to consensus despite constant disturbances. A distributed PI-controller for large-scale linear systems is proposed by Andreasson et al. (2014b). Chen et al. (2014) develop a distributed control law with integral action for distributed average tracking of Euler-Lagrange systems. A distributed PID control strategy for homogeneous and heterogeneous linear agents of order one is developed by Burbano Lombana and di Bernardo (2015).

Here, we generalize the class of agents and consider networks of high-order linear systems. While single and double-integrator agents have a wide range of applications, some cooperative control problems require more complex dynamical models of the individuals in the group. We present a procedure to establish integral action and design a suitable distributed control law such that the group achieves asymptotically exact output synchronization despite constant disturbances. Moreover, we present observer-based implementations of the novel control laws for the case that the agents do not have access to their (absolute or relative) states. We consider two cases: (i) each agent has access to its own output and (ii) each agent has only access to a relative output measurement to its neighboring agents. Moreover, we show that the proposed method can also be used in order to solve a cooperative distributed tracking problem. In this setup, each agent has its local reference signal and the group objective is that all agents shall synchronize to the average (or, more generally, weighted sum) of these local reference signals. Potential applications are, e.g., formation keeping tasks for mobile vehicles.

Section 4.1.1 contains the precise problem formulation. Section 4.1.2 presents the novel distributed control law with integral action. Section 4.1.3 contains its observer-based implementations. The cooperative tracking problem is addressed in Section 4.1.4. Numerical examples illustrate the results.

4.1.1 Problem Formulation

In the following, the agent models, the information structure, and the group objective of the multi-agent systems under consideration are introduced.

Agent Models We consider a group of $N > 1$ identical linear dynamical agents

$$\dot{x}_k = Ax_k + Bu_k + Pd_k \tag{4.1a}$$

$$y_k = Cx_k \tag{4.1b}$$

with state $x_k(t) \in \mathbb{R}^{n_x}$, control input $u_k(t) \in \mathbb{R}^{n_u}$, output $y_k(t) \in \mathbb{R}^{n_y}$, external disturbance $d_k(t) \in \mathbb{R}^{n_d}$, and matrices A, B, P, C of appropriate dimension, $k \in \mathcal{N}$.

Information Structure The same information structure as in Section 3.1 is assumed, i.e., the communication topology is described by a directed graph $\mathcal{G} = (\mathcal{V}, \mathcal{E}, A_{\mathcal{G}})$.

Group Objective The group objective is robust output synchronization, defined as follows.

Definition 4.1. The agents (4.1) are said to achieve *robust output synchronization*, if for all initial conditions $x_k(0) \in \mathbb{R}^{n_x}$ and all constant disturbances $d_k \in \mathbb{R}^{n_d}$, $k \in \mathcal{N}$, there exists a *synchronous output trajectory* $s : \mathbb{R}_0^+ \to \mathbb{R}^{n_y}$ s.t. for all $k \in \mathcal{N}$,

$$\lim_{t\to\infty} (y_k(t) - s(t)) = \mathbf{0}.$$

Problem 4.1 (Robust Output Synchronization Problem under Constant Disturbances)**.** Consider the multi-agent system (4.1) with an underlying graph $\mathcal{G}$. Find a distributed control law, such that the group achieves robust output synchronization despite constant external disturbances.

While solving Problem 4.1, emphasis is put on the characterization of the resulting synchronous output trajectories.

4.1.2 Robust Synchronization under Constant Disturbances

For classical single-loop regulation problems for linear dynamical systems subject to constant unknown external disturbances, the standard approach is to augment the original plant with integrators integrating the error and design a controller (e.g., LQR) for the augmented plant. The resulting controller guarantees asymptotic regulation with zero steady-state offset despite the disturbance. This procedure is described in the textbooks by Anderson and Moore (1990); Khalil (2002). In the following, we make use of the same idea in order to solve Problem 4.1. We propose distributed control laws with integral action which guarantee that the group achieves robust output synchronization despite constant disturbances.

The main idea is to augment each agent (4.1) with integrator states in a suitable way and construct an appropriate distributed control law such that the augmented systems achieve output synchronization. As will be shown in the sequel, this procedure will guarantee robust output synchronization despite constant external disturbances. Each agent (4.1) is augmented with integrator states $\zeta_k(t) \in \mathbb{R}^{n_y}$ that integrate the difference between neighboring outputs, i.e.,

$$\dot{\zeta}_k = \sum_{j=1}^{N} a_{kj}(y_j - y_k).$$

In view of the output synchronization objective, it is intuitive to integrate the output differences between neighboring agents. For notational convenience, we define the extended matrices

$$A_e = \begin{bmatrix} 0 & C \\ 0 & A \end{bmatrix}, \qquad B_e = \begin{bmatrix} 0 \\ B \end{bmatrix}, \qquad P_e = \begin{bmatrix} 0 \\ P \end{bmatrix}, \qquad C_e = \begin{bmatrix} 0 & C \end{bmatrix}$$

and formulate the extended agent model

$$\dot{z}_k = \begin{bmatrix} 0 & 0 \\ 0 & A \end{bmatrix} z_k + \begin{bmatrix} I_{n_y} \\ 0 \end{bmatrix} \sum_{j=1}^{N} a_{kj}(y_j - y_k) + B_e u_k + P_e d_k \tag{4.2a}$$

$$y_k = C_e z_k, \tag{4.2b}$$

where $z_k = [\zeta_k^\mathsf{T} \; x_k^\mathsf{T}]^\mathsf{T} \in \mathbb{R}^{n_y+n_x}$, for all $k \in \mathcal{N}$.

Theorem 4.1 (Diffusive Couplings with Integral Action). *Consider the multi-agent system* (4.1) *with constant disturbances $d_k \in \mathbb{R}^{n_d}$ for all $k \in \mathcal{N}$, underlying graph $\mathcal{G}$, and diffusive couplings with integral action*

$$\dot{\zeta}_k = \sum_{j=1}^{N} a_{kj}(y_j - y_k) \tag{4.3a}$$

$$u_k = K_\mathrm{P} \sum_{j=1}^{N} a_{kj}(x_j - x_k) + K_\mathrm{I}\zeta_k \tag{4.3b}$$

with proportional and integral coupling gain matrices $K_\mathrm{P} \in \mathbb{R}^{n_u \times n_x}$ and $K_\mathrm{I} \in \mathbb{R}^{n_u \times n_y}$. Suppose that the matrices

$$A_e - \lambda_k B_e K_e \tag{4.4}$$

are Hurwitz for all $k \in \mathcal{N}\backslash\{1\}$, where $K_e = [K_\mathrm{I} \; K_\mathrm{P}]$ and λ_k is the k-th eigenvalue of $L_\mathcal{G}$.

Then, the network (4.1), (4.3) *achieves robust output synchronization. Moreover,*

- *if* (4.3a) *is initialized with $\zeta_k(0) = \mathbf{0}$ for all $k \in \mathcal{N}$, then a synchronous output trajectory is*

$$s(t) = C\tilde{x}_1(t), \tag{4.5a}$$

$$\dot{\tilde{x}}_1 = A\tilde{x}_1 + \textstyle\sum_{k=1}^{N} p_k P d_k, \tag{4.5b}$$

$$\tilde{x}_1(0) = (p^\mathsf{T} \otimes I_{n_x})x(0), \tag{4.5c}$$

  where $p^\mathsf{T} = [p_1 \cdots p_N]$ satisfies $p^\mathsf{T} L_\mathcal{G} = \mathbf{0}^\mathsf{T}$, $p^\mathsf{T}\mathbf{1} = 1$ and $x(0)$ are the stacked initial conditions of the agents (4.1).

- *if $\mathcal{G}$ is connected, (A, B) is stabilizable, and*

$$\operatorname{rank} \begin{bmatrix} C & 0 \\ A & B \end{bmatrix} = n_x + n_y, \tag{4.6}$$

  then such a coupling gain matrix K_e exists.

Proof. See Appendix B.10. □

The diffusive couplings with integral action (4.3) are a generalization of the widely used static diffusive couplings (3.7) and guarantee exact output synchronization despite constant disturbances. Intuitively, the step from static diffusive couplings (3.7) to diffusive couplings with integral action (4.3) for networked systems is the counterpart to the step from static state feedback to state feedback for an integrator augmented plant in single-loop regulation problems.

Notable properties of the distributed control law (4.3) are the following:

- There is no need to communicate the controller states ζ_k among neighboring agents.
- The integrator states remain bounded for all times even if the synchronous output trajectory is not converging or unbounded.
- The synchronous output trajectory (4.5) of the group is a solution of the autonomous system (4.1) with initial condition depending on the initial conditions of the agents within the iSCC of the graph $\mathcal{G}$ (due to (4.5c) and Theorem A.27). Moreover, the synchronous output trajectory (4.5) is affected by the disturbances d_k acting on agents within the iSCC of $\mathcal{G}$.
- The proof of Theorem 4.1 reveals that the integrator states (4.3a) must be initialized appropriately in order not to affect $s(\cdot)$, where zero initial conditions are a feasible choice.
- The existence of suitable coupling gain matrices K_{P}, K_{I} is guaranteed if $\mathcal{G}$ is connected, (A, B) is stabilizable (analogously to the static diffusive coupling case, cf., Theorem 3.2), and if additionally the rank condition (4.6) is satisfied.
- Due to the rank condition (4.6), the input dimension n_u must be greater or equal to the output dimension n_y, i.e., $n_u \geq n_y$, and (4.1) must not have a transmission zero at zero.

With Theorem 4.1, we have recovered a condition for the coupling gain matrices which has the same form as the necessary and sufficient condition in Theorem 3.1, i.e., the coupling gain matrix $K_{\mathrm{e}} = [K_{\mathrm{I}}\ K_{\mathrm{P}}]$ has to stabilize $A_{\mathrm{e}} - \lambda_k B_{\mathrm{e}} K_{\mathrm{e}}$ simultaneously for all non-zero eigenvalues λ_k of $L_{\mathcal{G}}$, where $\mathcal{G}$ is connected. The derivation of this result in the proof of Theorem 4.1 involved a second particular similarity transformation. The advantage of this condition is the possibility to resort to the existence conditions and synthesis methods presented in Sections 3.1.2 and 3.1.3 for such K_{e}. Under the assumptions of Theorem 4.1, a feasible choice for K_{e} is

$$K_{\mathrm{e}} = cR^{-1}B_{\mathrm{e}}^{\mathsf{T}}\tilde{P}$$

with a scalar $c \geq (2\,\mathbf{Re}(\lambda_2))^{-1}$ and where $\tilde{P}$ is the unique positive semi-definite solution of the Algebraic Riccati Equation (ARE)

$$\tilde{P}A_{\mathrm{e}} + A_{\mathrm{e}}^{\mathsf{T}}\tilde{P} - \tilde{P}B_{\mathrm{e}}R^{-1}B_{\mathrm{e}}^{\mathsf{T}}\tilde{P} + Q = 0$$

for some positive definite matrices $Q \in \mathbb{R}^{(n_x+n_u)\times(n_x+n_u)}$ and $R \in \mathbb{R}^{n_u \times n_u}$, cf., Theorem 3.2. A sufficient condition in terms of LMIs is the following. Suppose that there exist matrices $X \in \mathbb{R}^{(n_x+n_u)\times(n_x+n_u)}$ and $Y \in \mathbb{R}^{n_u\times(n_x+n_u)}$ such that $X \succ 0$ and for $k = 2, \ldots, N$, the LMIs

$$XA_{\mathrm{e}}^{\mathsf{T}} + A_{\mathrm{e}}X - \lambda_k B_{\mathrm{e}}Y - \overline{\lambda_k}Y^{\mathsf{T}}B_{\mathrm{e}}^{\mathsf{T}} + 2\gamma X \prec 0 \tag{4.7}$$

are satisfied for some scalar $\gamma > 0$. Then, a feasible choice for K_{e} is

$$K_{\mathrm{e}} = YX^{-1},$$

cf., Theorem 3.5. The latter LMI based synthesis procedure allows to find a coupling gain K_{e} such that a desired exponential decay rate $\gamma > 0$ of the synchronization error is guaranteed. This can be generalized to pole placement in LMI regions. In particular, let $\mathcal{R} = \{z \in \mathbb{C} : L + zM + \bar{z}M^{\mathsf{T}} \prec 0\} \subset \mathbb{C}$ be an LMI region defined by the real matrices $L = L^{\mathsf{T}}$ and M. In order to include the pole placement constraint in the design, replace the LMI (4.7) by

$$L \otimes X + M \otimes (A_{\mathrm{e}}X - \lambda_k B_{\mathrm{e}}Y) + M^{\mathsf{T}} \otimes (A_{\mathrm{e}}X - \overline{\lambda_k}B_{\mathrm{e}}Y)^{\mathsf{T}} \prec 0, \tag{4.8}$$

cf., Theorem 3.7. Such pole placement constraints allow to include performance specifications such as a desired decay rate and damping ratio as shown in detail in Section 3.1.3.

4.1.3 Observer-based Distributed Control Laws

The diffusive couplings with integral action (4.3) are based on the states x_k and outputs y_k of the agents (4.1). In practical applications, the agent states may not be directly accessible. Therefore, we propose observer-based implementations of these protocols using only the agent outputs y_k. Firstly, we present a solution for the case when each agent has access to its own absolute output. Secondly, we present a solution which relies only on relative output measurements. The following detectability assumption is crucial in both cases.

Assumption 4.1. *The pair* $\left(\begin{bmatrix} A & P \\ 0 & 0 \end{bmatrix}, \begin{bmatrix} C & 0 \end{bmatrix}\right)$ *is detectable.*

Assumption 4.1 guarantees the existence of an observer gain matrix H such that the following matrix is Hurwitz:

$$\begin{bmatrix} A & P \\ 0 & 0 \end{bmatrix} - H \begin{bmatrix} C & 0 \end{bmatrix}. \tag{4.9}$$

Absolute Output Measurements

Suppose that each agent $k \in \mathcal{N}$ has access to its own, absolute output y_k. Then, under Assumption 4.1, we can construct local Luenberger-observers in order to estimate the state x_k and the disturbance d_k. This leads to the following observer-based implementation of the protocol (4.3).

Theorem 4.2 (Absolute Output Observer-based Diffusive Couplings with Integral Action). *Consider the multi-agent system* (4.1) *with constant disturbances* $d_k \in \mathbb{R}^{n_d}$ *for all* $k \in \mathcal{N}$ *and underlying graph* $\mathcal{G}$. *Suppose that there exists* K_e *such that the matrices* (4.4) *are Hurwitz for all* $k \in \mathcal{N}\backslash\{1\}$, *Assumption 4.1 is satisfied, and* H *is chosen such that* (4.9) *is Hurwitz. Then, the absolute output observer-based protocol*

$$\dot{\zeta}_k = \sum_{j=1}^{N} a_{kj}(y_j - y_k) \tag{4.10a}$$

$$\begin{bmatrix} \dot{\hat{x}}_k \\ \dot{\hat{d}}_k \end{bmatrix} = \begin{bmatrix} A & P \\ 0 & 0 \end{bmatrix} \begin{bmatrix} \hat{x}_k \\ \hat{d}_k \end{bmatrix} + \begin{bmatrix} B \\ 0 \end{bmatrix} u_k + H(y_k - C\hat{x}_k) \tag{4.10b}$$

$$u_k = K_P \sum_{j=1}^{N} a_{kj} \left(\hat{x}_j - \hat{x}_k\right) + K_I \zeta_k, \tag{4.10c}$$

guarantees that the systems (4.1) *achieve robust output synchronization. If* (4.10a) *is initialized with* $\zeta_k(0) = \mathbf{0}$ *for all* $k \in \mathcal{N}$, *then* $s(\cdot)$ *in* (4.5) *is a synchronous output trajectory.*

Proof. See Appendix B.11 □

Relative Output Measurements

Suppose that the agents do not have access to their absolute outputs y_k but only to a relative output measurement to their neighbors. The available relative measurement for agent k is

$$y_{\mathcal{N}_k} = \sum_{j=1}^{N} a_{kj}(y_j - y_k).$$

With this notation, (4.3a) becomes $\dot{\zeta}_k = y_{\mathcal{N}_k}$. Analogously, we define the relative variables

$$x_{\mathcal{N}_k} = \sum_{j=1}^{N} a_{kj}(x_j - x_k), \qquad d_{\mathcal{N}_k} = \sum_{j=1}^{N} a_{kj}(d_j - d_k), \qquad u_{\mathcal{N}_k} = \sum_{j=1}^{N} a_{kj}(u_j - u_k).$$

We construct an observer for the relative variables $x_{\mathcal{N}_k}$, $d_{\mathcal{N}_k}$, similarly to Li et al. (2010); Trentelman et al. (2013). The agents are assumed to have communication capabilities such that they can exchange controller states with neighboring agents.

Theorem 4.3 (Relative Output Observer-based Diffusive Couplings with Integral Action). *Consider the multi-agent system* (4.1) *with constant disturbances $d_k \in \mathbb{R}^{n_d}$ for all $k \in \mathcal{N}$ and underlying graph $\mathcal{G}$. Suppose that there exists K_e such that the matrices* (4.4) *are Hurwitz for all $k \in \mathcal{N}\backslash\{1\}$, Assumption 4.1 is satisfied and H is chosen such that* (4.9) *is Hurwitz. Then, the relative output observer-based protocol*

$$\dot{\zeta}_k = y_{\mathcal{N}_k} \tag{4.11a}$$

$$\begin{bmatrix} \dot{\hat{x}}_{\mathcal{N}_k} \\ \dot{\hat{d}}_{\mathcal{N}_k} \end{bmatrix} = \begin{bmatrix} A & P \\ 0 & 0 \end{bmatrix} \begin{bmatrix} \hat{x}_{\mathcal{N}_k} \\ \hat{d}_{\mathcal{N}_k} \end{bmatrix} + \begin{bmatrix} B \\ 0 \end{bmatrix} u_{\mathcal{N}_k} + H\left(y_{\mathcal{N}_k} - C\hat{x}_{\mathcal{N}_k}\right) \tag{4.11b}$$

$$u_k = K_P \hat{x}_{\mathcal{N}_k} + K_I \zeta_k, \tag{4.11c}$$

guarantees that the systems (4.1) *achieve robust output synchronization. If* (4.11a), (4.11b) *are initialized with $\zeta_k(0) = \mathbf{0}$, $\hat{x}_{\mathcal{N}_k}(0) = \mathbf{0}$, $\hat{d}_{\mathcal{N}_k}(0) = \mathbf{0}$, for all $k \in \mathcal{N}$, then $s(\cdot)$ in* (4.5) *is a synchronous output trajectory.*

Proof. See Appendix B.12 □

The variables ζ_k, $\hat{x}_{\mathcal{N}_k}$ and $\hat{d}_{\mathcal{N}_k}$ are local controller states of agent k. Besides the relative output measurement $y_{\mathcal{N}_k}$, agent k needs access to the variable $u_{\mathcal{N}_k}$ for the implementation of (4.11). Hence, neighboring agents must communicate the control inputs u_k so that they can compute $u_{\mathcal{N}_k}$. Since communication requires a certain amount of time, the question of robustness with respect to time delays arises. This question is addressed in the subsequent Example 4.1.

The proof of Theorem 4.3 reveals that the observers (4.11b) must be initialized such that the observer errors lie in a particular subspace in order to obtain (4.5) as synchronous output trajectory of the network. A feasible choice are zero initial conditions for all observers. Otherwise, this trajectory might be affected by the observer errors.

Example 4.1. Consider a group of $N = 5$ agents modeled by

$$\dot{x}_k = \begin{bmatrix} 0 & 5 & 0 \\ -5 & 0 & 1 \\ 0 & 0 & -5 \end{bmatrix} x_k + \begin{bmatrix} 0 \\ 0 \\ 1 \end{bmatrix} u_k + \begin{bmatrix} 0 \\ 1 \\ 0 \end{bmatrix} d_k$$

$$y_k = \begin{bmatrix} 1 & 0 & 0 \end{bmatrix} x_k,$$

and let the graph $\mathcal{G}$ be an undirected cycle. A distributed control law with integral action based on relative output measurements (4.11) is constructed according to Theorem 4.3. We obtain the gain matrices $K_P = [\,4.58\ 56.21\ 6.25\,]$, $K_I = 69.63$ from (4.8) with the LMI region $\mathcal{R} = S(2, 20, \pi/3)$, and the observer gain $H = [\,25.40\ 37.17\ {-0.76}\ 119.37\,]^{\mathsf{T}}$ by suitable pole placement. A simulation result for some random initial conditions and unit steps for the disturbance signals d_k of agents $k = 1, 3, 5$

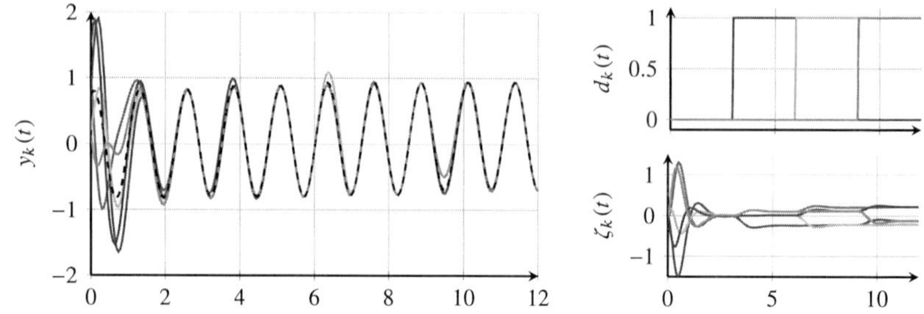

Figure 4.1: Robust output synchronization of the network from Example 4.1 under persistent disturbances.

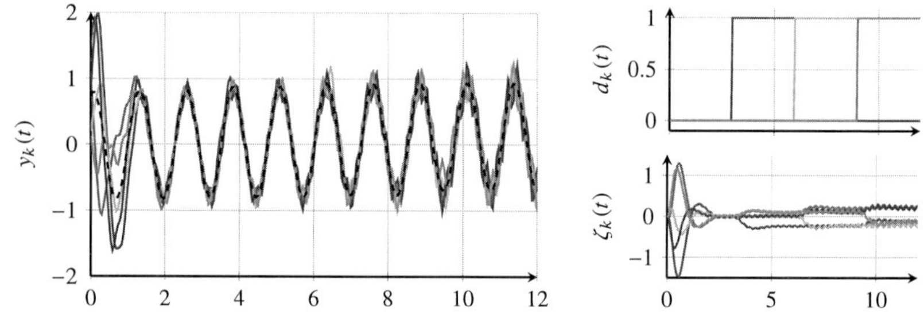

Figure 4.2: Robust output synchronization under persistent disturbances with a critical communication time delay of $\tau = 0.06$ in the network from Example 4.1.

at time $t = 3, 6, 9$, respectively, is shown in Fig. 4.1. The group achieves output synchronization as desired. The synchronous output trajectory (4.5) is plotted as dashed line.

The distributed control law (4.11) requires the sensing or communication with neighboring agents in order to obtain $y_{\mathcal{N}_k}$ and $u_{\mathcal{N}_k}$. In practice, information exchange over the network may be subject to a time delay $\tau > 0$. In this case, (4.11) must rely on the delayed quantities $y_{\mathcal{N}_k}(t) = \sum_{j=1}^{N} a_{kj}(y_j(t-\tau) - y_k(t-\tau))$ and $u_{\mathcal{N}_k}(t) = \sum_{j=1}^{N} a_{kj}(u_j(t-\tau) - u_k(t-\tau))$, i.e., the time delay enters the control-loop twice. A numerical evaluation reveals that the network under consideration remains stable for time delays $\tau \leq 0.05$. A simulation result for $\tau = 0.06$ is shown in Fig. 4.2. Only for such a large time delay, fast oscillations occur and the network becomes unstable. Concluding, the distributed control law (4.11) has a certain robustness with respect to time delays. Large time delays, however, need to be taken into account in the design since they may render the closed loop unstable. △

4.1.4 Cooperative Tracking of Constant Reference Signals

In this section, we show that distributed integral action allows to realize cooperative reference tracking for constant reference signals. As a preliminary step, we consider an isolated agent with dynamics

$$\dot{x}_k = Ax_k + Bu_k \tag{4.12a}$$
$$y_k = Cx_k, \tag{4.12b}$$

where (A, B) is stabilizable, with the asymptotic reference tracking objective $\lim_{t\to\infty} y_k(t) = r_k$ for a constant reference signal $r_k \in \mathbb{R}^{n_y}$. This problem is a special case of the output regulation problem reviewed in Section A.1.5 and Theorem A.20 yields the following result: There exists a control law of the form $u_k = -Fx_k + Vr_k$ that achieves internal stability of the closed loop and asymptotic output tracking, i.e., $\lim_{t\to\infty} y_k(t) = r_k$ for all initial conditions, if and only if there exists a solution Π, Γ to the regulator equations $A\Pi + B\Gamma = 0$, $C\Pi - I_{n_x} = 0$. If a solution to the latter exists, choose F such that $A - BF$ is Hurwitz and set $V = \Gamma + F\Pi$. The tracking control law is then

$$u_k = -Fx_k + Vr_k + v_k, \tag{4.13}$$

with a novel control input v_k for later use. The feed-forward filter matrix $V \in \mathbb{R}^{n_u \times n_y}$ given by

$$V = -(C(A - BF)^{-1}B)^{-1} \tag{4.14}$$

achieves unit steady-state gain. This leads to

$$\dot{x}_k = (A - BF)x_k + BVr_k + Bv_k \tag{4.15a}$$
$$y_k = Cx_k. \tag{4.15b}$$

By construction, all solutions of (4.15) satisfy $\lim_{t\to\infty} y_k(t) = r_k$ when $v_k(t) = \mathbf{0}$.

Here, we consider the cooperative reference tracking problem for a group of agents (4.12), each with a local external reference signal. The objective, formally defined below, is that the agent outputs synchronize and track the sum (or average) of all the external reference signals.

Definition 4.2. The group of agents (4.12) is said to achieve *cooperative reference tracking*, if for all initial conditions $x_k(0) \in \mathbb{R}^{n_x}$, $k \in \mathcal{N}$, and all constant reference signals $r_k \in \mathbb{R}^{n_y}$, $k \in \mathcal{N}_{\text{iSCC}}$, it holds that for all $k \in \mathcal{N}$,

$$\lim_{t\to\infty} y_k(t) = \bar{r},$$

where $\bar{r} = \sum_{j\in\mathcal{N}_{\text{iSCC}}} r_j$ is the sum of all reference signals acting on agents within the iSCC of $\mathcal{G}$.

In case there is only one reference signal $r_k \neq \mathbf{0}$ and all other reference signals are zero, i.e., $r_j = \mathbf{0}$, $j \neq k$, Definition 4.2 captures collective tracking of that reference by the outputs of all agents. In case of multiple reference signals to the group, the objective is to track the sum of all references to the group. By appropriate scaling of the individual reference signals, this problem formulation as well captures the distributed input averaging problem. The distributed input averaging problem was studied by Freeman et al. (2006) and has received considerable attention since then, cf., Zhu and Martínez (2010); Chen et al. (2012) and references therein. By rescaling with $1/N$, Definition 4.2 captures this problem and generalizes the solution to agents with general linear dynamics and directed graphs $\mathcal{G}$.

As we will show next, the cooperative reference tracking problem can be solved by the diffusive couplings with integral action developed in Section 4.1.2 and appropriate rescaling of the filters (4.14) for each agent. Observe that (4.15) equals (4.1) when replacing $P \leftarrow BV$, $A \leftarrow (A - BF)$, $d_k \leftarrow r_k$, and $u_k \leftarrow v_k$. The new input v_k is used in order to couple the agents

(4.15). Augmentation of the agents with integrator states (4.3a) and couplings of the form (4.3b) through v_k, i.e.,

$$v_k = K_P \sum_{j=1}^{N} a_{kj}(x_j - x_k) + K_I \zeta_k \tag{4.16}$$

guarantee that $y_k(t) - y_j(t) \to \mathbf{0}$ as $t \to \infty$ for all $j,k \in \mathcal{N}$. Furthermore, the synchronous output trajectory is $s(t) = C\tilde{x}_1(t)$, where $\tilde{x}_1(t)$ the solution of $\dot{\tilde{x}}_1 = (A - BF)\tilde{x}_1 + BV \sum_{k=1}^{N} p_k r_k$, analogously to (4.5). This shows that the reference signals affect only the synchronous motion of the group. By construction of V, it holds that $\lim_{t\to\infty} y_k(t) = \sum_{j=1}^{N} p_j r_j$ for all $k \in \mathcal{N}$. Recall that the normalized left eigenvector p^T of $L_\mathcal{G}$ corresponding to zero is non-negative and $p_k > 0$ if and only if $k \in \mathcal{N}_{\text{iSCC}}$ (Theorem A.27). Consequently, in order to achieve cooperative reference tracking, the filter V has to be rescaled individually for each agent such that the effects of the graph structure are compensated. This is achieved by

$$V_k = \frac{1}{p_k} V. \tag{4.17}$$

Corollary 4.4 (Cooperative Tracking of Constant Reference Signals)**.** *Consider a group of $N > 1$ controlled agents* (4.15) *with constant reference signals $r_k \in \mathbb{R}^{n_y}$, $k \in \mathcal{N}$, and underlying graph $\mathcal{G}$. Augment the agents with integrator states* (4.3a) *and construct couplings* (4.16) *according to Theorem 4.1. Furthermore, replace the filter V in* (4.15) *by* (4.17) *for all $k \in \mathcal{N}_{\text{iSCC}}$ and $V_k = 0$ for all $k \in \mathcal{N} \setminus \mathcal{N}_{\text{iSCC}}$.*

Then, the group achieves cooperative reference tracking.

Proof. The statement follows from the discussion above and the following observation. The filters V_k lead to $\dot{\tilde{x}}_1 = (A - BF)\tilde{x}_1 + B \sum_{k=1}^{N} V_k p_k r_k = (A - BF)\tilde{x}_1 + BV\bar{r}$. Hence, the synchronous output trajectory $s(t) = C\tilde{x}_1(t)$ satisfies $\lim_{t\to\infty} s(t) = \bar{r}$ for all initial conditions by design of the filter matrix V. □

We have seen that, in order to solve the cooperative tracking task, the filter V of each agent has to be rescaled, compared to the single-loop case, in order to compensate for the effects of cycles in the graph. Moreover, the group can only be coordinated through reference signals fed to nodes which lie within the iSCC of the graph. Corollary 4.4 is formulated for accessible states x_k. Observer-based implementations can be carried out analogously to Theorems 4.2 and 4.3.

Example 4.2. Consider a group of $N = 5$ agents with matrices A, B, C, and graph $\mathcal{G}$ as in Example 4.1. Each agent has a local reference signal r_k. The agents are locally stabilized by the LQR gain $F = [\,-20.74\ 39.93\ 6.15\,]$. We obtain the filter $V = 35.00$. For the undirected graph $\mathcal{G}$, it holds that $p = 1/N\mathbf{1}$, which leads to $V_k = 175.01$ for all $k \in \{1,\dots,5\}$. We assume that the agent states are accessible and employ couplings (4.16) with gains $K_P = [\,49.05\ 77.87\ 6.19\,]$, $K_I = 170.70$. K_e was obtained from (4.8) with the LMI region $\mathcal{R} = S(4,20,\pi/4)$. A simulation result for some random initial conditions and unit steps for the reference signals r_k of agents $k = 1,3,5$ at time $t = 3,6,9$, respectively, is shown in Fig. 4.3. The group achieves cooperative reference tracking, i.e., the synchronous output trajectory $s(t)$, plotted as dashed line, tracks $\bar{r}(t) = \sum_{k=1}^{5} d_k(t)$. △

4.2 Distributed Output Regulation

In Section 4.1, we moved from pure output synchronization to robust output synchronization and distributed tracking problems for multi-agent systems with exogenous input signals. For constant

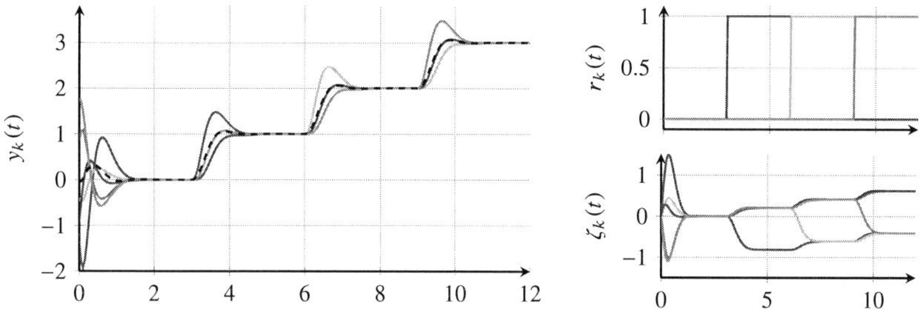

Figure 4.3: Cooperative reference tracking of the network from Example 4.2.

reference and disturbance signals, distributed control laws with integral action were developed for this purpose. In the present section, we move forward and enlarge both the class of external signals as well as the class of agent dynamics. The theoretical groundwork for the following developments is the output regulation theory for linear dynamical systems. A brief introduction and review of the essential result is provided in Section A.1.5. The present section is partially based on Seyboth et al. (2014a, 2016).

Xiang et al. (2009) and Huang (2011) suggested to formulate multi-agent coordination problems with external reference and disturbance signals as *synchronized output regulation* problem. Since then, a *cooperative output regulation* theory for a large class of practically relevant cooperative control problems has been under development. The problem setup considered by Xiang et al. (2009) and Huang (2011) consists of an autonomous exosystem and a group of identical linear agents, which are affected by the signal generated by the exosystem, and a tracking error for each agent which shall converge to zero. The problem becomes a cooperative control problem since, by assumption, not all agents have access to the external signal. In particular, the group is divided into a group of informed agents which are able to access or directly reconstruct the external signal, and uninformed agents which are dependent on information exchange with informed agents in order to solve their regulation task. The proposed distributed regulator consists of local feedback laws designed according to the classical output regulation theory and a distributed estimator for the external signal. It was shown by Hong et al. (2013) that the cooperative output regulation problem generalizes existing leader-follower control strategies. Wang et al. (2010) consider agents with non-identical and uncertain dynamics. The main limitation of the proposed solution is that the underlying communication graph is assumed to have no loops. This assumption was relaxed by Su et al. (2013) who, instead, assume the follower agents to have identical nominal dynamics. The work of Su and Huang (2012a,b) extends the results of Xiang et al. (2009); Huang (2011) and presents a solution to the cooperative output regulation problem with non-identical agents based on state feedback and output feedback. Therein, each agent is described by a generalized plant in which all matrices may be different for different agents. Cooperation among the agents is again required since the uninformed agents are not able to reconstruct the external signal locally. The solution proposed by Su and Huang (2012a) consists of three components: (1) local feedback laws which are constructed based on the classical output regulation theory for each agent, (2) local observers for the state of each agent, and (3) a distributed observer for the generalized disturbance. Further developments in this area focus on robust cooperative output regulation for uncertain agent dynamics (Su and Huang (2014); De Persis et al. (2012); Yu and Wang (2013)).

All studies mentioned above consider a single autonomous exosystem generating all reference and disturbance signals acting on the group. Meng et al. (2013) study a setup where each agent has an additional local exosystem that generates local reference signals but no disturbances are considered. Multiple exosystems are also considered by Liu et al. (2015).

In the present section, we present the distributed output regulation framework for heterogeneous linear multi-agent systems. In our formulation, we distinguish between global and local generalized disturbances affecting all agents or only individual agents in the group, respectively. In a preliminary step, we formulate the cooperative control problem as a single centralized output regulation problem. The solvability conditions for the centralized problem and its particular structure yield necessary and sufficient solvability conditions for the distributed regulation problem. Then, the distributed regulator is presented: The proposed distributed regulator is decentralized in the sense that there are no couplings based on the agent states or outputs besides the cooperative estimation protocol for the global generalized disturbance. It is inherent in the controller structure that local disturbances are rejected by the affected agent only and no other agent recognizes such a disturbance. This causes the limitation that the group is not able to react in a cooperative manner on local disturbances. In applications such as formation flight and vehicle platooning, keeping a desired formation typically has a higher priority than precisely following a given path with the formation center. As an example, a vehicle in a platoon is expected to slow down or accelerate in order to avoid collisions with its follower or predecessor when any of those experiences a disturbance. This fact is also highlighted by Bartels and Werner (2014) and motivates our further development of the distributed regulator. We introduce additional couplings among the agents in order to enable cooperation in transient phases. This extension of the distributed regulator can improve the cooperative behavior of the group significantly. We show that the transient state components (defined in Section A.1.5) of the agents are well suited for this purpose and we employ our synthesis method from Theorem 3.7. The novel distributed regulator ensures a cooperative reaction of the group on external disturbances. Finally, we discuss the problems arising when the generalized plants describing the individual agents are not decoupled. The cases of coupled agent states, measurement outputs, or regulation errors are addressed separately and corresponding solution approaches are shown.

Section 4.2.1 presents the distributed and centralized output regulation problems. The distributed regulator for general non-identical linear agents is presented in Section 4.2.2. In Section 4.2.3, the extension of the distributed regulator is derived, which guarantees an exponential decay of the synchronization error with a desired decay rate. The derivation is based on the assumption that the agents have identical dynamics. In Section 4.2.4, the assumption of identical agent dynamics is relaxed and it is shown how the coupling can be designed for similar but non-identical agents based on robust control methods. Numerical examples illustrate these results. Lastly, coupled distributed output regulation problems are addressed in Section 4.2.5.

4.2.1 Problem Formulation

In the following, the agent models, the information structure, and the group objective of the multi-agent systems under consideration are introduced.

Agent models The dynamics of the non-identical agents are described by linear state-space models. The index set of the agents is defined as $\mathcal{N} = \{1, ..., N\}$, where $N > 1$ is the number of agents in the group. The dynamics of the undisturbed agents are described by

$$\dot{x}_k = A_k x_k + B_k u_k$$

where $x_k(t) \in \mathbb{R}^{n_k^x}$ is the state and $u_k(t) \in \mathbb{R}^{n_k^u}$ is the control input of agent $k \in \mathcal{N}$. The coordination problem is formulated in terms of the generalized plant

$$\dot{x}_k = A_k x_k + B_k u_k + B_k^{d^g} d^g + B_k^{d^\ell} d_k^\ell \tag{4.18a}$$

$$y_k = C_k x_k + D_k u_k + D_k^{d^g} d^g + D_k^{d^\ell} d_k^\ell \tag{4.18b}$$

$$e_k = C_k^e x_k + D_k^e u_k + D_k^{ed^g} d^g + D_k^{ed^\ell} d_k^\ell \tag{4.18c}$$

where $y_k(t) \in \mathbb{R}^{n_k^y}$ is the measurement output of agent k and $d^g(t) \in \mathbb{R}^{n^{d^g}}$, $d_k^\ell(t) \in \mathbb{R}^{n_k^{d^\ell}}$ are external signals specified next. The regulation error $e_k(t) \in \mathbb{R}^{n_k^e}$ is defined such that asymptotic tracking and disturbance rejection is equivalent to $e_k(t) \to \mathbf{0}$ as $t \to \infty$ for all initial conditions.

External Signals We consider two types of external input signals which affect the group: a global signal that affects all agents and local signals that affect individual agents in the group. Each of these signals represents a generalized disturbance which may consist of reference signals and disturbances. While the problem formulation captures this general case, we may think of the global signal as a pure reference signal and include all disturbances into the local signals. We assume that all signals belong to a known family of signals. More precisely, the global signal $d^g(t), t \geq 0$, is a solution of the autonomous linear system,

$$\dot{d}^g = S^g d^g \tag{4.19}$$

called global exosystem, where $\sigma(S^g) \subset \mathbb{C}^0 \cup \mathbb{C}^+$. The local signal $d_k^\ell(t), t \geq 0$, acting on agent k is a solution of the local exosystem

$$\dot{d}_k^\ell = S_k^\ell d_k^\ell, \tag{4.20}$$

where $\sigma(S_k^\ell) \subset \mathbb{C}^0 \cup \mathbb{C}^+$ for all $k \in \mathcal{N}$.

Remark 4.1. The exosystems (4.19), (4.20) could be combined into a single large exosystem generating all disturbances acting on the group. In that sense, the present formulation does not enlarge the problem class compared to Su and Huang (2012a,c). The main benefit of taking the structure of the exosystem explicitly into account is the decomposition of the regulator equations and the reduction of the controller dimension, as we will see later in Lemma 4.5 and Theorem 4.6, respectively. ○

Information Structure The same information structure as in Section 3.1 is assumed, i.e., the communication topology is described by a directed graph $\mathcal{G} = (\mathcal{V}, \mathcal{E}, A_\mathcal{G})$.

The Distributed Output Regulation Problem

The group objective under consideration consists of two parts: asymptotic tracking of reference signals and asymptotic disturbance rejection. Typically, we are interested in synchronization problems which can be formulated as tracking problem of a common, global reference signal. Moreover, our focus will be on the cooperative behavior of the group in transient phases. The distributed output regulation problem is formally stated as follows.

Problem 4.2 (Distributed Output Regulation Problem). For each $k \in \mathcal{N}$, find a regulator

$$\dot{z}_k = A^c_{kk} z_k + B^c_{kk} y_k + \sum_{j \in \mathcal{N}_k} \left(A^c_{kj} z_j + B^c_{kj} y_j \right) \tag{4.21a}$$

$$u_k = C^c_{kk} z_k + \sum_{j \in \mathcal{N}_k} C^c_{kj} z_j \tag{4.21b}$$

with $z_k(t) \in \mathbb{R}^{n^z_k}$, such that the following two conditions are satisfied:

P1) If $d^g(0) = \mathbf{0}$, $d^\ell_k(0) = \mathbf{0}$, $k \in \mathcal{N}$, then, for all initial conditions $x_k(0)$, $z_k(0)$, it holds that

$$\lim_{t\to\infty} x_k(t) = \mathbf{0} \qquad \text{and} \qquad \lim_{t\to\infty} z_k(t) = \mathbf{0}.$$

P2) For all initial conditions $d^g(0) = d^g_0$, $d^\ell_k(0)$, $x_k(0)$, $z_k(0)$, $k \in \mathcal{N}$, it holds that

$$\lim_{t\to\infty} e_k(t) = \mathbf{0}.$$

Property **P1)** is referred to as internal stability while **P2)** is referred to as asymptotic regulation.

The Centralized Output Regulation Problem

Problem 4.2 is called Distributed Output Regulation Problem due to the structure imposed on the regulator (4.21). In case of a complete graph $\mathcal{G}$, i.e., all-to-all communication, (4.21) becomes a centralized dynamic output feedback controller of the form

$$\dot{z} = A^c z + B^c y \tag{4.22a}$$

$$u = C^c z, \tag{4.22b}$$

where z, y, u are the stack vectors of z_k, y_k, u_k, $k \in \mathcal{N}$, respectively. Let x, e be the stack vectors of x_k, e_k, $k \in \mathcal{N}$, and $d = [d^{g\mathsf{T}}\ d^{\ell\mathsf{T}}_1\ \cdots\ d^{\ell\mathsf{T}}_N]^\mathsf{T}$. Then, the overall cooperative control problem can be formulated as a single classical output regulation problem by aggregating all agents (4.18) and exosystems (4.19), (4.20) into one large generalized plant and one large exosystem as follows:

$$\dot{x} = \mathcal{A}x + \mathcal{B}u + \mathcal{B}_d d \tag{4.23a}$$

$$y = \mathcal{C}x + \mathcal{D}u + \mathcal{D}_d d \tag{4.23b}$$

$$e = \mathcal{C}_e x + \mathcal{D}_e u + \mathcal{D}_{ed} d \tag{4.23c}$$

and

$$\dot{d} = \mathcal{S}d, \tag{4.24}$$

with the matrices given by

$$\mathcal{A} = \begin{bmatrix} A_1 & & 0 \\ & \ddots & \\ 0 & & A_N \end{bmatrix}, \quad \mathcal{B} = \begin{bmatrix} B_1 & & 0 \\ & \ddots & \\ 0 & & B_N \end{bmatrix}, \quad \mathcal{B}_d = \begin{bmatrix} B^{d^g}_1 & B^{d^\ell}_1 & & 0 \\ \vdots & & \ddots & \\ B^{d^g}_N & 0 & & B^{d^\ell}_N \end{bmatrix},$$

$$\mathcal{C} = \begin{bmatrix} C_1 & & 0 \\ & \ddots & \\ 0 & & C_N \end{bmatrix}, \quad \mathcal{D} = \begin{bmatrix} D_1 & & 0 \\ & \ddots & \\ 0 & & D_N \end{bmatrix}, \quad \mathcal{D}_d = \begin{bmatrix} D^{d^g}_1 & D^{d^\ell}_1 & & 0 \\ \vdots & & \ddots & \\ D^{d^g}_N & 0 & & D^{d^\ell}_N \end{bmatrix},$$

$$\mathcal{C}_e = \begin{bmatrix} C^e_1 & & 0 \\ & \ddots & \\ 0 & & C^e_N \end{bmatrix}, \quad \mathcal{D}_e = \begin{bmatrix} D^e_1 & & 0 \\ & \ddots & \\ 0 & & D^e_N \end{bmatrix}, \quad \mathcal{D}_{ed} = \begin{bmatrix} D^{ed^g}_1 & D^{ed^\ell}_1 & & 0 \\ \vdots & & \ddots & \\ D^{ed^g}_N & 0 & & D^{ed^\ell}_N \end{bmatrix},$$

and

$$\mathcal{S} = \begin{bmatrix} S^g & & & 0 \\ & S_1^\ell & & \\ & & \ddots & \\ 0 & & & S_N^\ell \end{bmatrix}.$$

A distributed solution to the output regulation problem can only exist if the overall output regulation problem with (4.23), (4.24) has a centralized solution of the form (4.22). Hence, we study the necessary conditions for the solvability of the overall output regulation problem and exploit the structure in order to derive necessary conditions for the local output regulation problems. Theorem A.20 directly yields the following statement.

Let the pair $(\mathcal{A}, \mathcal{B})$ be stabilizable and the pair

$$\left(\begin{bmatrix} \mathcal{A} & \mathcal{B}_d \\ 0 & \mathcal{S} \end{bmatrix}, \begin{bmatrix} \mathcal{C} & \mathcal{D}_d \end{bmatrix} \right) \tag{4.25}$$

be detectable. Then, Problem 4.2 has a centralized solution (4.22), if and only if the regulator equation

$$\begin{bmatrix} \mathcal{A} & \mathcal{B} \\ \mathcal{C}_e & \mathcal{D}_e \end{bmatrix} \begin{bmatrix} \Pi \\ \Gamma \end{bmatrix} - \begin{bmatrix} \Pi \\ 0 \end{bmatrix} \mathcal{S} + \begin{bmatrix} \mathcal{B}_d \\ \mathcal{D}_{ed} \end{bmatrix} = 0 \tag{4.26}$$

is solvable with a solution Π, Γ. In case all these conditions are fulfilled, a centralized regulator can be constructed as follows. Choose F and L such that $\mathcal{A} - \mathcal{B}\mathsf{F}$ and

$$\begin{bmatrix} \mathcal{A} & \mathcal{B}_d \\ 0 & \mathcal{S} \end{bmatrix} - \mathsf{L} \begin{bmatrix} \mathcal{C} & \mathcal{D}_d \end{bmatrix}$$

are Hurwitz. Define $\mathsf{G} = \Gamma + \mathsf{F}\Pi$ where Π, Γ solve (4.26). Then, the output regulator is

$$\begin{aligned} \begin{bmatrix} \dot{\hat{x}} \\ \dot{\hat{d}} \end{bmatrix} &= \begin{bmatrix} \mathcal{A} & \mathcal{B}_d \\ 0 & \mathcal{S} \end{bmatrix} \begin{bmatrix} \hat{x} \\ \hat{d} \end{bmatrix} + \begin{bmatrix} \mathcal{B} \\ 0 \end{bmatrix} u + \mathsf{L}(y - \hat{y}) \\ u &= -\mathsf{F}\hat{x} + \mathsf{G}\hat{d}, \end{aligned}$$

where $\hat{y} = \mathcal{C}\hat{x} + \mathcal{D}u + \mathcal{D}_d\hat{d}$. The control law $u = -\mathsf{F}x + \mathsf{G}d$ is referred to as the full information control law. Since x and d are not directly accessible, the observer is constructed to obtain the estimates $\hat{x}$ and $\hat{d}$. This controller is of the form (4.22) with $z = [\hat{x}^\mathsf{T} \;\; \hat{d}^\mathsf{T}]^\mathsf{T}$.

It turns out that the solvability condition in terms of the centralized regulator equation (4.26) decomposes into a set of local regulator equations for the individual agents, as stated next.

Lemma 4.5. *The regulator equation* (4.26) *for the centralized output regulation problem is solvable, if and only if the local regulator equations*

$$\begin{bmatrix} A_k & B_k \\ C_k^e & D_k^e \end{bmatrix} \begin{bmatrix} \Pi_k^g \\ \Gamma_k^g \end{bmatrix} - \begin{bmatrix} \Pi_k^g \\ 0 \end{bmatrix} S^g + \begin{bmatrix} B_k^{d^g} \\ D_k^{ed^g} \end{bmatrix} = 0 \tag{4.27}$$

and

$$\begin{bmatrix} A_k & B_k \\ C_k^e & D_k^e \end{bmatrix} \begin{bmatrix} \Pi_k^\ell \\ \Gamma_k^\ell \end{bmatrix} - \begin{bmatrix} \Pi_k^\ell \\ 0 \end{bmatrix} S_k^\ell + \begin{bmatrix} B_k^{d^\ell} \\ D_k^{ed^\ell} \end{bmatrix} = 0 \tag{4.28}$$

have a solution $\Pi_k^g \in \mathbb{R}^{n_k^x \times n^{d^g}}$, $\Gamma_k^g \in \mathbb{R}^{n_k^u \times n^{d^g}}$, $\Pi_k^\ell \in \mathbb{R}^{n_k^x \times n_k^{d^\ell}}$, $\Gamma_k^\ell \in \mathbb{R}^{n_k^u \times n_k^{d^\ell}}$ *for all* $k \in \mathcal{N}$.

Proof. See Appendix B.13. □

Lemma 4.5 shows that, due to the structure of the overall generalized plant (4.23) and exosystem (4.24), the existence of a solution to (4.26) is equivalent to the existence of solutions to the local regulator equations (4.27), (4.28). Consequently, the solvability of (4.27), (4.28) is a necessary condition, not only for the existence of a distributed regulator but also for the existence of a centralized regulator with global information.

The following assumptions will be needed in the sequel.

Assumption 4.2. *The pair* (A_k, B_k) *is stabilizable for all* $k \in \mathcal{N}$.

Assumption 4.3. *The pair*

$$\left(\begin{bmatrix} A_k & B_k^{d^\ell} \\ 0 & S_k^\ell \end{bmatrix}, \begin{bmatrix} C_k & D_k^{d^\ell} \end{bmatrix}\right) \tag{4.29}$$

is detectable for all $k \in \mathcal{N}$.

Assumption 4.4. *The local regulator equations* (4.27) *and* (4.28) *have a solution for all* $k \in \mathcal{N}$.

Assumption 4.5. *The communication topology is described by a directed connected graph* $\mathcal{G}$ *and node* 1 *is the root of a spanning tree.*

Assumption 4.6. *Agent one has direct access to the signal* $d^g(t)$.

Remark 4.2. A discussion of each assumption is provided in order to point out their importance:

Assumption 4.2 is equivalent to stabilizability of the pair $(\mathcal{A}, \mathcal{B})$ due to the structure of (4.23). Moreover, it is required for **P1)** since the plant (4.18a) may be unstable.

Assumption 4.3 is necessarily satisfied if the pair (4.25) is detectable due to the structure of (4.23), (4.24). Note that detectability of (4.25) does *not* imply detectability of all pairs

$$\left(\begin{bmatrix} A_k & B_k^{d^g} & B_k^{d^\ell} \\ 0 & S^g & 0 \\ 0 & 0 & S_k^\ell \end{bmatrix}, \begin{bmatrix} C_k & D_k^{d^g} & D_k^{d^\ell} \end{bmatrix}\right). \tag{4.30}$$

The pair (4.25) is detectable if (4.30) is detectable for at least one agent $k \in \mathcal{N}$. This agent must be the root of a spanning tree in order to solve the distributed regulation problem. The agents have to cooperate in order to obtain an estimate of d^g.

Assumption 4.4 is motivated by Lemma 4.5. It is a necessary condition for the existence of a distributed (as well as a centralized) regulator solving Problem 4.2. Hence it causes no loss of generality. Equations (4.27), (4.28) can be rewritten as

$$A_k\Pi_k^g + B_k\Gamma_k^g - \Pi_k^g S^g + B_k^{d^g} = 0 \tag{4.31a}$$

$$C_k^e\Pi_k^g + D_k^e\Gamma_k^g + D_k^{ed^g} = 0 \tag{4.31b}$$

$$A_k\Pi_k^\ell + B_k\Gamma_k^\ell - \Pi_k^\ell S_k^\ell + B_k^{d^\ell} = 0 \tag{4.32a}$$

$$C_k^e\Pi_k^\ell + D_k^e\Gamma_k^\ell + D_k^{ed^\ell} = 0. \tag{4.32b}$$

A sufficient condition for the existence of a solution is the following (Huang, 2004, Thm. 1.9): The local regulator equations for agent $k \in \mathcal{N}$ are solvable, if for all $\lambda \in \sigma(S^g) \cup \sigma(S_k^\ell)$,

$$\operatorname{rank}\begin{bmatrix} A_k - \lambda I_{n_k^x} & B_k \\ C_k^e & D_k^e \end{bmatrix} = n_k^x + n_k^e.$$

Assumption 4.5 is a fundamental requirement for the construction of a distributed regulator. *Assumption 4.6* along with Assumption 4.3 guarantees detectability of (4.25). Assumption 4.6 can easily be relaxed to detectability of the pair (4.30) for agent $k = 1$ by construction of a local observer providing an estimate $\hat{d}_1^g$ of d^g. We work with the stricter assumption for ease of presentation. ○

4.2.2 The Distributed Regulator

The following result presents the general distributed regulator solving Problem 4.2. In its original form, the distributed regulator is due to Su and Huang (2012a). Here we present the refined formulation based on Seyboth et al. (2016), taking into account both global and local exosystems. The proof is included for a clearer understanding since Section 4.2.3 builds on insights gained in this proof.

Theorem 4.6 (Distributed Regulator). *Consider a group of $N > 1$ agents* (4.18) *with exosystems* (4.19), (4.20). *Suppose that Assumptions 4.2–4.6 are satisfied. Then, a distributed regulator solving Problem 4.2 can be constructed as follows:*

- *For all $k \in \mathcal{N}$, choose F_k such that $A_k - B_k F_k$ is Hurwitz.*
- *For all $k \in \mathcal{N}$, find a solution for* (4.27) *and* (4.28) *and set*

$$G_k^g = \Gamma_k^g + F_k \Pi_k^g \tag{4.33a}$$
$$G_k^\ell = \Gamma_k^\ell + F_k \Pi_k^\ell \tag{4.33b}$$

- *Let $\mathcal{G}_1$ be the graph $\mathcal{G}$ after deletion of all incoming edges to node 1. Set $\hat{d}_1^g = d^g$ and*

$$\dot{\hat{d}}_k^g = S^g \hat{d}_k^g + K \sum_{j \in \mathcal{N}_k} (\hat{d}_j^g - \hat{d}_k^g) \tag{4.34}$$

  for all $k \in \mathcal{N}\backslash\{1\}$, where K is chosen such that $S^g - \lambda_k K$ is stable for the non-zero eigenvalues λ_k, $k \in \mathcal{N}\backslash\{1\}$, of the Laplacian $L_{\mathcal{G}_1}$ of $\mathcal{G}_1$.

- *For all $k \in \mathcal{N}$, choose L_k such that*

$$\begin{bmatrix} A_k & B_k^{d^\ell} \\ 0 & S_k^\ell \end{bmatrix} - L_k \begin{bmatrix} C_k & D_k^{d^\ell} \end{bmatrix} \tag{4.35}$$

  is Hurwitz and construct the local observers

$$\begin{bmatrix} \dot{\hat{x}}_k \\ \dot{\hat{d}}_k^\ell \end{bmatrix} = \begin{bmatrix} A_k & B_k^{d^\ell} \\ 0 & S_k^\ell \end{bmatrix} \begin{bmatrix} \hat{x}_k \\ \hat{d}_k^\ell \end{bmatrix} + \begin{bmatrix} B_k & B_k^{d^g} \\ 0 & 0 \end{bmatrix} \begin{bmatrix} u_k \\ \hat{d}_k^g \end{bmatrix} + L_k (y_k - \hat{y}_k) \tag{4.36}$$

  where $\hat{y}_k = C_k \hat{x}_k + D_k u_k + D_k^{d^g} \hat{d}_k^g + D_k^{d^\ell} \hat{d}_k^\ell$.

Finally, the control law for each agent $k \in \mathcal{N}$ is given by

$$u_k = -F_k \hat{x}_k + G_k^g \hat{d}_k^g + G_k^\ell \hat{d}_k^\ell. \tag{4.37}$$

Proof. Define the observer errors

$$\epsilon_k^x = x_k - \hat{x}_k, \qquad \epsilon_k^{d^\ell} = d_k^\ell - \hat{d}_k^\ell, \qquad \epsilon_k^{d^g} = d^g - \hat{d}_k^g. \tag{4.38}$$

The observer errors ϵ_k^x and $\epsilon_k^{d^\ell}$ satisfy

$$\begin{bmatrix} \dot{\epsilon}_k^x \\ \dot{\epsilon}_k^{d^\ell} \end{bmatrix} = \begin{bmatrix} \dot{x}_k \\ \dot{d}_k^\ell \end{bmatrix} - \begin{bmatrix} \dot{\hat{x}}_k \\ \dot{\hat{d}}_k^\ell \end{bmatrix} = \left(\begin{bmatrix} A_k & B_k^{d^\ell} \\ 0 & S_k^\ell \end{bmatrix} - L_k \begin{bmatrix} C_k & D_k^{d^\ell} \end{bmatrix} \right) \begin{bmatrix} \epsilon_k^x \\ \epsilon_k^{d^\ell} \end{bmatrix} + \left(\begin{bmatrix} B_k^{d^g} \\ 0 \end{bmatrix} - L_k D_k^{d^g} \right) \epsilon_k^{d^g}.$$

The observer error $\epsilon_k^{d^g}$ satisfies $\epsilon_1^{d^g} = \mathbf{0}$ and for $k \in \mathcal{N} \setminus \{1\}$,

$$\dot{\epsilon}_k^{d^g} = \dot{d}^g - \dot{\hat{d}}_k^g = S^g \epsilon_k^{d^g} + K \sum_{j \in \mathcal{N}_k} (\epsilon_j^{d^g} - \epsilon_k^{d^g}).$$

By Assumption 4.5, $\mathcal{G}$ contains a spanning tree rooted at 1. Hence, a suitable gain matrix K in (4.34) exists by Theorem 3.2 and design methods are provided in Section 3.1.3. Consequently, for all $k \in \mathcal{N}$, $\epsilon_k^{d^g}(t) = d^g(t) - \hat{d}_k^g(t) \to \mathbf{0}$ as $t \to \infty$. Hence, it follows that for all $k \in \mathcal{N}$, $\epsilon_k^x(t) \to \mathbf{0}$ and $\epsilon_k^{d^\ell}(t) \to \mathbf{0}$ as $t \to \infty$ since (4.35) is Hurwitz by construction. Next, we define

$$\epsilon_k = x_k - \Pi_k^g d^g - \Pi_k^\ell d_k^\ell. \tag{4.39}$$

The variable ϵ_k is the transient state component of agent k. In particular, it is the state component in the complement of the subspace $\mathcal{V}_k^+ = \{(x_k, d^g, d_k^\ell) : x_k = \Pi_k^g d^g + \Pi_k^\ell d_k^\ell\}$. The equations (4.31a), (4.32a) express the fact that the subspace $\mathcal{V}_k^+$ is a controlled invariant subspace of the system

$$\begin{bmatrix} \dot{x}_k \\ \dot{d}^g \\ \dot{d}_k^\ell \end{bmatrix} = \begin{bmatrix} A_k & B_k^{d^g} & B_k^{d^\ell} \\ 0 & S^g & 0 \\ 0 & 0 & S_k^\ell \end{bmatrix} \begin{bmatrix} x_k \\ d^g \\ d_k^\ell \end{bmatrix} + \begin{bmatrix} B_k \\ 0 \\ 0 \end{bmatrix} u_k$$

and $\mathcal{V}_k^+$ is rendered invariant by $u_k = \Gamma_k^g d^g + \Gamma_k^\ell d_k^\ell$. Moreover, this subspace is in the null space of the regulation error map (4.18c) due to (4.31b), (4.32b). As we will show next, $\mathcal{V}_k^+$ is rendered asymptotically stable by the proposed distributed regulator, and $e_k(t)$ indeed converges to zero for all initial conditions. Using (4.18a), (4.19), (4.20), (4.33), (4.37), (4.38), and (4.39), we compute

$$\begin{aligned} \dot{\epsilon}_k &= \dot{x}_k - \Pi_k^g \dot{d}^g - \Pi_k^\ell \dot{d}_k^\ell \\ &= (A_k - B_k F_k)\epsilon_k + B_k F_k \epsilon_k^x - B_k G_k^g \epsilon_k^{d^g} - B_k G_k^\ell \epsilon_k^{d^\ell} \\ &\quad + (A_k \Pi_k^g + B_k \Gamma_k^g - \Pi_k^g S^g + B_k^{d^g}) d^g + (A_k \Pi_k^\ell + B_k \Gamma_k^\ell - \Pi_k^\ell S_k^\ell + B_k^{d^\ell}) d_k^\ell. \end{aligned}$$

Moreover, with (4.31a) and (4.32a), this reduces to

$$\dot{\epsilon}_k = (A_k - B_k F_k)\epsilon_k + B_k F_k \epsilon_k^x - B_k G_k^g \epsilon_k^{d^g} - B_k G_k^\ell \epsilon_k^{d^\ell}. \tag{4.40}$$

Since the observer errors ϵ_k^x, $\epsilon_k^{d^g}$, $\epsilon_k^{d^\ell}$ vanish asymptotically and $A_k - B_k F_k$ is Hurwitz by construction, we can conclude that $\epsilon_k(t) \to \mathbf{0}$ as $t \to \infty$ for all $k \in \mathcal{N}$. If $d^g(0) = \mathbf{0}$ and $d_k^\ell(0) = \mathbf{0}$ for all $k \in \mathcal{N}$, then $x_k(t) = \epsilon_k(t) \to \mathbf{0}$ as $t \to \infty$, and since the observer errors vanish asymptotically, also $\hat{x}_k(t) \to \mathbf{0}$, $\hat{d}_k^g(t) \to \mathbf{0}$, and $\hat{d}_k^\ell(t) \to \mathbf{0}$ as $t \to \infty$. Consequently, **P1)** is fulfilled. It remains to show that the regulation errors e_k converge to zero for all initial conditions. Using (4.18c), (4.33), (4.37), and (4.38), we compute

$$\begin{aligned} e_k &= C_k^e x_k + D_k^e u_k + D_k^{ed^g} d^g + D_k^{ed^\ell} d_k^\ell \\ &= (C_k^e - D_k^e F_k)\epsilon_k - D_k^e(-F_k \epsilon_k^x + G_k^g \epsilon_k^{d^g} + G_k^\ell \epsilon_k^{d^\ell}) \\ &\quad + (C_k^e \Pi_k^g + D_k^e \Gamma_k^g + D_k^{ed^g}) d^g + (C_k^e \Pi_k^\ell + D_k^e \Gamma_k^\ell + D_k^{ed^\ell}) d_k^\ell. \end{aligned}$$

Moreover, with (4.31b) and (4.32b), this reduces to

$$e_k = (C_k^e - D_k^e F_k)\epsilon_k - D_k^e(-F_k \epsilon_k^x + G_k^g \epsilon_k^{d^g} + G_k^\ell \epsilon_k^{d^\ell}). \tag{4.41}$$

Both the state component ϵ_k and the observer errors ϵ_k^x, $\epsilon_k^{d^g}$, $\epsilon_k^{d^\ell}$ vanish asymptotically. Hence, $e_k(t) \to \mathbf{0}$ as $t \to \infty$ for all $k \in \mathcal{N}$ and **P2)** is fulfilled. □

Remark 4.3. The dynamic diffusive couplings discussed in Remark 3.8 solving output synchronization problems for heterogeneous linear networks are a special case of the distributed regulator in Theorem 4.6. In particular, the distributed regulator reduces to the couplings in Remark 3.8 if there are no local generalized disturbance signals and if the global exosystem generates a common reference signal, i.e., the regulation errors are defined as $e_k = y_k - Rd^g$ for all $k \in \mathcal{N}$. From this point of view, the distributed regulator is a significant generalization of the dynamic diffusive couplings proposed by Wieland et al. (2011b). ○

The distributed regulator realizes a two-level controller architecture. The virtual network (4.34) is established on a higher *network level* and guarantees that all agents have access to synchronized reference signals. The decoupled output regulation control laws (4.37) on a lower *agent level* achieve internal stability, disturbance rejection, and asymptotic tracking of the synchronized and local reference signals for each agent.

The example presented next serves two purposes. On the one hand, it shall illustrate the distributed regulator presented in Theorem 4.6 by application to a concrete multi-agent system. On the other hand, it shall reveal a limitation inherent to the two-level architecture and motivate the developments in the next section. Note that the distributed regulator allows for heterogeneous agent dynamics, even though the following example involves identical agents.

Example 4.3 (Vehicle Platooning)**.** We consider a group of $N = 5$ vehicles in a platoon, each modeled as a double-integrator system of the form $\ddot{s}_k = u_k + w_k$, where $s_k(t) \in \mathbb{R}$ is the vehicle position, $\dot{s}_k = v_k$ its velocity, $u_k(t) \in \mathbb{R}$ its control input, and $w_k(t) \in \mathbb{R}$ is a local disturbance acting on vehicle k, e.g., due to a mass change, gear shift, or other external influence. There is a ramp shaped global reference signal $s_0(t) \in \mathbb{R}$, $\dot{s}_0 = v_0$, $v_0 \in \mathbb{R}$, for the position of the platoon, i.e., for the first vehicle. Moreover, there are constant local reference signals $r_k(t) \in \mathbb{R}$ commanding the position of each vehicle with respect to the global reference. This scenario is illustrated in Fig. 4.4 and captured by the generalized plants

$$\begin{aligned} \dot{x}_k &= \begin{bmatrix} 0 & 1 \\ 0 & 0 \end{bmatrix} x_k + \begin{bmatrix} 0 \\ 1 \end{bmatrix} u_k + \begin{bmatrix} 0 & 0 \\ 0 & 0 \end{bmatrix} d^g + \begin{bmatrix} 0 & 0 \\ 0 & -1 \end{bmatrix} d_k^\ell \\ y_k &= \begin{bmatrix} 0 & 0 \\ 1 & 0 \\ 0 & 1 \end{bmatrix} x_k + \begin{bmatrix} 0 & 0 \\ 0 & 0 \\ 0 & 0 \end{bmatrix} u_k + \begin{bmatrix} 0 & 0 \\ 0 & 0 \\ 0 & 0 \end{bmatrix} d^g + \begin{bmatrix} 1 & 0 \\ 0 & 0 \\ 0 & 0 \end{bmatrix} d_k^\ell \\ e_k &= \begin{bmatrix} 1 & 0 \\ 0 & 1 \end{bmatrix} x_k + \begin{bmatrix} 0 & 0 \\ 0 & 0 \end{bmatrix} u_k + \begin{bmatrix} -1 & 0 \\ 0 & -1 \end{bmatrix} d^g + \begin{bmatrix} 0 & -1 \\ 0 & 0 \end{bmatrix} d_k^\ell \end{aligned}$$

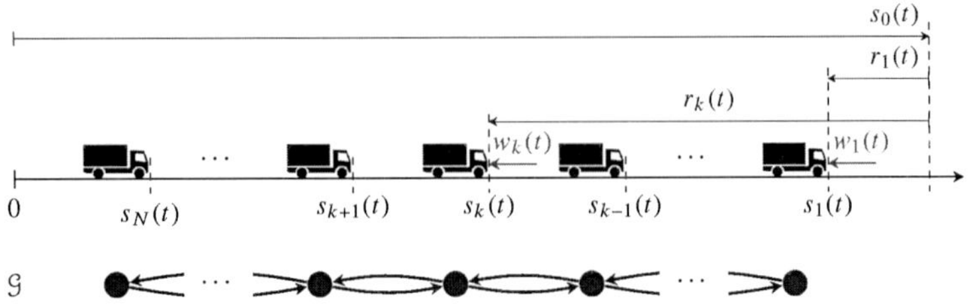

Figure 4.4: Illustration of the vehicle platoon from Example 4.3.

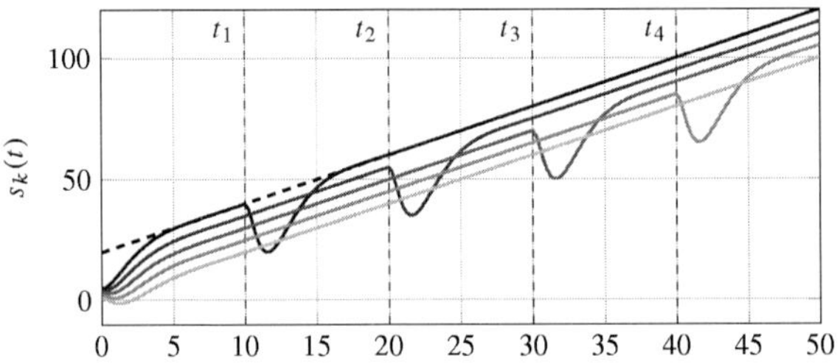

Figure 4.5: Simulation results for the vehicle platoon from Example 4.3 with the distributed regulator in Theorem 4.6. The plot shows the reference $s_0(t)$ (dashed) and the vehicle positions $s_k(t)$ for $k = 1,\ldots,5$ (black–light gray).

with $x_k = [s_k \;\; v_k]^\mathsf{T}$ for all $k \in \mathcal{N}$. The generalized disturbances $d^g = [s_0 \;\; v_0]^\mathsf{T}$ and $d_k^\ell = [r_k \;\; w_k]^\mathsf{T}$ are generated by the exosystems

$$\dot{d}^g = \begin{bmatrix} 0 & 0 \\ 0 & 1 \end{bmatrix} d^g, \qquad \dot{d}_k^\ell = \begin{bmatrix} 0 & 0 \\ 0 & 0 \end{bmatrix} d_k^\ell, \quad k \in \mathcal{N}.$$

Each vehicle can communicate with its follower and predecessor in the platoon, i.e., $\mathcal{G}$ is an undirected line graph as shown in Fig. 4.4, and only agent 1 has direct access to d^g. All assumptions of Theorem 4.6 are satisfied and a distributed regulator is constructed as follows. We find $F_k = [\,0.7071 \;\; 1.3836\,]$ via LQR design. Equations (4.27), (4.28) are solved by the matrices

$$\Pi_k^g = \begin{bmatrix} 1 & 0 \\ 0 & 1 \end{bmatrix}, \qquad \Gamma_k^g = \begin{bmatrix} 0 & 0 \end{bmatrix}, \qquad \Pi_k^\ell = \begin{bmatrix} 1 & 0 \\ 0 & 0 \end{bmatrix}, \qquad \Gamma_k^\ell = \begin{bmatrix} 0 & 1 \end{bmatrix},$$

which yield $G_k^g = \Gamma_k^g + F_k\Pi_k^g$ and $G_k^\ell = \Gamma_k^\ell + F_k\Pi_k^\ell$. Each agent knows its local reference r_k but the local disturbance w_k needs to be estimated. For that purpose, local observers analogous to (4.36) with state $[\hat{x}_k^\mathsf{T} \;\; \hat{w}_k]^\mathsf{T}$ are constructed, where $L_k = \left[\begin{smallmatrix} 4.0999 & -0.1622 & -0.6699 \\ 0.9999 & 8.2001 & 16.8003 \end{smallmatrix}\right]^\mathsf{T}$ places the poles at $\{-4.0, -4.1, -4.2\}$. The coupling gain K is designed via Theorem 3.7 for $\eta = 2.5$ with the LMI region $S(2.5, 75, \pi/4)$, which yields $K = \left[\begin{smallmatrix} -20.9938 & -1.6601 \\ -0.0261 & -21.0682 \end{smallmatrix}\right]$.

For the simulations, we set $s_0(0) = 20$, $v_0 = 2$, and $r_k = -5(k-1)$. In order to assess the behavior of the closed loop, particularly in transient phases, the step signals $w_1(t) = 100$ for $t \geq t_1$, $w_2(t) = 100$ for $t \geq t_2$, $w_3(t) = 100$ for $t \geq t_3$, $w_4(t) = 100$ for $t \geq t_4$ are applied to the vehicles. The distributed regulator (4.37) leads to the simulation result shown in Fig. 4.5. The distributed regulation problem is solved, i.e., the reference signals are tracked and the disturbances are rejected asymptotically.

This example illustrates an inherent limitation of the control scheme: The platoon does not react cooperatively on local disturbances acting on individual vehicles. The steps in the constant disturbances are rejected by the local controllers but the other vehicles have no information about the disturbance and cannot adjust their actions accordingly. A desirable and cooperative reaction to the local disturbances would be that the other vehicles in the platoon slow down or accelerate in order to maintain the desired inter-vehicle distances during the transient phase. For safety reasons, maintaining the desired distances is of higher importance than maintaining the desired velocity. Such performance requirements cannot be taken into account explicitly with the distributed regulator described in Theorem 4.6. △

4.2.3 Transient Synchronization: Identical Agents

The distributed regulator presented in Section 4.2.2 solves the distributed output regulation problem and realizes a cooperative behavior of the group. However, as we will see next, the cooperative behavior can be improved significantly by a suitable extension of the distributed regulator. As motivated in the introduction and in Example 4.3, we realize a cooperative reaction to local disturbances, such that a suitably defined synchronization error is kept small in transient phases. We propose a novel distributed regulator with additional couplings among the agents *on the agent level* stabilizing the synchronization error and guaranteeing a desired performance. In a first step, we consider groups of agents with identical dynamics. In particular, we impose the following assumption.

Assumption 4.7. *There exist matrices A, B, C^e, D^e, F such that for all $k \in \mathcal{N}$, it holds that $A_k = A$, $B_k = B$, $C^e_k = C^e$, $D^e_k = D^e$, $F_k = F$.*

Note that not all matrices in (4.18) are required to be identical for the individual agents. The measurement output maps (4.18b) are allowed to be non-identical, and the local exosystems (4.20) as well as the generalized disturbance input matrices in (4.18a), (4.18c), can still be non-identical. This leaves great flexibility in the problem formulation despite Assumption 4.7. Moreover, Assumption 4.7 will be relaxed to a certain extent in the next section.

From the proof of Theorem 4.6, we see that the disagreement of the group can be quantified based on the transient state components ϵ_k defined in (4.39). According to (4.41), $e_k = (C^e - D^e F)\epsilon_k$ under Assumption 4.7 and without observer errors. Since $C^e - D^e F$ is identical for all $k \in \mathcal{N}$, agreement of ϵ_k corresponds to agreement of the regulation errors e_k. For each agent k, we define the transient synchronization error

$$\epsilon^s_k = \epsilon_k - \frac{1}{N}\sum_{j=1}^{N} \epsilon_j. \tag{4.42}$$

In the following, we propose a distributed regulator which solves Problem 4.2 and, additionally, exponentially stabilizes the synchronization errors ϵ^s_k with a certain desired decay rate $\gamma > 0$. In order to guarantee that this decay rate can be achieved, we refine Assumption 4.2 as follows.

Assumption 4.8. *For a given $\gamma > 0$, the pair $(A_k + \gamma I_{n^x_k}, B_k)$ is stabilizable for all $k \in \mathcal{N}$.*

We start with the full information case. This allows to present the main idea in a clear way and will be instrumental in the proof of the output feedback case.

Lemma 4.7 (Distributed Full Information Regulator). *Consider a group of $N > 1$ agents* (4.18) *with exosystems* (4.19), (4.20). *Suppose that Assumptions 4.4, 4.5, 4.7, and 4.8 are satisfied. Then, a distributed full-information regulator solving Problem 4.2 and additionally achieving exponential stability of the synchronization errors ϵ_k^s with decay rate $\gamma > 0$ can be constructed as follows:*

- *Choose F such that $A - BF$ is Hurwitz.*
- *For all $k \in \mathcal{N}$, find a solution for* (4.31) *and* (4.32) *and set $G_k^g = \Gamma_k^g + F\Pi_k^g$, $G_k^\ell = \Gamma_k^\ell + F\Pi_k^\ell$.*
- *Choose H such that for all $k \in \mathcal{N}\backslash\{1\}$,*

$$\max\{\mathbf{Re}(\mu) : \mu \in \sigma(A - BF - \lambda_k BH)\} < -\gamma, \tag{4.43}$$

  i.e., the maximal real part of all eigenvalues of $A - BF - \lambda_k BH$ is smaller than $-\gamma$.

Finally, the control law for each agent $k \in \mathcal{N}$ is given by

$$u_k = -Fx_k + G_k^g d^g + G_k^\ell d_k^\ell + H \sum_{j \in \mathcal{N}_k} \left(\epsilon_j - \epsilon_k\right). \tag{4.44}$$

Proof. See Appendix B.14. □

The novel control law (4.44) solves Problem 4.2 and additionally enforces synchronization of the regulation errors with a desired decay rate and therefore has the desired effect, that is, a cooperative reaction of the group on disturbances acting on individual agents. The gain matrices F and H allow to tune the local disturbance rejection of each agent and the synchronization of the group separately. Theorem 3.7 serves as a design method for H.

The full information control law (4.44) is impractical when the agents do not have direct access to their state and generalized disturbance signals. The following theorem shows that (4.44) can be implemented based on observers, analogously to (4.37) in Theorem 4.6. In order to guarantee that the desired decay rate can be achieved, we refine Assumption 4.3.

Assumption 4.9. *For a given $\eta > 0$, the pair*

$$\left(\begin{bmatrix} A_k & B_k^{d^\ell} \\ 0 & S_k^\ell \end{bmatrix} + \eta I_{(n_k^x + n_k^{d^\ell})}, \begin{bmatrix} C_k & D_k^{d^\ell} \end{bmatrix}\right),$$

is detectable for all $k \in \mathcal{N}$.

Theorem 4.8 (Distributed Output Feedback Regulator). *Consider a group of $N > 1$ agents* (4.18) *with exosystems* (4.19), (4.20). *Suppose that Assumptions 4.4, 4.5, 4.6, 4.7, 4.8, and 4.9 are satisfied. Construct a distributed regulator as in Theorem 4.6 and choose L_k and K such that the corresponding systems are exponentially stable with decay rate $\eta > \gamma > 0$. Choose H as in* (4.43). *Then, the control law*

$$u_k = -F\hat{x}_k + G_k^g \hat{d}_k^g + G_k^\ell \hat{d}_k^\ell + H \sum_{j \in \mathcal{N}_k} \left(\hat{\epsilon}_j - \hat{\epsilon}_k\right) \tag{4.45}$$

with $\hat{\epsilon}_k = \hat{x}_k - \Pi_k^g \hat{d}_k^g - \Pi_k^\ell \hat{d}_k^\ell$, $k \in \mathcal{N}$, solves Problem 4.2 and additionally achieves exponential stability of the synchronization errors ϵ_k^s with decay rate $\gamma > 0$.

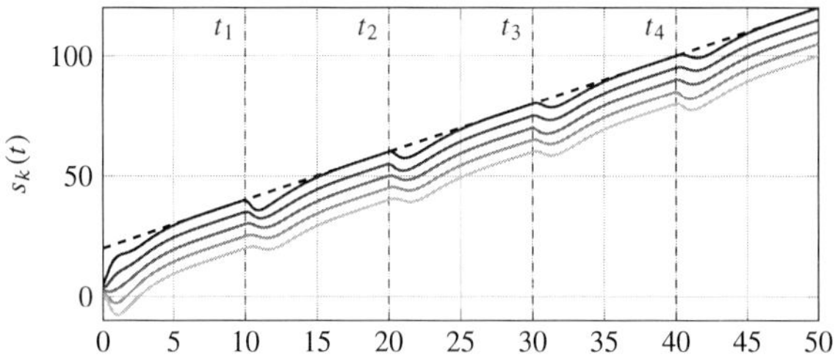

Figure 4.6: Simulation result for the vehicle platoon from Example 4.4 with the distributed regulator in Theorem 4.8. The plot shows the reference $s_0(t)$ (dashed) and the vehicle positions $s_k(t)$ for $k = 1,...,5$ (black–light gray).

Proof. See Appendix B.15. □

Example 4.4 (Vehicle Platooning (continued from Example 4.3)). The distributed regulator for the vehicle platoon considered in Example 4.3 is extended by a coupling term based on the transient state components as in (4.45) in order to enable a cooperative reaction on local external disturbances. The coupling gain matrix $H = [\,-14.6573\;\; -7.0411\,]$ was obtained via Theorem 3.7 for $\gamma = 2$ with the pole placement region $S(2,25,\pi/4)$. The corresponding simulation result in Fig. 4.6 shows the resulting behavior of the closed-loop system. The comparison with Fig. 4.5 on page 70 clearly shows the benefit of the additional coupling: the effect of the disturbances on the inter-vehicle distances is indeed rejected much more efficiently. The vehicles react cooperatively and maintain a small synchronization error. △

As another illustrative example, we present a synchronization problem for two-mass-spring systems under disturbances. A pure synchronization problem without external disturbances or reference signals for the same type of systems was studied by Lewis et al. (2014).

Example 4.5 (Two-Mass-Spring Systems). We consider a set of $N = 5$ two-mass-spring systems as illustrated in Fig. 4.7. Each system can be controlled through a force $u_k(t) \in \mathbb{R}$ acting on the first mass and is disturbed by a constant external force $d_k^\ell(t) \in \mathbb{R}$ acting on the second mass. The control objective is that the positions $s_2(t) \in \mathbb{R}$ of mass two of all systems synchronize and asymptotically track a sinusoidal reference signal. The external signals are generated by $\dot{d}_k^\ell = 0$ and

$$\dot{d}^g = \begin{bmatrix} 0 & -\omega \\ \omega & 0 \end{bmatrix} d^g$$

with $\omega = \pi/2$. The agent state $x_k(t) \in \mathbb{R}^4$ is defined as $x_k = [s_{1,k}\; v_{1,k}\; s_{2,k}\; v_{2,k}]^\mathsf{T}$, i.e., the stack of position and velocity of mass one and position and velocity of mass two, for each $k \in \mathcal{N}$. This

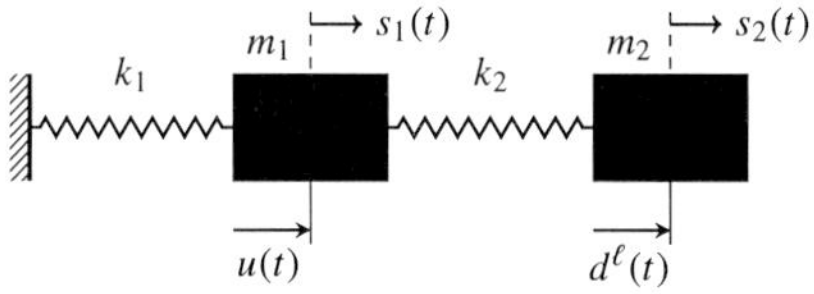

Figure 4.7: Illustration of the two-mass-spring system from Example 4.5.

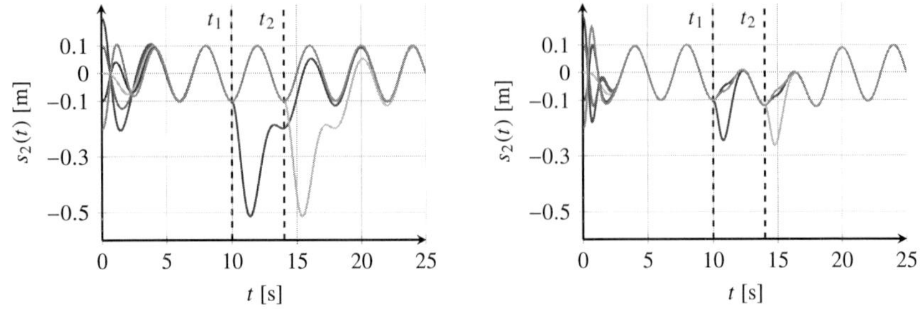

(a) Distributed regulator in Theorem 4.6.

(b) Distributed regulator in Theorem 4.8.

Figure 4.8: Simulation results for the two-mass-spring-systems from Example 4.5.

leads to the generalized plant description

$$\dot{x}_k = \begin{bmatrix} 0 & 1 & 0 & 0 \\ -\frac{k_1+k_2}{m_1} & 0 & \frac{k_2}{m_1} & 0 \\ 0 & 0 & 0 & 1 \\ \frac{k_2}{m_2} & 0 & -\frac{k_2}{m_2} & 0 \end{bmatrix} x_k + \begin{bmatrix} 0 \\ \frac{1}{m_1} \\ 0 \\ 0 \end{bmatrix} u_k + \begin{bmatrix} 0 \\ 0 \\ 0 \\ \frac{1}{m_2} \end{bmatrix} d_k^{\ell} \tag{4.46a}$$

$$y_k = \begin{bmatrix} 1 & 0 & 0 & 0 \\ 0 & 0 & 1 & 0 \end{bmatrix} x_k \tag{4.46b}$$

$$e_k = \begin{bmatrix} 0 & 0 & 1 & 0 \end{bmatrix} x_k - \begin{bmatrix} 1 & 0 \end{bmatrix} d^g \tag{4.46c}$$

for each system $k \in \mathcal{N}$. By assumption, system one has access to the global reference signal $d^g(t)$. The information structure of the multi-agent system is defined by an undirected cycle graph $\mathcal{G}$. The masses and spring constants are $m_1 = m_2 = 0.75$kg and $k_1 = k_2 = 2$N/m for all systems. The construction according to Theorem 4.6 yields the stabilizing control gain $F_k = [\, 5.7836 \;\, 5.3549 \;\, -4.9551 \;\, 2.0273 \,]$ as LQR with weights $Q = \operatorname{diag}(2,20,2,20)$, $R = 1$, the feed-forward gain matrices $G_k^g = [\, -6.3624 \;\, -3.8129 \,]$, $G_k^{\ell} = 4.8918$, and the coupling gain matrix $K = \left[\begin{smallmatrix} -15.8040 & 0 \\ 0 & -15.8040 \end{smallmatrix}\right]$ for (4.34) via Theorem 3.5 with $\gamma = 3$. The observer gain matrices obtained via pole placement in the interval $[-10,-12]$ are $L_k = \left[\begin{smallmatrix} 21.9978 & 115.3948 & 2.1536 & 50.1671 & -195.9574 \\ 0.0167 & 2.8513 & 33.0022 & 359.4153 & -990.4791 \end{smallmatrix}\right]^{\mathsf{T}}$ for each of the two-mass-spring systems $k \in \mathcal{N}$.

Fig. 4.8(a) shows a numerical simulation result for the closed-loop system. The initial positions of the second mass of all systems are chosen in the interval ± 0.2m, all other initial conditions are set to zero. Two unit step signals are applied as local external disturbances to agents $k = 1,3$ at times $t_1 = 10$s, $t_2 = 14$s, respectively. As expected, the distributed regulator successfully rejects these disturbances and all agents synchronize to the global reference signal of amplitude 0.1m.

Next, the distributed regulator is extended by a coupling term based on the transient state components as in (4.45) in order to enable a cooperative reaction on local external disturbances. The coupling gain matrix $H = [\, 28.6174 \;\, 2.9285 \;\, 14.4063 \;\, 31.5896 \,]$ is obtained via Theorem 3.7 for the pole placement region $S(2.5, 12, \pi/3)$. The corresponding simulation result in Fig. 4.8(b) shows the resulting behavior of the closed-loop system. The comparison with Fig. 4.8(a) clearly shows the benefit of the additional coupling: the effect of the disturbances on the synchronization error is indeed rejected much more efficiently. △

As an additional application example, the distributed output regulation framework is used for solving a motion coordination problem for a group of nonholonomic mobile robots in Section 5.1. This application example further demonstrates the flexibility of the problem class, the systematic construction of the distributed regulator, as well as the benefit of transient synchronization.

4.2.4 Transient Synchronization: Non-identical Agents

In this section, we aim at relaxing Assumption 4.7. In case of non-identical agents, an analogue coupling term as in (4.44) leads to a diffusively coupled network of non-identical stable linear systems in (B.17). In this case, it is hard to find a suitable coupling gain H. In case of mismatching state dimensions n_k^x such couplings cannot be realized. We exclude this case and consider heterogeneous networks consisting of non-identical perturbed versions of a common nominal agent. We suggest to design the couplings for the nominal dynamics and evaluate the design for the heterogeneous network a posteriori. Moreover, we show how to compute the size of tolerable stable additive perturbations in terms of their $\mathcal{H}_\infty$ norm, which yields a stability guarantee for a whole class of non-identical agents. For ease of presentation, we discuss only the full information case in the following. The output feedback implementation can be carried out analogously to Theorem 4.8.

Suppose that we implement a control law of the form (4.45) in a heterogeneous group, i.e.,

$$u_k = -F_k x_k + G_k^g d^g + G_k^\ell d_k^\ell + H \sum_{j \in \mathcal{N}_k} \left(\epsilon_j - \epsilon_k \right), \tag{4.47}$$

where F_k, G_k^g, G_k^ℓ are designed as in Theorem 4.6 and H is a coupling gain to be specified in the following. Then, analogously to (B.17), the error variables ϵ_k obey the dynamics

$$\dot{\epsilon}_k = (A_k - B_k F_k)\epsilon_k + B_k \nu_k, \tag{4.48}$$

where $A_k - B_k F_k$ is Hurwitz and $\nu_k = H \sum_{j \in \mathcal{N}_k} \left(\epsilon_j - \epsilon_k \right)$. The key assumption in the following is that the systems (4.48) have similar dynamics.

Design of the Couplings for Nominal Dynamics

The following heuristic procedure may be used in order to design H in (4.47):

Firstly, define a nominal system $\dot{z}_k = \tilde{A} z_k + \tilde{B} \nu_k$ for (4.48), where $\tilde{A}$ is Hurwitz. We use the average matrices

$$\tilde{A} = \frac{1}{N} \sum_{k=1}^{N} (A_k - B_k F_k), \qquad \tilde{B} = \frac{1}{N} \sum_{k=1}^{N} B_k.$$

Alternatively, if this does not yield $\tilde{A}$ Hurwitz, one can choose a particular agent $j \in \mathcal{N}$ as the nominal system by setting $\tilde{A} = A_j - B_j F_j$, $\tilde{B} = B_j$.

Secondly, design the coupling gain H for the nominal system, analogously to Lemma 4.7, such that for all $k \in \mathcal{N} \setminus \{1\}$, $\max\{\mathbf{Re}(\mu) : \mu \in \sigma(\tilde{A} - \lambda_k \tilde{B} H)\} < -\gamma$, using Theorem 3.7.

Thirdly, implement (4.47) in the group of non-identical agents and evaluate the design. In particular, it has to be checked whether the network (4.48) with $\nu_k = H \sum_{j \in \mathcal{N}_k} (\epsilon_j - \epsilon_k)$, $k \in \mathcal{N}$, is asymptotically stable and has a satisfactory performance.

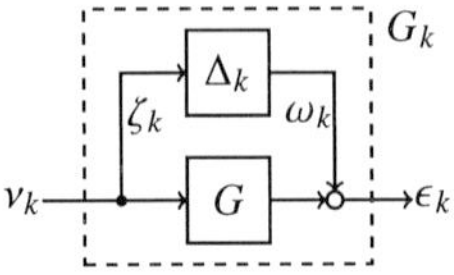

Figure 4.9: Decomposition of $G_k(s) = G(s) + \Delta_k(s)$.

Robustness Analysis

Besides evaluating the design for a particular set of non-identical agents, it is possible to quantify the tolerable heterogeneity in the agent dynamics for a given coupling gain H via a robustness analysis as follows. The nominal transfer function matrix from ν_k to z_k is given by

$$G(s) = (sI_{n^x} - \tilde{A})^{-1}\tilde{B}. \tag{4.49}$$

The transfer function matrix from ν_k to ϵ_k is $G_k(s) = (sI_{n^x} - (A_k - B_k F_k))^{-1}B_k$, for $k \in \mathcal{N}$. Then, $G_k(s)$ can be expressed as perturbed version of the nominal G with an additive uncertainty $\Delta_k(s) = G_k(s) - G(s)$. Disconnecting Δ_k from the block diagram in Fig. 4.9 leads to the description

$$\begin{bmatrix}\zeta_k\\ \epsilon_k\end{bmatrix} = \begin{bmatrix}0 & I_{n^x}\\ I_{n^x} & G\end{bmatrix}\begin{bmatrix}\omega_k\\ \nu_k\end{bmatrix}, \qquad \omega_k = \Delta_k\zeta_k. \tag{4.50}$$

Lemma 4.9. *Let the graph $\mathcal{G}$ be undirected and connected. Suppose that $G_k(s)$ can be expressed as $G_k(s) = G(s) + \Delta_k(s)$ for some real-rational strictly proper stable $\Delta_k(s)$, $k \in \mathcal{N}$. Suppose that there exist a scalar $\eta > 0$ and positive definite matrices $X_k \succ 0$ such that*

$$\begin{bmatrix}(\tilde{A} - \lambda_k\tilde{B}H)^{\mathsf{T}}X_k + X_k(\tilde{A} - \lambda_k\tilde{B}H) & -\lambda_k X_k\tilde{B}H & -\lambda_k H^{\mathsf{T}}\\ -\lambda_k H^{\mathsf{T}}\tilde{B}^{\mathsf{T}}X_k & -\eta I_{n^x} & -\lambda_k H^{\mathsf{T}}\\ -\lambda_k H & -\lambda_k H & -\eta I_{n^u}\end{bmatrix} \prec 0 \tag{4.51}$$

for $k = 2, \ldots, N$ and $\|\Delta_k\|_\infty < 1/\eta$ for all $k \in \mathcal{N}$. Then, the network (4.48) *with coupling term $\nu_k = H\sum_{j\in\mathcal{N}_k}\left(\epsilon_j - \epsilon_k\right)$, $k \in \mathcal{N}$, is asymptotically stable.*

Proof. See Appendix B.16. □

For a given coupling gain H and a nominal system (4.49), Lemma 4.9 allows to compute a bound on the $\mathcal{H}_\infty$ norm of tolerable uncertainties Δ_k. Since (4.51) is an LMI, $\eta > 0$ can be minimized via SDP. Stability of the network is guaranteed for all Δ_k with $\|\Delta_k\|_\infty < 1/\eta$. Note that Lemma 4.9 provides a robust stability result but does not guarantee the desired performance of the heterogeneous network in terms of the exponential decay rate. Nevertheless, the additional coupling term can be expected to improve the convergence speed of the synchronization error significantly, also in groups of non-identical agents.

Example 4.6 (Two-Mass-Spring Systems (continued))**.** We again consider the two-mass-spring systems studied in Example 4.5. Suppose that the spring constants of the systems are non-identical and vary by ±5%. In particular, the pairs (k_1, k_2) for agents 1–5 are $(2.0, 2.0)$, $(2.1, 2.1)$, $(2.1, 1.9)$, $(1.9, 1.9)$, $(1.9, 2.1)$ N/m. The procedure outlined above leads to the coupling gain matrix $H = [\,9.7102\ \ 1.1861\ \ -1.4417\ \ 6.9029\,]$ via Theorem 3.7 for the pole placement region $S(2.5, 12, \pi/3)$. We obtain $\max_{k\in\mathcal{N}} \|\Delta_k\|_\infty = 0.0226$ and the robustness analysis with Lemma 4.9 yields $1/\eta = 0.0229$ when minimizing η. △

4.2.5 Coupled States, Measurement Outputs, and Regulation Errors

A crucial property of the multi-agent system model with agents described by (4.18) is that not only the agent states are physically decoupled, but also the measurement outputs y_k and regulation errors e_k of the individual agents are decoupled. This decoupled structure is essential for the design of local observers (4.36) and output regulation control laws (4.37) for each agent. Cooperation among the agents is only required in order to spread the global external signal d^g over the entire network, since not all agents have direct access.

We have seen that the description (4.18) captures a wide range of practically relevant cooperative control problems and distributed control laws can be constructed as described in Theorems 4.6, 4.8. It is desirable to generalize the distributed output regulation framework to scenarios without the decoupled structure. Depending on the application, couplings among the individual agents in the physical agent states (4.18a) or in the measurement outputs (4.18b) or in the regulation errors (4.18c) may be inevitable. Each of these cases is discussed separately in the present section.

Coupled Agent States

When considering physical couplings among the agent states (4.18a), we are no longer faced with a multi-agent system consisting of autonomous agents but rather with a large-scale linear dynamical system consisting of physically coupled subsystems. The study of such large-scale systems and the development of tailored analysis and controller synthesis methods is the subject of the field of decentralized control, cf., Bakule (2008). This research field became particularly active in the 1980s and has been boosted again in recent years due to the rapid growth of communication networks and the emergence of LMIs as a powerful computational tool (Zecevic and Šiljak (2010)). The decentralized output regulation problem (also known as the servomechanism problem) for large-scale linear dynamical systems was studied within the field of decentralized control and solved by Davison (1976) and Vaz and Davison (1989).

Coupled Measurement Outputs

The most general form of the generalized plant with coupled measurement outputs is (4.18a), (4.18c), and

$$y_k = C_k x_k + D_k^{d^g} d^g + D_k^{d^\ell} d_k^\ell + D_k u_k + \sum_{j\in\mathcal{N}_k} (C_{kj} x_j + D_{kj}^{d^\ell} d_j^\ell). \tag{4.52}$$

Note that for simplicity, we assume that there are no direct feed-through terms from u_j, $j \in \mathcal{N}_k$, to y_k. Coupled measurement outputs are of great practical importance since they naturally arise when using distance measurements among mobile vehicles. For this reason, relative sensing constraints have received considerable attention in the field of distributed and cooperative control (see, e.g., Wu and Allgöwer (2012); Grip et al. (2012); Wu et al. (2014)). This situation complicates the design of local observers replacing (4.36) significantly. It is positive that the structure of the regulator equations is unaffected by the couplings in (4.52) and Lemma 4.5 still holds. Therefore, the problem at hand reduces to the design of a distributed estimator which reconstructs $\hat{x}_k$ and $\hat{d}_k^\ell$ locally for each agent based on the measurement (4.52) and, if required, communication with neighboring agents. If agent k had an estimate $\hat{x}_k$ of its state and an estimate $\hat{d}_k^\ell$ of its local external signal available, the regulation task could be solved by the same control law (4.37) as in Theorem 4.6 due to the separation principle. A straight-forward but rather conservative way of designing the distributed estimator is the following result.

Theorem 4.10 (Distributed Regulator for Coupled Measurement Outputs). *Consider a group of* $N > 1$ *agents* (4.18a), (4.18c) *with exosystems* (4.19), (4.20) *and coupled measurement outputs* (4.52). *Suppose that Assumptions 4.2–4.6 are satisfied. Moreover, suppose that there exist matrices* $P_k \in \mathbb{R}^{(n_k^x+n_k^{d^\ell})\times(n_k^x+n_k^{d^\ell})}$ *and* $Y_k \in \mathbb{R}^{(n_k^x+n_k^{d^\ell})\times n_k^y}$, $k \in \mathcal{N}$, *such that*

$$P \succ 0 \tag{4.53a}$$

$$P\mathcal{A} + \mathcal{A}^\mathsf{T} P - Y\mathcal{C} - \mathcal{C}^\mathsf{T} Y^\mathsf{T} \prec 0, \tag{4.53b}$$

where $P = \operatorname{diag}(P_1, ..., P_N)$, $Y = \operatorname{diag}(Y_1, ..., Y_N)$, *and*

$$\mathcal{A} = \begin{bmatrix} \begin{matrix} A_1 & B_1^{d^\ell} \\ 0 & S_1^\ell \end{matrix} & & 0 \\ & \ddots & \\ 0 & & \begin{matrix} A_N & B_N^{d^\ell} \\ 0 & S_N^\ell \end{matrix} \end{bmatrix}, \mathcal{C} = \begin{bmatrix} C_1 & D_1^{d^\ell} & C_{12} & D_{12}^{d^\ell} & \cdots & & C_{1N} & D_{1N}^{d^\ell} \\ \vdots & & & & & & & \vdots \\ C_{N1} & D_{N1}^{d^\ell} & \cdots & & C_{N(N-1)} & D_{N(N-1)}^{d^\ell} & C_N & D_N^{d^\ell} \end{bmatrix}$$

with $C_{kj} = 0$, $D_{kj}^{d^\ell} = 0$ *if* $j \notin \mathcal{N}_k$.

Then, a distributed regulator solving Problem 4.2 can be constructed analogously to Theorem 4.6, where the local observers (4.36) *are replaced by the distributed state estimator*

$$\begin{bmatrix} \dot{\hat{x}}_k \\ \dot{\hat{d}}_k^\ell \end{bmatrix} = \begin{bmatrix} A_k & B_k^{d^\ell} \\ 0 & S_k^\ell \end{bmatrix} \begin{bmatrix} \hat{x}_k \\ \hat{d}_k^\ell \end{bmatrix} + \begin{bmatrix} B_k & B_k^{d^g} \\ 0 & 0 \end{bmatrix} \begin{bmatrix} u_k \\ \hat{d}_k^g \end{bmatrix} + L_k\left(y_k - \hat{y}_k\right) \tag{4.54a}$$

$$\hat{y}_k = C_k\hat{x}_k + D_k^{d^g}\hat{d}_k^g + D_k^{d^\ell}\hat{d}_k^\ell + D_k u_k + \sum_{j\in\mathcal{N}_k}(C_{kj}\hat{x}_j + D_{kj}^{d^\ell}\hat{d}_j^\ell) \tag{4.54b}$$

with observer gain matrices $L_k = P_k^{-1} Y_k$ *for* $k \in \mathcal{N}$.

Proof. See Appendix B.17. □

The conservatism of Theorem 4.10 is introduced by restricting the matrix P to be block diagonal. This constraint is imposed in order to obtain a block diagonal observer gain matrix $L = \operatorname{diag}(L_1, ..., L_N)$. The design of L is centralized and results in LMIs (4.53) of high order for large networks. The implementation of the distributed state estimator (4.54) requires the exchange of state estimates among neighboring agents with coupled measurement outputs. The same graph is assumed for the coupled measurement outputs and for possible information exchange among the agents. In essence, in case of coupled measurement outputs, the distributed regulation problem requires the solution of a distributed estimation problem. There are sophisticated synthesis methods for distributed estimators, most notably the work of Listmann et al. (2011); Ugrinovskii (2011b) and Wu et al. (2014). The combination of these methods with the distributed output regulation framework is a promising topic of future research.

Coupled Regulation Errors

The formulation of cooperative control tasks may naturally lead to a problem description with coupled regulation errors (4.18c). A compelling example for such a scenario is a vehicle platoon in which the desired *relative distances* among subsequent vehicles are commanded by external signals. The distance regulation objective for vehicle k with position $s_k(t)$ is then $s_k(t) - s_{k-1}(t) + r_k(t) \to 0$ as $t \to \infty$, where $r_k(t)$ is the corresponding desired inter-vehicle distance. Formulating this problem as distributed output regulation problem leads to generalized plants with coupled regulation errors e_k involving differences between local and neighboring agent states. The purpose of the global exosystem (4.19) was to define the group objective for the decoupled agents (4.18) through a common global reference signal for all agents. Alternatively, typical group objectives such as output agreement or formation keeping can be captured by coupled regulation errors. Therefore, we do not consider a global exosystem but only a local exosystems for each agent in the following. In order to simplify the notation, the superscript $\cdot^{\ell}$ is omitted so that the local exosystems (4.20) read $\dot{d}_k = S_k d_k$, $d_k(t) \in \mathbb{R}^{n_k^d}$. Consequently, the most general form of the generalized plant description for agent $k \in \mathcal{N}$ with coupled regulation errors is given by

$$\dot{x}_k = A_k x_k + B_k u_k + B_k^d d_k \tag{4.55a}$$
$$y_k = C_k x_k + D_k u_k + D_k^d d_k \tag{4.55b}$$
$$e_k = C_k^e x_k + D_k^e u_k + D_k^{ed} d_k + \sum_{j\in\mathcal{N}_k} (C_{kj}^e x_j + D_{kj}^e u_j + D_{kj}^{ed} d_j). \tag{4.55c}$$

Due to the couplings in (4.55c), the regulator equation (4.26) for the centralized output regulation problem can no longer be decomposed as in Lemma 4.5. When formulating the centralized output regulation problem based on (4.55), analogously to (4.23), we see that the matrices $\mathcal{S}$, $\mathcal{A}$, $\mathcal{B}$, $\mathcal{B}_d$, $\mathcal{C}$, $\mathcal{D}$, $\mathcal{D}_d$ are block diagonal. The matrices $\mathcal{C}_e$, $\mathcal{D}_e$, $\mathcal{D}_{ed}$, however, no longer admit this structure. Instead, we obtain

$$\mathcal{C}_e = \begin{bmatrix} C_1^e & C_{12}^e & \cdots & C_{1N}^e \\ \vdots & & & \vdots \\ C_{N1}^e & \cdots & C_{N(N-1)}^e & C_N^e \end{bmatrix}$$

and $\mathcal{D}_e$, $\mathcal{D}_{ed}$ analogously, where $C_{kj}^e = 0$, $D_{kj}^e = 0$, $D_{kj}^{ed} = 0$ whenever $j \notin \mathcal{N}_k$. Recalling (4.26),

$$\mathcal{A}\Pi + \mathcal{B}\Gamma - \Pi\mathcal{S} + \mathcal{B}_d = 0$$
$$\mathcal{C}_e\Pi + \mathcal{D}_e\Gamma + \mathcal{D}_{ed} = 0.$$

While Π and Γ can be chosen block diagonal for the former equation, there is no hope that a solution to the latter equation is in general block diagonal. Since the structure of Π, Γ determines the structure of the resulting output regulation control law, we cannot hope for a distributed regulator solving Problem 4.2 for the agents (4.55). In special cases, however, distributed solutions may exist. An example is the aforementioned vehicle platooning problem, which is worked out below. It is worth noting that state estimation for (4.55) can still be done by decoupled local observers as in Theorem 4.6 as long as the measurement outputs y_k are decoupled and the corresponding observability condition (Assumption 4.3) is satisfied.

Example 4.7 (Vehicle Platooning (revisited)). We consider the platooning problem for a string of $N > 1$ vehicles illustrated in Fig. 4.10. Each vehicle k in the platoon is modeled with double-integrator dynamics, control input $u_k(t) \in \mathbb{R}$, measurement output $y_k(t) \in \mathbb{R}^{n_k^y}$, and is subject to an external local disturbance $w_k(t) \in \mathbb{R}$. In contrast to Examples 4.3, 4.4, the reference signals $r_k(t) \in \mathbb{R}$ for vehicles $2, ..., N$ define the desired relative distances to the next preceding vehicle and not to the leading vehicle. Each vehicle measures its velocity and its distance to the preceding vehicle by local sensors. The leading vehicle measures its absolute position and velocity. The reference signal commanded to the leading vehicle is its desired absolute position, evolving as a

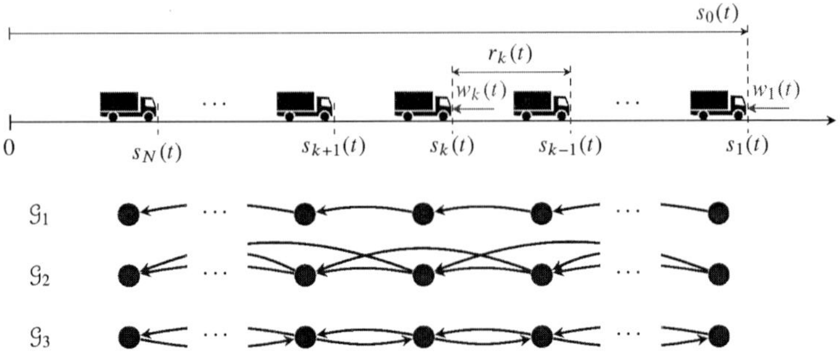

Figure 4.10: Illustration of the vehicle platoon from Example 4.7 with three communication graphs.

ramp function of time, and its velocity. Each vehicle may communicate with neighboring vehicles in the string according to the graphs $\mathcal{G}_1$, $\mathcal{G}_2$, $\mathcal{G}_3$ shown in Fig. 4.10, which will be investigated individually. This setup leads to a distributed output regulation problem with coupled regulation errors as in (4.55). In particular, the generalized plant descriptions for the leading vehicle one and the following vehicles $k = 2, ..., N$ are

$$
\begin{aligned}
\dot{d}_1 &= \begin{bmatrix} 0 & 1 & 0 \\ 0 & 0 & 0 \\ 0 & 0 & 0 \end{bmatrix} d_1 & \dot{d}_k &= \mathbf{0} \\
\dot{x}_1 &= \begin{bmatrix} 0 & 1 \\ 0 & 0 \end{bmatrix} x_1 + \begin{bmatrix} 0 \\ 1 \end{bmatrix} u_1 + \begin{bmatrix} 0 & 0 & 0 \\ 0 & 0 & -1 \end{bmatrix} d_1 & \dot{x}_k &= \begin{bmatrix} 0 & 1 \\ 0 & 0 \end{bmatrix} x_k + \begin{bmatrix} 0 \\ 1 \end{bmatrix} u_k + \begin{bmatrix} 0 & 0 \\ 0 & -1 \end{bmatrix} d_k \\
y_1 &= \begin{bmatrix} 0 & 0 \\ 0 & 0 \\ 1 & 0 \\ 0 & 1 \end{bmatrix} x_1 + \begin{bmatrix} 1 & 0 & 0 \\ 0 & 1 & 0 \\ 0 & 0 & 0 \\ 0 & 0 & 0 \end{bmatrix} d_1 & y_k &= \begin{bmatrix} 0 & 0 \\ -1 & 0 \\ 0 & 1 \end{bmatrix} x_k + \begin{bmatrix} 0 & 0 \\ 1 & 0 \\ 0 & 0 \end{bmatrix} x_{k-1} + \begin{bmatrix} 1 & 0 \\ 0 & 0 \\ 0 & 0 \end{bmatrix} d_k \\
e_1 &= \begin{bmatrix} 1 & 0 \end{bmatrix} x_1 + \begin{bmatrix} -1 & 0 & 0 \end{bmatrix} d_1 & e_k &= \begin{bmatrix} 1 & 0 \end{bmatrix} x_k + \begin{bmatrix} -1 & 0 \end{bmatrix} x_{k-1} + \begin{bmatrix} 1 & 0 \end{bmatrix} d_k
\end{aligned}
$$

where $d_1 = [s_0 \; v_0 \; w_1]^\mathsf{T}$ and $d_k = [r_k \; w_k]^\mathsf{T}$. The state $x_k = [s_k \; v_k]^\mathsf{T}$ comprises vehicle k's position $s_k(t) \in \mathbb{R}$ and velocity $v_k(t) \in \mathbb{R}$. The formulation of the centralized output regulation problem leads to the matrices

$$
\mathcal{C}_e = \left[\begin{array}{cc:cc:cc:cc:cc}
1 & 0 & 0 & 0 & \cdots & & & & & \\ \hdashline
-1 & 0 & 1 & 0 & 0 & 0 & \cdots & & & \\ \hdashline
0 & 0 & -1 & 0 & 1 & 0 & 0 & 0 & \cdots & \\ \hdashline
 & & & & \ddots & & \ddots & & & \\ \hdashline
 & & & \cdots & 0 & 0 & -1 & 0 & 1 & 0
\end{array}\right], \quad
\mathcal{D}_{ed} = \left[\begin{array}{ccc:cc:cc:cc}
-1 & 0 & 0 & 0 & 0 & \cdots & & & \\ \hdashline
0 & 0 & 0 & 1 & 0 & 0 & 0 & \cdots & \\ \hdashline
 & & \cdots & 0 & 0 & 1 & 0 & 0 & 0 \; \cdots \\ \hdashline
 & & & & & & \ddots & & \\ \hdashline
 & & & & & \cdots & 0 & 0 & 1 \;\; 0
\end{array}\right].
$$

It can be verified that the corresponding regulator equation (4.26) admits the solution

$$\Pi = \left[\begin{array}{ccc|cc|cc|cc} 1 & 0 & 0 & 0 & 0 & \cdots & & & \\ 0 & 1 & 0 & 0 & 0 & \cdots & & & \\ \hline 1 & 0 & 0 & -1 & 0 & 0 & 0 & \cdots & \\ 0 & 1 & 0 & 0 & 0 & 0 & 0 & \cdots & \\ \hline & & & & & \ddots & & & \\ \hline 1 & 0 & 0 & -1 & 0 & \cdots & & -1 & 0 \\ 0 & 1 & 0 & 0 & 0 & \cdots & & 0 & 0 \end{array}\right], \quad \Gamma = \left[\begin{array}{ccc|cc|cc|cc} 0 & 0 & 1 & 0 & 0 & \cdots & & & \\ \hline 0 & 0 & 0 & 0 & 1 & 0 & 0 & \cdots & \\ \hline & & \cdots & 0 & 0 & 0 & 1 & 0 & 0 \; \cdots \\ \hline & & & & & & & \ddots & \\ \hline & & & & & \cdots & & 0 & 0 \; 0 \; 1 \end{array}\right].$$

A centralized output regulation control law has the form $u = -\mathsf{F}x + \mathsf{G}d$, where F is such that $\mathcal{A} - \mathcal{B}\mathsf{F}$ is Hurwitz and $\mathsf{G} = \Gamma + \mathsf{F}\Pi$. The challenge regarding a distributed implementation of the control law is to find a sparse control gain matrix F respecting the graph topology, such that the gain matrix G inherits the desired sparsity pattern as well. Suppose that $K = [k_1 \; k_2] \in \mathbb{R}^{1\times 2}$ is such that $A_k - B_k K$ is Hurwitz. Then, $\mathsf{F} = I_N \otimes K$ would render $\mathcal{A} - \mathcal{B}\mathsf{F}$ Hurwitz. However, due to the structure of Π, the resulting matrix $\mathsf{G} = \Gamma + \mathsf{F}\Pi$ would be lower block triangular and would hence violate the graph topology. It turns out that a suitable distributed control gain can be constructed as follows. Let $\mathcal{G}$ be the underlying connected graph and define $\tilde{L}_{\mathcal{G}} = L_{\mathcal{G}} + \mathrm{diag}(\alpha, 0, ..., 0)$ for some $\alpha > 0$. Then, $\sigma(\tilde{L}_{\mathcal{G}}) \subset \mathbb{C}^+$, i.e., all its eigenvalues are positive (Qu, 2009, Cor. 4.33). Based on the results of Section 3.1.3, it is easy to find K such that $\mathsf{F} = (\tilde{L}_{\mathcal{G}} \otimes K)$ renders $\mathcal{A} - \mathcal{B}\mathsf{F}$ Hurwitz (by substituting $\mathbf{Re}(\lambda_2(L_{\mathcal{G}}))$ with $\mathbf{Re}(\lambda_1(\tilde{L}_{\mathcal{G}})) > 0$ in Theorem 3.2). This structure of F results in $\mathsf{G} = \Gamma + \mathsf{F}\Pi$ with a desired sparsity pattern if $\mathcal{G}$ is chosen appropriately. The control input for vehicle k is the k-th row of $u = -(\tilde{L}_{\mathcal{G}} \otimes K)x + (\Gamma + (\tilde{L}_{\mathcal{G}} \otimes K)\Pi)d$. We obtain

$$\begin{aligned} u_1 &= -\alpha K x_1 + K \textstyle\sum_{j\in\mathcal{N}_1}(x_j - x_1) + \alpha k_1 s_0 + \alpha k_2 v_0 - k_1 \sum_{i=2}^{N} \sum_{j=i}^{N} l_{1j} r_i + w_1 \\ u_k &= K \textstyle\sum_{j\in\mathcal{N}_k}(x_j - x_k) - k_1 \sum_{i=2}^{N} \sum_{j=i}^{N} l_{kj} r_i + w_k \qquad k = 2, ..., N, \end{aligned}$$

where l_{kj} is the kj-entry of the graph Laplacian $L_{\mathcal{G}}$. Obviously, the control law does not inherit the sparsity structure of $L_{\mathcal{G}}$ for arbitrary graphs, but it does for several common choices. For the three communication graphs $\mathcal{G}_1$, $\mathcal{G}_2$, $\mathcal{G}_3$ shown in Fig. 4.10, the control laws above specialize to

$$\begin{aligned} \mathcal{G}_1: \quad & u_1 = -\alpha K x_1 + \alpha k_1 s_0 + \alpha k_2 v_0 + w_1 \\ & u_k = K(x_{k-1} - x_k) - k_1 r_k + w_k \qquad k = 2, ..., N, \\ \mathcal{G}_2: \quad & u_1 = -\alpha K x_1 + \alpha k_1 s_0 + \alpha k_2 v_0 + w_1 \\ & u_2 = K(x_1 - x_2) - k_1 r_2 + w_2 \\ & u_k = K(x_{k-1} - x_k) + K(x_{k-2} - x_k) - k_1 r_{k-1} - 2k_1 r_k + w_k \qquad k = 3, ..., N, \\ \mathcal{G}_3: \quad & u_1 = -\alpha K x_1 + K(x_2 - x_1) + \alpha k_1 s_0 + \alpha k_2 v_0 + k_1 r_2 + w_1 \\ & u_k = K(x_{k-1} - x_k) + K(x_{k+1} - x_k) - k_1 r_k + k_1 r_{k+1} + w_k \qquad k = 2, ..., N-1, \\ & u_N = K(x_{N-1} - x_N) - k_1 r_N + w_N. \end{aligned}$$

By assumption, each vehicle k is capable of measuring its own velocity $v_k(t) = \dot{s}_k(t)$ and the distance to its preceding vehicle $s_{k-1}(t) - s_k(t)$. The distance $s_k(t) - s_{k+1}(t)$ to the following vehicle and the distance $s_{k-2}(t) - s_k(t)$ to the second vehicle in front can be obtained through additional sensors or through communication, if required. The leading vehicle has access to its absolute position $s_1(t)$. Moreover, all vehicles have access to their reference r_k. The only quantities that need to be estimated are the local disturbances $w_k(t)$. This can easily be done by means of the local estimators

$$\begin{bmatrix} \dot{\hat{v}}_k \\ \dot{\hat{w}}_k \end{bmatrix} = \begin{bmatrix} 0 & -1 \\ 0 & 0 \end{bmatrix} \begin{bmatrix} \hat{v}_k \\ \hat{w}_k \end{bmatrix} + \begin{bmatrix} 1 \\ 0 \end{bmatrix} u_k + L(v_k - \hat{v}_k),$$

where L is chosen such that the observer error dynamics are asymptotically stable.

For a string of $N = 10$ vehicles and $\alpha = 1$, we obtained $K = [\,2.9515 \;\; 2.6021\,]$ for both $\mathcal{G}_1$ and $\mathcal{G}_2$, and $K = [\,24.1589 \;\; 46.0216\,]$ for $\mathcal{G}_3$, and the observer gain matrix $L = [\,13.4099 \;\; 44.6835\,]^\mathsf{T}$. In order to illustrate the behavior of the vehicle string under the distributed control law, the following external input signals are applied. The initial velocities of all vehicles are set to zero. At time $t_1 = 10$, a step disturbance $w_3(t) = 100$, $t \geq t_1$, acts on vehicle 3. The reference signal $s_0(t)$ is a ramp function of slope $v_0(t) = 3$ for $0 \leq t < t_2 = 20$, switching to $v_0(t) = 6$ for $t \geq t_2$. The inter-vehicle distances are initialized with $r_k(t) = 2$ for all following vehicles. At time $t_3 = 30$, the spacing between vehicles 2–3 is increased to $r_3(t) = 6$ for $t \geq t_3$. Simulation results for the three graphs $\mathcal{G}_1$, $\mathcal{G}_2$, and $\mathcal{G}_3$ are shown in Fig. 4.11. As expected, each of the distributed control laws solves the output regulation problem with coupled regulation errors.

Fig. 4.11 reveals that the distributed control law based on graph $\mathcal{G}_1$ leads to string instability, i.e., the regulation errors are amplified along the vehicle string. The notion of string stability was introduced by Swaroop and Hedrick (1996). Since then, the vehicle platooning problem has received considerable attention, with a focus on the design of controllers ensuring that disturbances are attenuated along the vehicle string, cf., Ploeg et al. (2014) and references therein. For graph $\mathcal{G}_2$, the behavior of the string is much better and for graph $\mathcal{G}_3$, no string instability is observed. This demonstrates that more and particularly bi-directional information exchange among the vehicles, which can be realized via car-to-car communication, is beneficial for the platooning task. △

The distributed output regulation problem in Example 4.7 results in matrices Π, Γ with a particular structure. Along with an appropriate choice of the control gain matrix F, this structure indeed leads to distributed control laws for the vehicle platoon. A generalization of this result is presented in the following theorem.

Theorem 4.11. *Consider a group of agents* (4.55) *with identical exosystems* $\dot{d}_k = Sd_k$, $d_k(t) \in \mathbb{R}^{n_d}$, $\sigma(S) \subset \mathbb{C}^0 \cup \mathbb{C}^+$, *and suppose that*

i) all agents have identical dynamics, i.e., $A_k = A$, $B_k = B$, $B_k^d = B_d$ *for all* $k \in \mathcal{N}$,

ii) the pair (A, B) *is stabilizable,*

iii) the pair $\left(\begin{bmatrix} A & B_d \\ 0 & S \end{bmatrix}, \begin{bmatrix} C_k & D_k^d \end{bmatrix}\right)$ *is detectable for all* $k \in \mathcal{N}$,

iv) the regulation errors are coupled in a chain structure according to

$$\begin{aligned} e_1 &= C_e x_1 + D_{ed} d_1 \\ e_k &= C_e(x_k - x_{k-1}) + D_{ed} d_k, \qquad k \in \mathcal{N}\backslash\{1\}, \end{aligned}$$

v) there exist matrices Π_1, Γ_1 *such that* $A\Pi_1 - \Pi_1 S = 0$, $B\Gamma_1 + B_d = 0$, *and* $C_e\Pi_1 + D_{ed} = 0$.

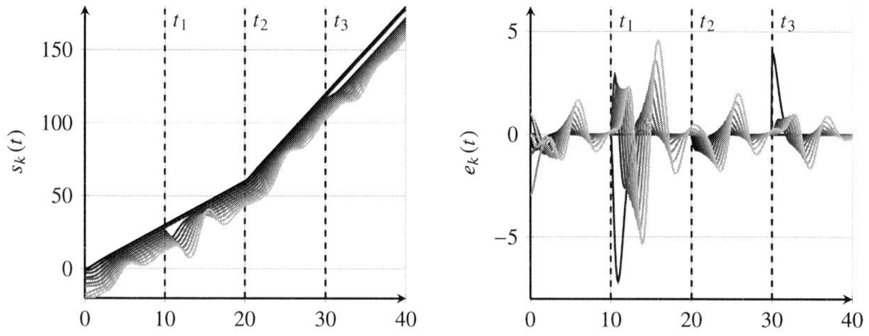

(a) Vehicle positions (left) and regulation errors (right) under graph $\mathcal{G}_1$.

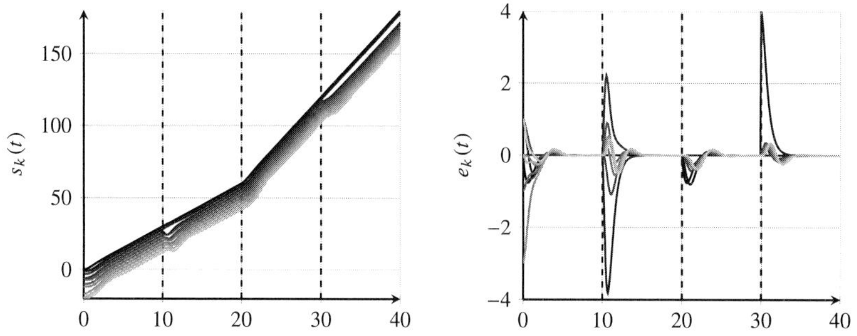

(b) Vehicle positions (left) and regulation errors (right) under graph $\mathcal{G}_2$.

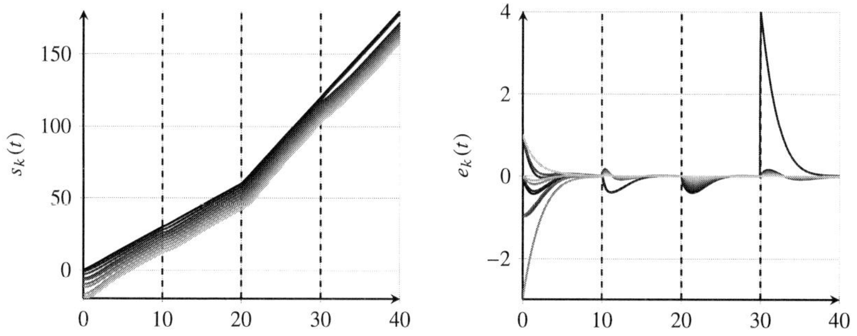

(c) Vehicle positions (left) and regulation errors (right) under graph $\mathcal{G}_3$.

Figure 4.11: Simulation results for the vehicles 1–10 (black–light gray) in the platoon from Example 4.7 and external step signals $w_3(t) = 100$, $v_0(t) = 6$, $r_3(t) = 6$ applied at times $t \geq t_1, t_2, t_3$, respectively. While regulation is achieved in all cases, string instability is observed in (a).

Then, the matrices

$$\Pi = \begin{bmatrix} 1 & & 0 \\ \vdots & \ddots & \\ 1 & \cdots & 1 \end{bmatrix} \otimes \Pi_1, \qquad \Gamma = I_N \otimes \Gamma_1,$$

solve the corresponding centralized regulator equations (4.26).

Let $L_{\mathcal{G}}$ be the Laplacian matrix of an undirected line graph and define $\tilde{L}_{\mathcal{G}} = L_{\mathcal{G}} + \mathrm{diag}(\alpha, 0, ..., 0)$ for some $\alpha > 0$. Choose F such that $A - \tilde{\lambda}_k BF$ is Hurwitz for all eigenvalues $\tilde{\lambda}_k$ of $\tilde{L}_{\mathcal{G}}$, $k \in \mathcal{N}$. Then, the regulator

$$u = -(\tilde{L}_{\mathcal{G}} \otimes F)\hat{x} + (\Gamma + (\tilde{L}_{\mathcal{G}} \otimes F)\Pi)\hat{d}$$

with local observers of the form (4.36) *solves the output regulation problem and is distributed.*

Proof. See Appendix B.18. □

The crucial structural properties of the regulation problem in Theorem 4.11 are identical agent dynamics, coupled regulation errors with a chain structure (which is a topological constraint), and solvability of the output regulation problem for agent $k = 1$. More precisely, condition *v)* states that the subspace $\mathrm{im}(\Pi_1)$ is A-invariant (and the agent dynamics restricted to $\mathrm{im}(\Pi_1)$ are described by S) and $\mathrm{im}(B_d) \subseteq \mathrm{im}(B)$.

Summarizing, the assumption of decoupled regulation errors as in (4.18) facilitates the construction of a distributed regulator significantly, cf., Theorem 4.6. The reason is that the regulation problem can in essence be decomposed into individual local regulation problems for each agent, cf., Lemma 4.5. In the presence of coupled regulation errors, such a decomposition is in general no longer possible and the resulting regulators are in general centralized and may not admit a distributed realization. However, as shown in Theorem 4.11, distributed regulators may be found for problems involving coupled regulation errors in special cases. The vehicle platooning problem in Example 4.7 is a corresponding application example of practical relevance.

Remark 4.4. The problem of robust output synchronization under constant disturbances addressed in Section 4.1 may be formulated as a distributed output regulation problem as well. In this case, the output synchronization objective would lead to coupled regulation errors such as $e_k = y_k - 1/d_k \sum_{j \in \mathcal{N}_k} y_j$, where d_k is the degree of node k. As a viable solution for such problems, we presented distributed control laws with integral action in Section 4.1. ○

4.3 Summary and Discussion

In Chapter 4, we move beyond consensus and synchronization and focus on multi-agent coordination problems involving exogenous input signals in order to arrive at more realistic problem statements. The exogenous input signals comprise both external reference signals and disturbances and may affect either all agents in the group (global reference signals and global disturbances) or only individual agents (local reference signals and local disturbances). Effective distributed control laws are developed for a variety of such scenarios.

In Section 4.1, we develop a solution to the robust output synchronization problem for groups of identical linear agents under persistent external disturbances. As our main contribution, we bring the classical approach of control with integral action forward to the field of cooperative control and present an appropriate novel distributed control law with integral action. Two observer-based implementations of the distributed control law are presented for the cases where only absolute or relative output measurements are available. The resulting synchronous output trajectory is characterized in all cases. Moreover, we show that the novel distributed control law with integral action additionally provides a solution to a cooperative tracking problem with multiple local reference signals. Numerical examples illustrate the effectiveness of the novel control laws. Prior results on distributed PI-controllers for single and double-integrator networks are contained in our results as special cases.

Section 4.2 presents the distributed output regulation framework as a powerful method to formulate and systematically solve a wide range of practical multi-agent coordination problems. We present a distributed regulator solving the regulation problem in presence of both global and local exogenous input signals and discuss necessary and sufficient conditions for its existence. As a major contribution, we extend the distributed regulator in order to improve and tune the synchronization error dynamics of the group. In case of identical agent dynamics, a novel coupling term based on the transient state components of each agent allows to impose a desired exponential decay rate on the synchronization error among agents, which leads to a significant improvement of the cooperative behavior of the group in transient phases. Under the novel control law, the group is able to react cooperatively on external disturbances acting on individual agents. Several numerical examples emphasize the importance of a cooperative reaction on disturbances and illustrate the design procedure and effectiveness of the novel distributed regulator with transient synchronization. Moreover, we outline a procedure to establish such couplings in groups of non-identical agents with similar but non-identical dynamics. Finally, we generalize the problem formulation and discuss distributed output regulation problems with coupled agent states, coupled measurement outputs, and coupled regulation errors. Each of these cases may well arise in practical applications. Problems with coupled agent states fall into the field of decentralized control which was particularly active in the 1980s. Coupled measurement outputs lead to distributed estimation problems which are an active research area in their own right. Modern distributed estimation schemes can be integrated in the distributed output regulation framework in order to solve such problems. Coupled regulation errors complicate the distributed regulation problem significantly. Due to the more complex structure of the regulator equations, it is in general not guaranteed that a centralized regulator can be realized in a distributed fashion. Nevertheless, we identify a class of problems with coupled regulation errors which is solvable via distributed control. A corresponding compelling application example of practical relevance is the vehicle platooning problem, which is worked out in detail.

In the next chapter, we shift the focus from linear multi-agent systems to motion coordination problems for groups of nonholonomic vehicles and address typical formation control tasks of practical interest. For their solution, we utilize distributed control design methods developed in the present chapter and additionally employ new theoretic concepts.

Chapter 5

Motion Coordination for Groups of Nonholonomic Vehicles

After the thorough study of state and output synchronization problems in Chapter 3 and Chapter 4, we now turn our attention to formation control problems for teams of mobile vehicles as a major application domain of distributed control methods. The development of formation control algorithms for groups of autonomous vehicles has received considerable attention over the past decade. An overview is provided in the recent survey article by Oh et al. (2015). A majority of the existing formation control algorithms were developed for agents with single-integrator or double-integrator dynamics (Olfati-Saber and Murray (2004); Ren and Beard (2008); Mesbahi and Egerstedt (2010)). Distributed control techniques for networks of general linear systems were the subject of Chapter 3 and Chapter 4 of the present thesis. Certainly, there are motion coordination problems which require more complex and nonlinear models for the agents in the group. Oftentimes, autonomous vehicles need to be modeled as nonholonomic systems, cf., Qu (2009). A suitable description for the kinematics of vehicles such as unmanned aerial vehicles (UAVs) or mobile robots with tank drives instead of omni-directional drives is the unicycle model.

In the present chapter, we focus on two motion coordination scenarios for groups of nonholonomic vehicles. In Chapter 5.1, we propose a formation control architecture for a group of mobile robots. The robots shall form a time-varying formation, follow a desired path with the formation center, and maintain the formation despite external disturbances. We show that the distributed output regulation framework can be made applicable to this coordination scenario by means of the feedback linearization technique and nonlinear disturbance estimation. In Chapter 5.2, we consider heterogeneous groups of unicycle-type vehicles with constant and non-identical velocities. This scenario is of particular practical relevance in the context of UAV coordination. After analyzing the effect of the mismatching velocities on the motion coordination problem, we develop smooth nonlinear control laws which steer the vehicles to a variety of desirable circular formations.

5.1 Mobile Robots

In this section, we pose and solve an advanced motion coordination problem for a group of nonholonomic mobile robots. Previous work on this problem was done by Lawton et al. (2003) and Ren and Atkins (2007). In our formulation, the formation shape needs not to be static but can change dynamically over time. Moreover, we take external disturbances (forces and momenta) explicitly into account, which may act on individual robots. The development of the motion coordination algorithm tailored to this scenario involves the feedback linearization technique and a nonlinear observer for the local disturbances, additionally to the output regulation framework presented in Section 4.2. The purpose of these developments is to demonstrate the applicability of the distributed output regulation framework to this more complex cooperative control with nonlinear dynamical agents. More importantly, it further illustrates this framework, including the transient synchronization, which allows to improve the cooperative behavior in presence of local external disturbances.

5.1.1 Problem Formulation

In the following, the agent models, information structure, and the group objective are introduced.

Agent Models The dynamical model of the nonholonomic mobile robots is shown in Fig. 5.1. The coordinates $x_k(t), y_k(t) \in \mathbb{R}$ describe the position of robot k in the plane, $v_k(t) \in \mathbb{R}$ is its

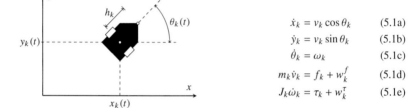

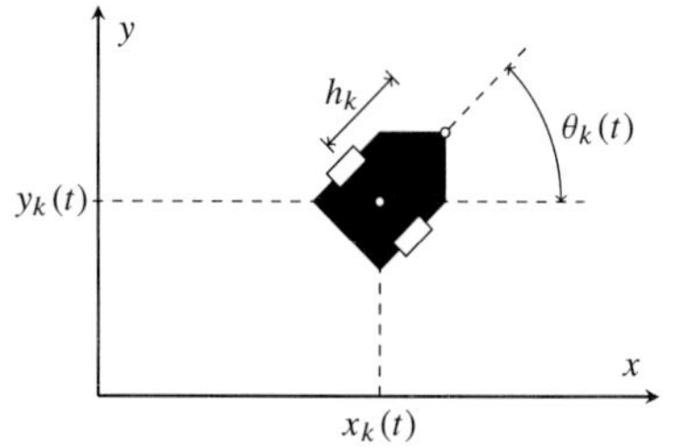

$$
\begin{aligned}
\dot{x}_k &= v_k \cos\theta_k && (5.1a)\\
\dot{y}_k &= v_k \sin\theta_k && (5.1b)\\
\dot{\theta}_k &= \omega_k && (5.1c)\\
m_k \dot{v}_k &= f_k + w_k^f && (5.1d)\\
J_k \dot{\omega}_k &= \tau_k + w_k^\tau && (5.1e)
\end{aligned}
$$

Figure 5.1: Model of the nonholonomic mobile robot.

longitudinal velocity, $\theta_k(t) \in \mathbb{R}$ is its orientation, $\omega_k(t) \in \mathbb{R}$ is its angular speed, $m_k > 0$ its mass and $J_k > 0$ its moment of inertia. The force $f_k(t) \in \mathbb{R}$ and torque $\tau_k(t) \in \mathbb{R}$ serve as control inputs. The exogenous inputs $w_k^f \in \mathbb{R}$ and $w_k^\tau \in \mathbb{R}$ represent external force and torque disturbances acting on the robot, which are assumed to be constant. The constant $h_k > 0$ is the distance between the robot center and its "hand" position with coordinates

$$
\begin{aligned}
x_k^{\mathrm{h}} &= x_k + h_k \cos\theta_k && (5.2a)\\
y_k^{\mathrm{h}} &= y_k + h_k \sin\theta_k. && (5.2b)
\end{aligned}
$$

Information Structure The same information structure as in Section 3.1 is assumed, i.e., the communication topology is described by a directed graph $\mathcal{G} = (\mathcal{V}, \mathcal{E}, A_{\mathcal{G}})$. For illustration, a group of $N = 6$ robots as shown in Fig. 5.2 is considered in the sequel and the underlying graph $\mathcal{G}$ is assumed to be an undirected cycle graph. Besides the communication capabilities, it is assumed that each robot has access to its own position (x_k, y_k), hand position $(x_k^{\mathrm{h}}, y_k^{\mathrm{h}})$, velocity v_k, orientation θ_k, and ω_k through on-board sensors.

Group Objective The group objective under consideration is the following. The mobile robots shall reach a predefined formation and the formation center shall track a given path in the plane. While the formation center point tracks a desired reference $r^g(t)$, each agent keeps a desired relative distance $r_k^\ell(t)$ from that center point, i.e., the objective is

$$
\lim_{t\to\infty} \left(\begin{bmatrix} x_k^{\mathrm{h}}(t) \\ y_k^{\mathrm{h}}(t) \end{bmatrix} - r^g(t) - r_k^\ell(t) \right) = \mathbf{0}.
$$

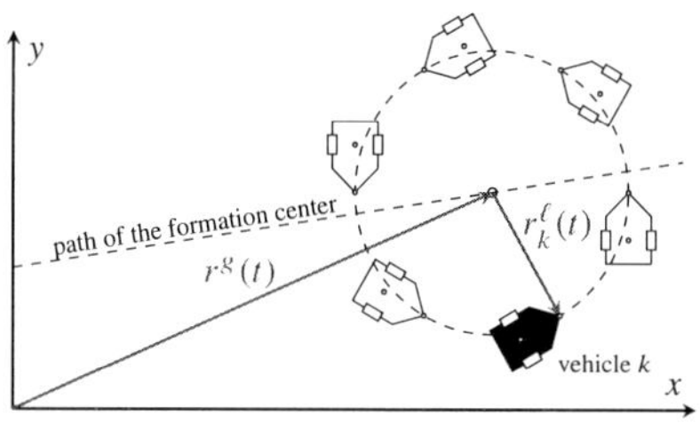

Figure 5.2: Illustration of the motion coordination task for the nonholonomic mobile robots.

for all $k \in \mathcal{N}$. Fig. 5.2 illustrates this scenario. The global reference signal $r^g(t)$ describes a straight line in the plane. The local reference signals $r^\ell_k(t)$ are such that each agent performs a circular motion around the formation center point. Such motion coordination tasks appear, e.g., in patrolling and surveillance scenarios.

5.1.2 Motion Coordination Algorithm

As proposed by Lawton et al. (2003), we resort to the feedback linearization technique in order to convert the coordination problem for the robots (5.1) into a coordination problem for double-integrator agents. Then, we employ the distributed output regulation framework and extend it by a suitable nonlinear estimator in order to compensate for the local disturbances.

Feedback Linearization

The system dynamics (5.1) (without disturbances) can be linearized by feedback as follows. Let $R(\theta_k)$ and $S_k(\theta_k)$ be the rotation matrix and a scaled rotation matrix

$$R(\theta_k) = \begin{bmatrix} \cos\theta_k & -\sin\theta_k \\ \sin\theta_k & \cos\theta_k \end{bmatrix}, \qquad S_k(\theta_k) = R(\theta_k)\begin{bmatrix} \frac{1}{m_k} & 0 \\ 0 & \frac{h_k}{J_k} \end{bmatrix}.$$

Both $R(\theta_k)$ and $S_k(\theta_k)$ are invertible and $R(\theta_k)^{-1} = R(-\theta_k)$. Applied to (5.1), the feedback

$$\begin{bmatrix} f_k \\ \tau_k \end{bmatrix} = S_k(\theta_k)^{-1}\begin{bmatrix} v_k\omega_k \sin\theta_k + h_k\omega_k^2\cos\theta_k + \bar{f}_k \\ -v_k\omega_k\cos\theta_k + h_k\omega_k^2\sin\theta_k + \bar{\tau}_k \end{bmatrix} \tag{5.3}$$

with the new control inputs $\bar{f}_k(t) \in \mathbb{R}$, $\bar{\tau}_k(t) \in \mathbb{R}$ leads to the linearized and decoupled model

$$\dot{\xi}^{\mathrm{h}}_k = \left(\begin{bmatrix} 0 & 1 \\ 0 & 0 \end{bmatrix} \otimes I_2\right)\xi^{\mathrm{h}}_k + \left(\begin{bmatrix} 0 \\ 1 \end{bmatrix} \otimes I_2\right)\begin{bmatrix} \bar{f}_k \\ \bar{\tau}_k \end{bmatrix} + \left(\begin{bmatrix} 0 \\ 1 \end{bmatrix} \otimes I_2\right) S_k(\theta_k)\begin{bmatrix} w^f_k \\ w^\tau_k \end{bmatrix},$$

where we defined the agent state $\xi^{\mathrm{h}}_k = [x^{\mathrm{h}}_k \;\; y^{\mathrm{h}}_k \;\; \dot{x}^{\mathrm{h}}_k \;\; \dot{y}^{\mathrm{h}}_k]^{\mathsf{T}}$. Besides the disturbances w^f_k and w^τ_k entering nonlinearly, the linearized system consists of two decoupled double-integrators.

Distributed Output Regulation

For the agents with double-integrator dynamics, the problem at hand can be formulated as distributed output regulation problem with the global and local exosystems

$$\dot{d}^g = \left(\begin{bmatrix} 0 & 1 \\ 0 & 0 \end{bmatrix} \otimes I_2\right) d^g, \qquad \dot{d}^\ell_k = \begin{bmatrix} 0 & \alpha \\ -\alpha & 0 \end{bmatrix} d^\ell_k$$

for some $\alpha > 0$, generating the global reference signal $r^g = ([1\ 0] \otimes I_2)d^g$ and the local reference signals $r_k^\ell = d_k^\ell$ for $k \in \mathcal{N}$. After the feedback linearization the robot hand positions obey double-integrator dynamics. We define the control input $u_k = [\bar{f}_k \ \ \bar{\tau}_k]^\mathsf{T}$ and, for now, we neglect the local disturbances. This leads to the agent dynamics

$$\dot{\xi}_k^\mathrm{h} = \left(\begin{bmatrix} 0 & 1 \\ 0 & 0 \end{bmatrix} \otimes I_2\right) \xi_k^\mathrm{h} + \left(\begin{bmatrix} 0 \\ 1 \end{bmatrix} \otimes I_2\right) u_k.$$

The regulation error e_k defining the group objective reads

$$e_k = ([1\ 0] \otimes I_2)\, \xi_k^\mathrm{h} - ([1\ 0] \otimes I_2)\, d^g - d_k^\ell.$$

For this scenario, a distributed regulator can be constructed as described in Theorem 4.6 and Theorem 4.8. A stabilizing feedback for the agents is $u_k = -F\xi_k^\mathrm{h}$ with $F = [\, 0.1000\ 0.4583 \,] \otimes I_2$ (LQR with weights $Q = I_4$, $R = 100 I_2$). The regulator equations (4.27) and (4.28) are solved by $\Pi^g = I_4$, $\Gamma^g = 0$, $\Pi^\ell = [I_2\ \Omega(\alpha)^\mathsf{T}]^\mathsf{T}$, $\Gamma^\ell = -\alpha^2 I_2$ with the skew-symmetric map $\Omega : \mathbb{R} \to \mathbb{R}^{2\times 2}$ defined by

$$\Omega(\alpha) = \begin{bmatrix} 0 & \alpha \\ -\alpha & 0 \end{bmatrix}.$$

The gain matrices (4.33) are $G^g = F$ and $G^\ell = -\alpha^2 I_2 + F[I_2\ \Omega(\alpha)^\mathsf{T}]^\mathsf{T}$. A suitable gain matrix K for the distributed observer (4.34) is $K = \left[\begin{smallmatrix} 11.9504 & 1.5317 \\ 0.9511 & 12.1072 \end{smallmatrix}\right] \otimes I_2$, obtained with Theorem 3.7 and region $\mathcal{R} = S(3, 50, \pi/3)$. The underlying graph $\mathcal{G}$ is an undirected cycle with $N = 6$ nodes. Since each agent has direct access to its local reference signal $r_k^\ell(t)$, no local observers (4.36) are needed. The last component of the distributed regulator are couplings based on the transient state components of each agent, as proposed in Section 4.2.3. The coupling gain matrix H is synthesized with Theorem 3.7 and region $S(2, 15, \pi/4)$, which yields $H = [\, 8.3190\ 4.0836 \,] \otimes I_2$.

Local Disturbance Estimation and Compensation

The distributed regulator constructed above needs to be extended such that the local disturbances are estimated and rejected. For this purpose, a nonlinear observer for the disturbances is constructed. It is based on an augmented agent model including the dynamics of the disturbances. For the augmented system, a Luenberger-type observer with observer gain matrix scheduled by the agent orientation is constructed in the following. We define the augmented agent state vector $z_k = [x_k^\mathrm{h}\ y_k^\mathrm{h}\ \dot{x}_k^\mathrm{h}\ \dot{y}_k^\mathrm{h}\ w_k^f\ w_k^\tau]^\mathsf{T}$, which obeys the dynamics

$$\dot{z}_k = A_k(\theta_k) z_k + B \begin{bmatrix} \bar{f}_k \\ \bar{\tau}_k \end{bmatrix} \quad \text{with} \quad A_k(\theta_k) = \begin{bmatrix} 0 & I_2 & 0 \\ 0 & 0 & S_k(\theta_k) \\ 0 & 0 & 0 \end{bmatrix}, \quad B = \begin{bmatrix} 0 \\ I_2 \\ 0 \end{bmatrix}.$$

A Luenberger-type observer for this system takes the form

$$\dot{\hat{z}}_k = A_k(\theta_k)\hat{z}_k + B \begin{bmatrix} \bar{f}_k \\ \bar{\tau}_k \end{bmatrix} + L_k(\theta_k) \left(\begin{bmatrix} x_k^\mathrm{h} \\ y_k^\mathrm{h} \end{bmatrix} - \begin{bmatrix} \hat{x}_k^\mathrm{h} \\ \hat{y}_k^\mathrm{h} \end{bmatrix} \right), \tag{5.4}$$

where $L_k : \mathbb{R} \to \mathbb{R}^{6\times 2}$ is a continuous bounded map to be determined. The resulting dynamics of the observer error $e_k = \hat{z}_k - z_k$ can be computed as $\dot{e}_k = (A_k(\theta_k) - L_k(\theta_k)C)e_k$ with $C = [I_2\ 0\ 0]$.

The nonlinear change of coordinates $\tilde{e}_k = \mathrm{diag}(I_2, I_2, S_k(\theta_k))e_k$ leads to the transformed dynamics $\dot{\tilde{e}}_k = \mathrm{diag}(I_2, I_2, S_k(\theta_k))\dot{e}_k + \mathrm{diag}(0, 0, \dot{S}_k(\theta_k))e_k$, i.e.,

$$\dot{\tilde{e}}_k = \left(\begin{bmatrix} 0 & I_2 & 0 \\ 0 & 0 & I_2 \\ 0 & 0 & \dot{R}(\theta_k)R(\theta_k)^{-1} \end{bmatrix} - \begin{bmatrix} I_2 & 0 & 0 \\ 0 & I_2 & 0 \\ 0 & 0 & S_k(\theta_k) \end{bmatrix} L_k(\theta_k)C \right) \tilde{e}_k.$$

Exponential stability of $e_k = \mathbf{0}$ is equivalent to exponential stability of $\tilde{e}_k = \mathbf{0}$. We choose the orientation-dependent observer gain matrix $L_k(\theta_k)$ as

$$L_k(\theta_k) = \mathrm{diag}(I_2, I_2, S_k(\theta_k)^{-1})L \tag{5.5}$$

for some constant matrix $L \in \mathbb{R}^{6\times 2}$ to be determined. With $\dot{R}(\theta_k)R(\theta_k)^{-1} = \Omega(\omega_k)$, this yields

$$\dot{\tilde{e}}_k = \left(\begin{bmatrix} 0 & I_2 & 0 \\ 0 & 0 & I_2 \\ 0 & 0 & \Omega(\omega_k) \end{bmatrix} - LC \right) \tilde{e}_k. \tag{5.6}$$

Through this transformation, the observer design problem reduces to finding a constant gain matrix L such that the error dynamics (5.6) are exponentially stable for all $\omega_k(t) \in \mathbb{R}$. In order to obtain a bounded observer gain, we assume that the angular velocity is bounded, i.e., there are constants $\underline{\omega}, \overline{\omega} \in \mathbb{R}$ such that

$$\underline{\omega} \leq \omega_k(t) \leq \overline{\omega}$$

for all $t \geq 0$ and all $k \in \mathcal{N}$. For given bounds $\underline{\omega}, \overline{\omega}$, the error dynamics can be described as a polytopic LPV system with parameter $\omega_k(t) \in \mathcal{P} = [\underline{\omega}, \overline{\omega}]$. A stabilizing gain matrix L can hence be found via SDP based on Lemma A.23 (exploiting duality). For the bounds $\underline{\omega} = -2\pi\,\mathrm{rad/s^2}$ and $\overline{\omega} = 2\pi\,\mathrm{rad/s^2}$, we obtained $L = ([\,24.4140\ 274.9224\ 542.9262\,] \otimes I_2)^\mathsf{T}$. Note that the gain increases for larger intervals $[\underline{\omega}, \overline{\omega}]$.

With the disturbance estimates $\hat{w}_k^f$ and $\hat{w}_k^\tau$ provided by the observer (5.4), the feedback

$$\begin{bmatrix} f_k \\ \tau_k \end{bmatrix} = S_k(\theta_k)^{-1} \begin{bmatrix} v_k\omega_k \sin\theta_k + h_k\omega_k^2 \cos\theta_k + \bar{f}_k \\ -v_k\omega_k \cos\theta_k + h_k\omega_k^2 \sin\theta_k + \bar{\tau}_k \end{bmatrix} - \begin{bmatrix} \hat{w}_k^f \\ \hat{w}_k^\tau \end{bmatrix} \tag{5.7}$$

asymptotically achieves feedback linearization and, simultaneously, rejection of the local disturbances. The following example summarizes the motion coordination algorithm for the group of mobile robots and presents numerical simulation results.

Example 5.1. The motion coordination algorithm consists of the following components: For each agent, a nonlinear observer (5.4) reconstructs the disturbances $\hat{w}_k^f$, $\hat{w}_k^\tau$. The feedback (5.7) compensates these disturbances and linearizes the system dynamics. The motion coordination problem is solved by means of a distributed regulator through the new control inputs $\bar{f}_k$, $\bar{\tau}_k$. The distributed regulator, designed according to Theorem 4.8, is

$$\begin{bmatrix} \bar{f}_k \\ \bar{\tau}_k \end{bmatrix} = -F\xi_k^h + G^g \hat{d}_k^g + G^\ell d_k^\ell + H \sum_{j=1}^{N} a_{kj}(\hat{\epsilon}_j - \hat{\epsilon}_k)$$

with the transient state component $\hat{\epsilon}_k = \xi_k^\mathrm{h} - \Pi^g \hat{d}_k^g - \Pi^\ell d_k^\ell$. By assumption, both $[x_k^\mathrm{h}\ y_k^\mathrm{h}]^\mathsf{T}$ and $[\dot{x}_k^\mathrm{h}\ \dot{y}_k^\mathrm{h}]^\mathsf{T} = R(\theta_k)[v_k\ h_k\omega_k]^\mathsf{T}$ can be measured or computed, i.e., ξ_k^h is available. The estimate $\hat{d}_k^g$ is obtained via a distributed observer of the form (4.34).

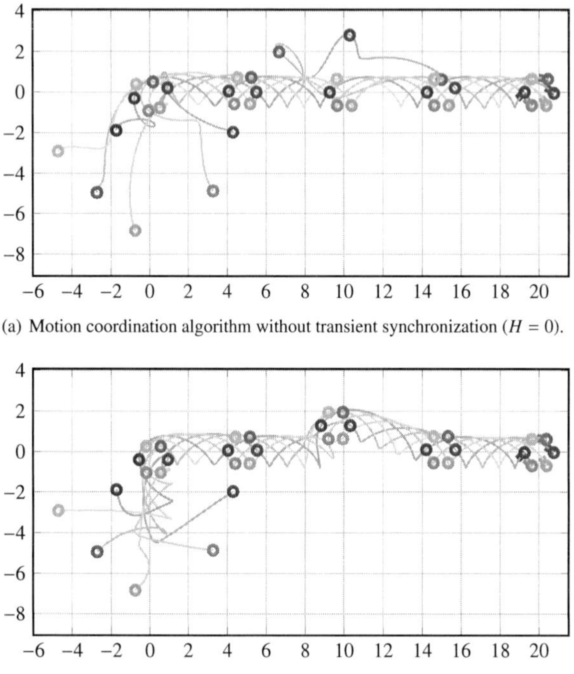

(a) Motion coordination algorithm without transient synchronization ($H = 0$).

(b) Motion coordination algorithm with transient synchronization.

Figure 5.3: Motion of the group of mobile robots from Example 5.1 in the plane. The thick circles mark formation snapshots at the time instances $t = 0$ s, 10 s, 20 s, . . . , 50 s.

The cooperative behavior of the group running this motion coordination algorithm is illustrated by two numerical simulation results in Fig. 5.3. Typical physical parameters are $m_k = 10$ kg and $J_k = 0.15$ kg m^2 (Ren and Atkins (2007)). In this example, the formation center point shall rest at (0 m, 0 m) until $t = 10$ s and travel linearly from (0 m, 0 m) to (20 m, 0 m) during $t \in [10\text{ s}, 50\text{ s}]$. In the formation, each agent shall circle around the formation center point with a distance of 0.75 m and angular speed $\alpha = \pi/5$ rad/s. An equal spacing is achieved by suitable initial conditions for the local exosystems. At $t = 27$ s, a step disturbance of 20 N and 2 Nm acts on both vehicles one and two, i.e., $w_1^f(t) = w_2^f(t) = 20$ N and $w_1^\tau(t) = w_2^\tau(t) = 2$ Nm for $t \geq 27$ s. Fig. 5.3 shows two simulation results. In both cases, the motion coordination task is fulfilled. In Fig. 5.3(a), the coupling gain H is set to zero. Due to the local disturbance acting on two robots, these robots deviate significantly from the formation. Asymptotically, the disturbances are rejected and the robots return to the desired formation. Fig. 5.3(b) demonstrates the effect of the additional coupling term. In this case, the robots remain in a close formation during the transient phase and maintain the desired relative distances, while the formation as a whole deviates from the reference path. This behavior has clear advantages for practical applications. The distributed regulator (Theorem 4.8) allows for a trade-off between these two behaviors by appropriately designing the control gains F and H. △

5.2 Unicycle-type Vehicles with Constant Velocities

This section is dedicated to formation control problems for unicycle-type vehicles with constant and non-identical velocities. In contrast to Section 5.1, the focus is on the kinematic equations (5.1a)–(5.1c) of the unicycle-type vehicles while the dynamics (5.1d)–(5.1e) are disregarded. The following literature review is intended to provide the required background information and to motivate our contributions. This section is based on Seyboth et al. (2014b).

Formation control of unicycle-type vehicles with continuous feedback laws is studied by, e.g., Lin et al. (2005); Marshall et al. (2006); Zheng et al. (2009), where it is assumed that both the orientation and the velocity of each unicycle can be controlled. Furthermore, Chen and Zhang (2011) consider fully controllable unicycles and propose a distributed control law, which achieves a collective circular motion over jointly connected communication graphs. Chen and Zhang (2013) present an extension to heterogeneous groups, where the heterogeneity is assumed in the angular velocities while the forward speed is used as control input. Rendezvous problems for fully controllable unicycle-type vehicles based on limited information are studied by Yu et al. (2012) and Zheng and Sun (2013). Oftentimes, however, the velocity of a vehicle cannot be controlled in practice, e.g., when using UAVs with limited capabilities. This motivated the study of coordination and formation control problems for unicycle-type vehicles with constant velocities, cf., Justh and Krishnaprasad (2004); Marshall et al. (2004); Klein and Morgansen (2006); Lalish et al. (2007); Sepulchre et al. (2007, 2008); Ceccarelli et al. (2008). Justh and Krishnaprasad (2004) construct a two-vehicle formation control law, which achieves global convergence to a desired configuration. An extension to N vehicles is discussed based on simulations, but no rigorous convergence analysis was done. Marshall et al. (2004) study the cyclic pursuit problem, in which each agent follows its designated predecessor in a cyclic fashion. It is shown that the pursuit rule leads to regular polygon formations. Klein and Morgansen (2006); Lalish et al. (2007) pursue a different approach, termed oscillatory control, in order to control unicycles with fixed speed. An inner loop controller allows to adjust the effective forward speed of the vehicle by steering the vehicle on a sinusoidal line along a desired path. The outer control loop can be designed as if the vehicle was a fully controllable unicycle. The control scheme is based on multiple approximations, which makes a rigorous convergence analysis hard. The seminal articles by Sepulchre et al. (2007, 2008) present a study of circular formation stabilization and stabilization of desired phase arrangements in a group of particles moving in the plane, i.e., in a group of unicycles with unit speed. The results of Sepulchre et al. (2007) are based on all-to-all communication in the group, which is relaxed to limited communication by Sepulchre et al. (2008). Ceccarelli et al. (2008) take sensory limitation in terms of a limited visibility region into account as an additional practical issue. The research in this area was strongly influenced by the study of synchronization phenomena in physics and mathematics (Strogatz (2000); Kuramoto (2003)), which has a long history and where the goal is to describe and understand synchronizing behavior in dynamical networks consisting of individual dynamical subsystems with a certain interconnection structure.

Summarizing, the forward velocity of each vehicle is assumed to be controllable by Lin et al. (2005); Marshall et al. (2006); Zheng et al. (2009); Chen and Zhang (2011, 2013); Yu et al. (2012); Zheng and Sun (2013), whereas all results of Marshall et al. (2004); Justh and Krishnaprasad (2004); Klein and Morgansen (2006); Lalish et al. (2007); Sepulchre et al. (2007, 2008); Ceccarelli et al. (2008) are based on the assumption that the velocities of all unicycles in the group are constant and identical. It was pointed out by Sepulchre et al. (2007) that the assumption of identical agents is fundamental to the symmetry properties of the closed-loop system and also facilitates the stability analysis. However, as already pointed out, this assumption may not be satisfied in practice. This motivates the work of Sinha and Ghose (2007) and van der Walle et al. (2008). Sinha and Ghose (2007) address the cyclic pursuit problem with heterogeneous velocities. The possible equilibrium configurations are computed and discussed, but no stability analysis is provided. The work of van der Walle et al. (2008) is focused on a particular formation control problem for mobile agents modeled as unicycles with non-identical fixed velocities. A triangular formation of unicycles with non-identical fixed speeds is considered and a spiral motion of the formation is studied. It turns out that the formation starts to rotate around the formation center, which is unwanted and has to be compensated for. The proposed control law is a switching controller which steers each agent to a sequence of way points which are generated along the desired spiral path. Concluding, unicycle formations with heterogeneous constant velocities have received very little attention despite their practical relevance.

With this motivation, we study motion coordination problems for groups of unicycle-type vehicles with non-identical speed. As a fundamental coordination task, we focus on the asymptotic stabilization of circular formations in which all vehicles move collectively on circles in the plane. We discuss the limitations caused by the heterogeneity and present possible circular motions in such groups. The agents can traverse circles around a common center point only if either different radii or different angular frequencies for each agent are permitted. In case of different radii and identical angular frequencies, the orientation of all agents can be coordinated and synchronized as well. We propose suitable static and smooth control laws that guarantee (global) convergence to such circular formations.

The problem formulation is presented and the effects of non-identical velocities on the formation control task are discussed in Section 5.2.1. The main results are presented in Sections 5.2.2 and 5.2.3. The stabilization problem of circular formations in heterogeneous groups is addressed and it is shown how circular formations with synchronized or balanced phase configurations can be achieved. All results are illustrated by numerical examples.

5.2.1 Problem Formulation

In the following, some additional notation, the agent models, the information structure, and the group objective of the multi-agent systems under consideration are introduced.

Additional Notation In the following, stack vectors are boldface for convenience, i.e., $\mathbf{r} = [r_1 \ \cdots \ r_N]^\mathsf{T} \in \mathbb{C}^N$ for $r_k \in \mathbb{C}$, $k \in \mathcal{N}$. For $\mathbf{z}_1, \mathbf{z}_2 \in \mathbb{C}^N$, we define the scalar product $\langle \mathbf{z}_1, \mathbf{z}_2 \rangle = \mathbf{Re}(\bar{\mathbf{z}}_1^\mathsf{T} \mathbf{z}_2)$, i.e., the real part of the standard scalar product over $\mathbb{C}^N$. It holds that $\langle \mathbf{z}_1, \mathbf{z}_2 \rangle = \langle \mathbf{z}_2, \mathbf{z}_1 \rangle$. Note that $\langle \mathbf{z}, \mathbf{z} \rangle = \mathbf{Re}(\bar{\mathbf{z}}^\mathsf{T} \mathbf{z}) = \|z\|^2$, i.e., the squared Euclidean norm of $\mathbf{z}$. Let $D \subseteq \mathbb{R}^N$ and $f : D \to \mathbb{R}$ be a differentiable map. Then, the gradient ∇f and the Hessian matrix H_f of f are defined by $\nabla f = \text{grad} f = [\frac{\partial f}{\partial x_1} \ \cdots \ \frac{\partial f}{\partial x_N}]^\mathsf{T}$ and $[H_f]_{jk} = \frac{\partial^2 f}{\partial x_j \partial x_k}$ element-wise, respectively.

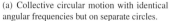

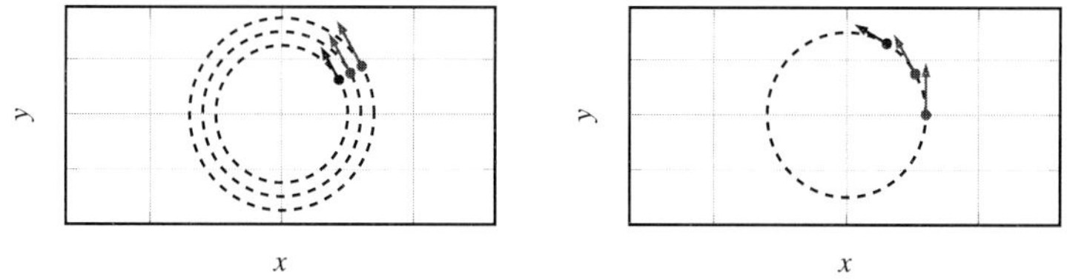

(a) Collective circular motion with identical angular frequencies but on separate circles.

(b) Collective circular motion on identical circles but with non-identical angular frequencies.

Figure 5.4: Two possible collective circular motion patterns. The orientation of each agent is depicted as arrow and the length of each arrow is proportional to the agent's forward velocity.

Agent Models We consider a formation control problem for a group of $N > 1$ mobile agents of unicycle-type with constant and non-identical velocities. The kinematic equations of agent k are

$$\dot{x}_k = v_k \cos(\theta_k) \tag{5.8a}$$
$$\dot{y}_k = v_k \sin(\theta_k) \tag{5.8b}$$
$$\dot{\theta}_k = u_k, \tag{5.8c}$$

where $x_k(t) \in \mathbb{R}$ and $y_k(t) \in \mathbb{R}$ are the coordinates of agent k in the plane and $\theta_k(t) \in \mathbb{S}$ is its orientation, also referred to as the heading angle or phase angle, for $k \in \mathcal{N}$. Each agent has a positive fixed cruising speed $v_k > 0$. The orientation of each agent can be controlled through its input $u_k(t) \in \mathbb{R}$.

With the complex coordinate $r_k(t) = x_k(t) + \mathbf{i}y_k(t) = |r_k(t)|e^{\mathbf{i}\phi_k(t)} \in \mathbb{C}$, where $\phi_k(t) = \arg(r_k(t))$, the model (5.8) can be rewritten as

$$\dot{r}_k = v_k e^{\mathbf{i}\theta_k} \tag{5.9a}$$
$$\dot{\theta}_k = u_k. \tag{5.9b}$$

It is assumed that each agent is aware of its own velocity v_k as well as its position $r_k(t)$ and orientation $\theta_k(t)$ in a global coordinate frame. With the terminology of Oh et al. (2015), our approach is displacement-based control (with couplings among the agents and no individual pre-defined target positions) and partly position-based control (since each agent senses its own position and orientation with respect to a global coordinate frame).

Effects of Non-identical Velocities The so-called *collective circular motion* is a motion of the group, in which all agents move on a common circle of radius ρ_0 and with constant angular frequency ω_0. Such a collective circular motion is impossible for a group of unicycles with non-identical velocities. It is only possible to achieve either a movement of all agents on the same circle but with non-identical angular frequencies, or a movement with identical angular frequency (and possibly with synchronized phase angles), but on circles of different radii with a common center point. Both configurations are shown in Fig. 5.4. Our goal is to find smooth control laws that steer the group of unicycles with non-identical fixed velocities to configurations as shown in Fig. 5.4. These configurations may serve as basis for typical practical applications such as surveillance, exploration, or target circumnavigation (cf., Deghat et al. (2014)). In formations according to Fig. 5.4(b), it is assumed that the agents move on different altitudes such that there can be no collisions.

A suitable measure of synchrony in networks of coupled oscillators is the so-called *order parameter* p_θ, defined as the average of all agents phasors $e^{\mathrm{i}\theta_k}$, i.e.,

$$p_\theta = \frac{1}{N}\sum_{k=1}^{N} e^{\mathrm{i}\theta_k}. \tag{5.10}$$

The absolute value $|p_\theta|$ of the complex order parameter is a measure of phase coherence and the argument $\arg(p_\theta)$ is the phase average, cf., Strogatz (2000); Kuramoto (2003); Sepulchre et al. (2007). A second important quantity is the average linear momentum $\dot{R}$ of the group, i.e., the time derivative of the formation center point $R = 1/N\sum_{k=1}^{N} r_k$. With (5.9a), the average linear momentum is given by

$$\dot{R} = \frac{1}{N}\sum_{k=1}^{N} v_k e^{\mathrm{i}\theta_k}. \tag{5.11}$$

In case of identical unit velocities $v_1 = \ldots = v_N = 1$, the average linear momentum $\dot{R}$ and the order parameter p_θ coincide. In particular, when $p_\theta = \dot{R} = 0$, then the average position of all agents rests on a fixed point and the phase angles are balanced.

In case of non-identical velocities, the situation is different. The average linear momentum $\dot{R}$ does not coincide with p_θ. The absolute value of $\dot{R}$ is non-negative and its maximum is the average velocity $v_{\text{avg}} = \frac{1}{N}\sum_{k=1}^{N} v_k$. The maximal value $|\dot{R}| = v_{\text{avg}}$ is achieved if and only if all phase angles θ_k, $k \in \mathcal{N}$, are identical, i.e., if all vehicles have a common orientation. Hence, in case of non-identical velocities, both the order parameter (5.10) and the linear momentum (5.11) are suitable measures of synchrony. The value $|\dot{R}| = 0$, however, cannot always be achieved. If there is an agent with velocity larger than the sum of the velocities of all other agents, then the velocity of this agent dominates the whole group and there is no configuration with $|\dot{R}| = 0$. To be precise, $\dot{R} = 0$ is impossible if $v_{\max} > (\sum_{k=1}^{N} v_k) - v_{\max}$, where $v_{\max} = \max_{k\in\mathcal{N}} v_k$ is the maximal velocity.

Information Structure The agents have communication capabilities such that they can exchange information with other agents in the group. For most of the following developments, it is assumed that information exchange is possible among any two agents in the group, i.e., $\mathcal{G}$ is a complete (all-to-all) graph. The limited communication case with general (undirected) graphs $\mathcal{G}$ is addressed at the end of the Section 5.2.3.

Group Objectives In the following, we develop motion coordination algorithms for groups of $N > 1$ agents (5.9). Four distinct cooperative control problems can be defined:

Problem 5.1 (Phase Agreement)**.** Find a smooth control law u_k, $k \in \mathcal{N}$, such that for all $j, k \in \mathcal{N}$,

$$\lim_{t\to\infty}(\theta_k(t) - \theta_j(t)) = 0.$$

In words, phase agreement (or synchronization) is achieved if the phase differences between all agents vanish asymptotically, as in Fig. 5.4(a).

Problem 5.2 (Phase Balancing)**.** Find a smooth control law u_k, $k \in \mathcal{N}$, such that

$$\lim_{t\to\infty} p_\theta = 0.$$

In words, phase balancing is achieved if the sum of the phasors $e^{\mathrm{i}\theta_k}$ of all agents converges to zero. There is a variety of balanced configurations, as discussed in detail in Sepulchre et al. (2007). Of particular interest is the splay state, a balanced configuration in which the difference between the phase angles of two consecutive agents is $2\pi/N$, i.e., the phase angles are uniformly spaced around the circle.

Problem 5.3 (Stabilization of the Average Linear Momentum). Find a smooth control law u_k, $k \in \mathcal{N}$, such that

$$\lim_{t\to\infty} \dot{R} = 0.$$

Problem 5.4 (Stabilization of Circular Formations). Find a smooth control law u_k, $k \in \mathcal{N}$, such that the steering rate $\dot{\theta}_k(t)$ of every agent converges to a constant $\omega_k \neq 0$, and the center points of the resulting circular trajectories are identical.

Two examples for circular formations are depicted in Fig. 5.4. In the following, we present solutions to these problems as well as to combinations thereof.

5.2.2 Coordination of the Vehicle Orientations

Before turning to circular formations, we address Problem 5.1, Problem 5.2, and Problem 5.3.

Phase Agreement and Balancing

Since the dynamics of the phase angles (5.9b) are not affected by the non-identical velocities, the same control law as in the homogeneous case treated by Sepulchre et al. (2007) provides a solution to Problem 5.1 and Problem 5.2 in the heterogeneous case.

Theorem 5.1 (Phase Agreement or Balancing). *Consider a group of $N > 1$ unicycles (5.9) with non-identical velocities $v_k > 0$, $k \in \mathcal{N}$, and consider the control law*

$$u_k = \frac{\gamma}{N}\sum_{j=1}^{N} \sin(\theta_j - \theta_k). \tag{5.12}$$

Then, for $\gamma > 0$, the synchronous state $\theta_1 = \ldots = \theta_N$ is asymptotically stable and all other equilibria are unstable. For $\gamma < 0$, the balanced set on which $p_\theta = 0$ is asymptotically stable and all other equilibria are unstable.

Proof. See Appendix B.19. □

The control law (5.12) is a gradient control law for the potential function

$$U_{p_\theta} = \frac{N}{2}|p_\theta|^2. \tag{5.13}$$

Depending on the sign of the gain γ, it stabilizes the maxima or minima of U_{p_θ}, which corresponds to phase agreement or balancing, respectively.

Stabilization of the Average Linear Momentum

In contrast to the homogeneous case, the balanced state with $p_\theta = 0$ does not coincide with $\dot{R} = 0$ in the heterogeneous case. Hence, control law (5.12) can not be used in order to solve Problem 5.3. As discussed in Section 5.2.1, stabilization of the average linear momentum $\dot{R}$ is only possible, if $v_{\max} \leq \frac{1}{2}\sum_{k=1}^{N} v_k$. Therefore, we impose the following assumption.

Assumption 5.1. *No individual agent has a velocity larger than the sum of the velocities of all other agents in the group, i.e.,*

$$v_{\max} \leq \frac{1}{2}\sum_{k=1}^{N} v_k,$$

where $v_{\max} = \max_{k\in\mathcal{N}} v_k$ *is the maximal velocity of all agents in the group.*

A suitable control law solving Problem 5.1 and Problem 5.3 is given in the following theorem.

Theorem 5.2 (Stabilization of the Average Linear Momentum). *Consider a group of* $N > 1$ *unicycles* (5.9) *with non-identical velocities* $v_k > 0$, $k \in \mathcal{N}$, *and suppose that Assumption 5.1 is satisfied. Consider the control law*

$$u_k = \frac{\gamma}{N}\sum_{j=1}^{N} v_j v_k \sin(\theta_j - \theta_k). \tag{5.14}$$

Then, for $\gamma > 0$, *the synchronous state* $\theta_1 = \ldots = \theta_N$ *is asymptotically stable and all other equilibria are unstable. For* $\gamma < 0$, *the set on which the linear momentum is constant, i.e.,* $\dot{R} = 0$, *is asymptotically stable and all other equilibria are unstable.*

Proof. See Appendix B.20. □

The control law (5.14) is a gradient control law for the potential function

$$U_{\dot{R}} = \frac{N}{2}|\dot{R}|^2. \tag{5.15}$$

Depending on the sign of the gain γ, it stabilizes the maxima or minima of $U_{\dot{R}}$, which corresponds to phase agreement or stabilization of the average linear momentum, respectively.

Summarizing, both (5.12) and (5.14) asymptotically stabilize the synchronous state $\theta_1 = \ldots = \theta_N$ for $\gamma > 0$. In contrast to the homogeneous case, it is in general impossible to achieve $p_\theta = 0$ and $\dot{R} = 0$ at the same time. The state with balanced phase angles, i.e., $p_\theta = 0$, is asymptotically stabilized by (5.12) with $\gamma < 0$. The set on which $|\dot{R}|$ takes its minimum is asymptotically stabilized by (5.14) with $\gamma < 0$ and if Assumption 5.1 is satisfied, $\dot{R}$ vanishes and the average position of all agents converges to a fixed point. The control law (5.14) depends explicitly on the vehicle velocities. Interestingly, in contrast to (5.12), the coupling among two vehicles is weighted by the product of their respective velocities.

Corollary 5.3. *Consider a group of* $N > 1$ *unicycles* (5.9) *with non-identical velocities* $v_k > 0$, $k \in \mathcal{N}$, *and consider the control law*

$$u_k = \omega_0 + \frac{\gamma}{N}\sum_{j=1}^{N} \sin(\theta_j - \theta_k) \tag{5.16}$$

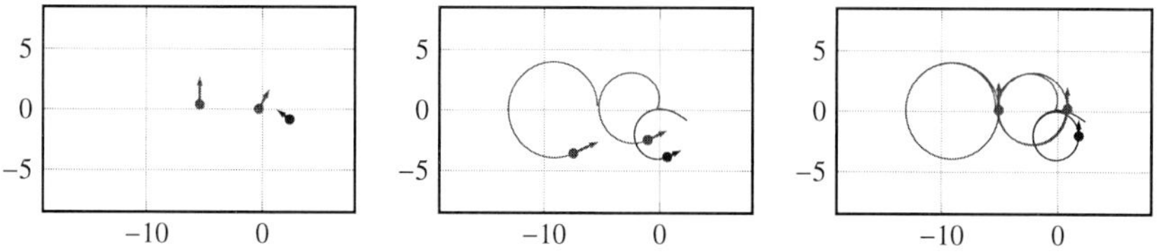

(a) $t = 0$. (b) $t = 10$. (c) $t = 50$.

Figure 5.5: Phase synchronization, i.e., stabilization of $p_\theta = 1$ with control law (5.16) in Example 5.2.

with $\gamma \neq 0$. Then, each agent k moves on a circle of radius $\rho_k = v_k/|\omega_0|$. For $\gamma > 0$, the synchronous state $\theta_1 = \ldots = \theta_N$ is asymptotically stable and all other equilibria are unstable. For $\gamma < 0$, the balanced set on which $p_\theta = 0$ is asymptotically stable and all other equilibria are unstable.

Moreover, suppose that Assumption 5.1 is satisfied and consider the control law

$$u_k = \omega_0 + \frac{\gamma}{N} \sum_{j=1}^{N} v_j v_k \sin(\theta_j - \theta_k) \tag{5.17}$$

with $\gamma \neq 0$. Then, each agent k moves on a circle of radius $\rho_k = v_k/|\omega_0|$. For $\gamma > 0$, the synchronous state $\theta_1 = \ldots = \theta_N$ is asymptotically stable and all other equilibria are unstable. For $\gamma < 0$, the set on which the linear momentum is constant, i.e., $\dot{R} = 0$, is asymptotically stable and all other equilibria are unstable.

Proof. The corollary follows from stating the results of Theorems 5.1 and 5.2 in a rotating coordinate frame, i.e., we consider the model $\dot{\theta}_k = \omega_0 + u_k$ instead of (5.9b). Since $\langle \nabla U_{p_\theta}, \mathbf{1} \rangle = 0$ and $\langle \nabla U_{\dot{R}}, \mathbf{1} \rangle = 0$, the stability analysis holds unchanged. Since in steady-state $u_k = \omega_0$ for all $k \in \mathcal{N}$, all agents converge to circular movements of radius $\rho_k = v_k/|\omega_0|$. □

Example 5.2. A simulation is performed for a group of $N = 3$ unicycles (5.8) with velocities $v_1 = 1$, $v_2 = 1.5$, $v_3 = 2$ and control laws (5.16) and (5.17). First, control law (5.16) with $\omega_0 = 0.5$ and $\gamma = 0.6$ is applied. The motion of the agents in the plane is shown in Fig. 5.5. As expected, the agents reach a phase agreement and move on circles. Second, the control law (5.17) with $\omega_0 = 0.5$ and $\gamma = -0.6$ is applied. The motion of the agents in the plane is shown in Fig. 5.6. As expected, the group center point R, marked as $\oplus$, converges to a fixed point in the plane while the agents move on circles. $\wedge$

These results show how the vehicle orientations θ_k of the agents can be coordinated such that they reach an agreement or a balanced configuration. Furthermore, the novel control law (5.14) stabilizes the average linear momentum $\dot{R} = 0$, i.e., it stabilizes the center point of the group to a fixed point in the plane. As noted in Corollary 5.3, the controllers can be augmented with a constant angular frequency ω_0 in order to achieve the same phase configurations while moving on circles.

5.2.3 Stabilization of Circular Formations

In a first step, we propose a control law that solves Problem 5.4, i.e., that achieves a circular formation of the group according to Fig. 5.4(a). Circular formations of the alternative form

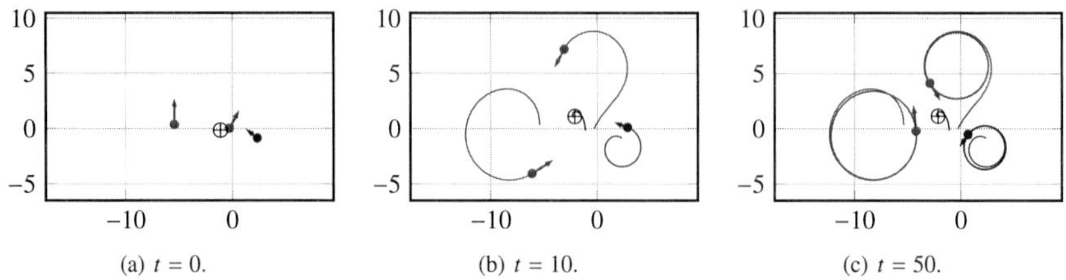

Figure 5.6: Stabilization of the average linear momentum in Example 5.2, i.e., stabilization of $\dot{R} = 0$, with control law (5.17). The group center point R is marked by $\oplus$.

Fig. 5.4(b) will be treated thereafter. The following theorem presents a suitable control law for the stabilization of circular formations in heterogeneous groups. At first, it is assumed that all agents may exchange information with each other. The limited communication case with an underlying communication graph will be addressed in the last paragraph of the present section.

Theorem 5.4 (Circular Formation Stabilization). *Consider a group of $N > 1$ unicycles* (5.9) *with non-identical velocities $v_k > 0$, $k \in \mathcal{N}$, and consider the control law*

$$u_k = \gamma \frac{\partial U_{\dot{R}}}{\partial \theta_k} + \omega_0 \left(1 + \gamma \langle \tilde{r}_k, \dot{r}_k \rangle\right), \tag{5.18}$$

with $\gamma > 0$ and $\omega_0 \neq 0$, and where $\tilde{r}_k = r_k - R$. Then, all solutions converge to a circular formation in which each agent k moves on a circle of radius $\rho_k = v_k/|\omega_0|$ and all circles have a fixed common center point.

Proof. See Appendix B.21. □

Theorem 5.4 is a global convergence result. The control law (5.18) guarantees that a circular formation according to Fig. 5.4(a) is achieved for arbitrary initial conditions and non-identical velocities. In contrast to the control law proposed by Sepulchre et al. (2007), (5.18) involves the gradient of $U_{\dot{R}}$ instead of U_{p_θ} and hence takes into account the velocities of the agents. The control law (5.18) can be expressed in real variables using (B.24) and (B.27). This yields

$$u_k = \omega_0 - \frac{\gamma}{N} \sum_{j=1}^{N} v_j v_k \sin(\theta_k - \theta_j) + \omega_0 \gamma v_k |r_k| \cos\left(\phi_k - \theta_k\right) - \omega_0 \frac{\gamma}{N} \sum_{j=1}^{N} |r_j| v_k \cos\left(\phi_j - \theta_k\right). \tag{5.19}$$

Example 5.3. A simulation is performed for a group of $N = 3$ unicycles (5.8) with velocities $v_1 = 1$, $v_2 = 1.5$, $v_3 = 2$, and control law (5.19) with $\omega_0 = 0.6$ and $\gamma = 0.5$. The motion of the agents in the plane is shown in Fig. 5.7. As expected, the agents move on circles around a common center point with different radii $\rho_k = v_k/|\omega_0|$ according to the respective velocities. The final configuration of the phase angles is not predefined and depends on the initial conditions. △

Circular Formations with Synchronized or Balanced Phase

The next step is combining the previous results. We show that Problems 5.1 and 5.4 or Problems 5.2 and 5.4 can be solved at the same time as a circular formation can be achieved with a

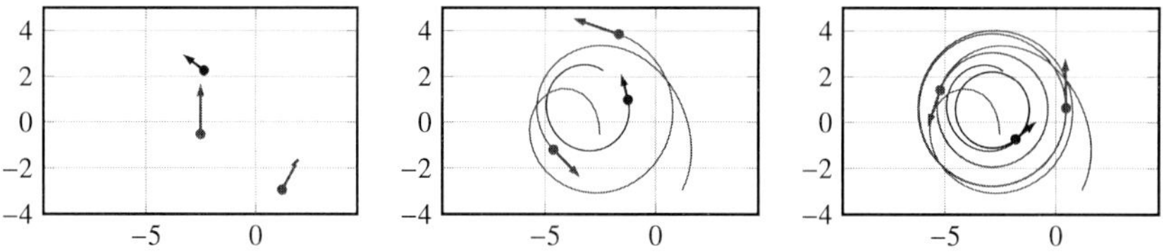

(a) $t = 0$.	(b) $t = 10$.	(c) $t = 50$.

Figure 5.7: Stabilization of a collective circular motion with control law (5.19) in Example 5.3.

desired synchronized or balanced configuration of the phase angles. As a first step, we observe that the control law (5.18) can be augmented with the gradient of a suitable potential U that is minimized by the desired phase configuration. The following result extends (Sepulchre et al., 2007, Thm. 3) to groups with non-identical velocities.

Theorem 5.5 (Stabilization of Phase Configurations)**.** *Consider a group of $N > 1$ unicycles* (5.9) *with non-identical velocities $v_k > 0$, $k \in \mathcal{N}$. Let $U(\boldsymbol{\theta})$ be a smooth phase potential that satisfies $\langle \nabla U, \mathbf{1}\rangle = 0$. Then, the control law*

$$u_k = \gamma \frac{\partial U_{\dot{R}}}{\partial \theta_k} + \omega_0 \left(1 + \gamma \langle \tilde{r}_k, \dot{r}_k \rangle\right) - \frac{\partial U}{\partial \theta_k}, \tag{5.20}$$

with $\omega_0 \neq 0$ guarantees that all solutions converge to a circular formation where all agents k move on circles of radius $\rho_k = v_k/|\omega_0|$ with fixed common center point and direction given by the sign of ω_0. Furthermore, the phase arrangement lies in the critical set of U. Every (local) minimum of U defines an asymptotically stable phase configuration and all other equilibria are unstable.

Proof. See Appendix B.22. □

Theorem 5.5 presents a control law that guarantees convergence of the group to a circular formation and, at the same time, allows to choose a desired phase configuration. Of particular interest are the synchronized phase configuration (phase agreement) and the balanced phase configuration. The control law (5.20) is based on (5.18) which takes into account the non-identical velocities of the agents. A synchronized phase configuration is achieved if the phase angles of all agents coincide, i.e., $\theta_1 = \ldots = \theta_N$. A balanced phase configuration is achieved if the phasors $e^{\mathrm{i}\theta_k}$ add to zero, i.e., $p_\theta = 0$. Here, we focus on the splay state, a balanced configuration in which the difference between the phase angles of two consecutive agents is $2\pi/N$, i.e., the phase angles are uniformly spaced around the circle.

Sepulchre et al. (2007) construct suitable phase potentials in order to stabilize certain phase patterns, including the synchronized and the splay configurations. The potentials for these two cases are described next. A synchronized phase configuration is achieved if and only if $|p_\theta| = 1$, which corresponds to the global minimum of the potential

$$U^{\text{sync}} = K U_{p_\theta} = K \frac{N}{2} |p_\theta|^2 \tag{5.21}$$

with $K < 0$. Note that for $K > 0$, the global minimum of (5.21) corresponds to the balanced phase configuration with $|p_\theta| = 0$. Further, let $p_{m\theta}$ be the m-th moment of the phase distribution on the circle

$$p_{m\theta} = \frac{1}{mN}\sum_{k=1}^{N} e^{\mathrm{i}m\theta_k}$$

and define the corresponding potentials

$$U_m = \frac{N}{2}|p_{m\theta}|^2.$$

The unique minimum of U_m is achieved by $|p_{m\theta}| = 0$, which corresponds to balancing modulo $2\pi/m$. The splay state is the global minimum of the potential

$$U^{\text{splay}} = \sum_{m=1}^{\lfloor \frac{N}{2} \rfloor} K_m U_m \tag{5.22}$$

with $K_m > 0$ for $m = 1, ..., \lfloor N/2 \rfloor$. For more details, we refer to Sepulchre et al. (2007).

Theorem 5.6 (Circular Formations with Phase Synchronization and Balancing)**.** *Consider a group of N unicycles* (5.9) *with non-identical velocities* $v_k > 0$, $k \in \mathcal{N}$. *The control law* (5.20) *with potential* $U = U^{\text{sync}}$ *asymptotically stabilizes a circular formation with synchronized phase configuration. The control law* (5.20) *with potential* $U = U^{\text{splay}}$ *asymptotically stabilizes a circular formation with phase configuration in the splay state.*

Proof. The statement follows from Theorem 5.5 and the construction of the potentials U^{sync} and U^{splay}. □

In real coordinates, the control law (5.20) with potentials U^{splay} and U^{sync} can be computed with the gradient of U_m,

$$\frac{\partial U_m}{\partial \theta_k} = \left\langle p_{m\theta}, \mathbf{i}e^{\mathrm{i}m\theta_k} \right\rangle = -\frac{1}{mN}\sum_{j=1}^{N} \sin\left(m(\theta_k - \theta_j)\right),$$

and the gradient of U^{sync} in (5.12).

Example 5.4. A simulation is performed for a group of $N = 3$ unicycles (5.8) with velocities $v_1 = 1$, $v_2 = 1.5$, $v_3 = 2$, and control law (5.20) with $U = U^{\text{sync}}$, $K = -0.5$, $\omega_0 = 0.6$, $\gamma = 0.5$. The motion of the agents in the plane is shown in Fig. 5.8. As expected, the agents move on circles around a common center point but with different radii $\rho_k = v_k/\omega_0$ and achieve a synchronized phase configuration. In the same setup, control law (5.20) with $U = U^{\text{splay}}$, $K_1 = 0.5$, $\omega_0 = 0.6$, $\gamma = 0.5$ leads to the result in Fig. 5.9. As expected, the agents move on circles around a common center point with different radii $\rho_k = v_k/\omega_0$ and the phase angles achieve a splay state. △

Remark 5.1. The formation center point is not prescribed but depends on the initial conditions of the agents. In order to achieve a circular motion around a given point $R_0 \in \mathbb{C}$, following the idea of (Sepulchre et al., 2007, Cor. 2), the following variation of the control law (5.20) can be used:

$$u_k = \omega_0 \left(1 + \gamma \langle r_k - R_0, \dot{r}_k \rangle\right) - \frac{\partial U}{\partial \theta_k}. \tag{5.23}$$

Further details on the derivation are provided in Seyboth et al. (2014b). ○

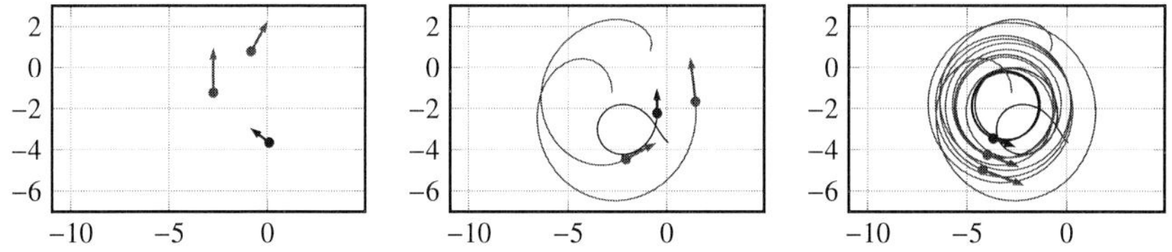

(a) $t = 0$. (b) $t = 10$. (c) $t = 50$.

Figure 5.8: Circular formation with synchronized phase angles with control law (5.20) and $U = U^{\text{sync}}$ in Example 5.4.

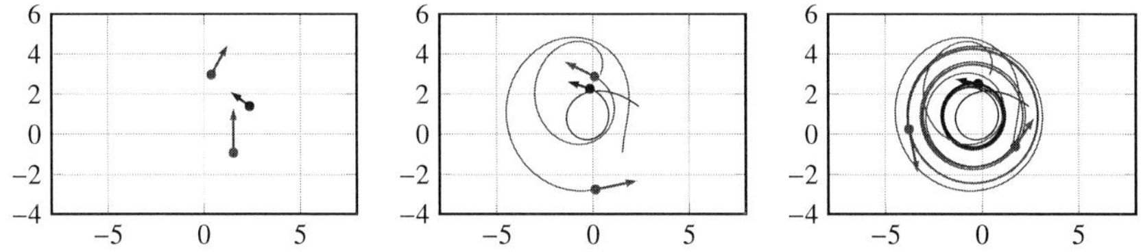

(a) $t = 0$. (b) $t = 10$. (c) $t = 50$.

Figure 5.9: Circular formation with balanced phase angles with control law (5.20) and $U = U^{\text{splay}}$ in Example 5.4.

Unknown Neighbor Velocities

In order to implement control law (5.18), all agents have to receive or sense the position and orientation of all other agents in the group. Moreover, (5.18) depends on the velocity v_k of agent k itself as well as on the velocities v_j of all other agents in the group. The next result shows that a circular formation can also be achieved if each agent k is aware of its own velocity v_k but is not aware of the velocities v_j of the other agents in the group.

Theorem 5.7 (Ignoring the Neighbors' Velocities). *Consider a group of $N > 1$ unicycles* (5.9) *with non-identical velocities $v_k > 0$, $k \in \mathcal{N}$, and consider the control law*

$$u_k = \gamma \frac{\partial U_{p\theta}}{\partial \theta_k} + \omega_0 \left(1 + \gamma \langle \tilde{r}_k, \dot{r}_k \rangle\right), \tag{5.24}$$

with $\gamma > 0$ and $\omega_0 \neq 0$. Then, all solutions converge to a circular formation in which each agent k moves on a circle of radius $\rho_k = v_k / |\omega_0|$ and all circles have a fixed common center point.

Proof. See Appendix B.23. □

In real variables, control law (5.24) is given by

$$u_k = \omega_0 - \frac{\gamma}{N} \sum_{j=1}^{N} \sin(\theta_k - \theta_j) + \omega_0 \gamma v_k |r_k| \cos\left(\phi_k - \theta_k\right) - \omega_0 \frac{\gamma}{N} \sum_{j=1}^{N} |r_j| v_k \cos(\phi_j - \theta_k).$$

This result shows that an agent does not need to know the velocities of the other agents in the network in order to stabilize a circular formation. In order to compute control law (5.24), each

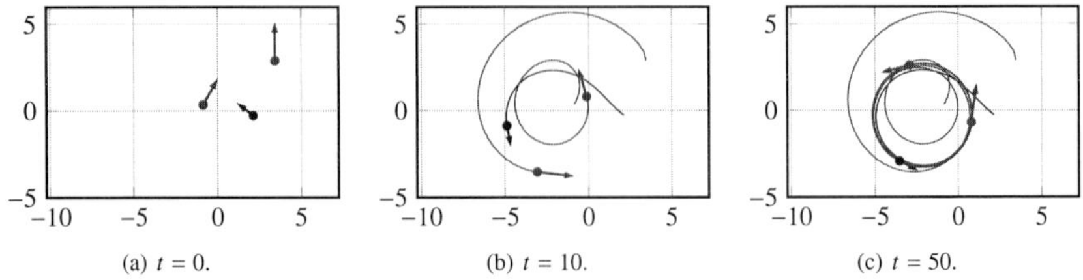

Figure 5.10: Stabilization of a collective circular motion on a common circle with control law (5.25) in Example 5.5.

agent k only has to be aware of its own velocity v_k. However, a drawback is that the final configuration of the phase angles is determined by the potential $U(\boldsymbol{\theta}) = \gamma(U_{\dot{R}}(\boldsymbol{\theta}) - U_{p_\theta}(\boldsymbol{\theta}))$. In particular, it follows from Theorem 5.5 that the resulting phase arrangement lies in the critical set of U and the (local) minima of U are asymptotically stable phase configurations. Hence the control law (5.24) does in general not lead to a synchronous or balanced phase configuration.

Collective Circular Motion on a Common Circle

We have shown how circular formations as shown in Fig. 5.4(a) can be realized in heterogeneous groups. Next, we propose a control law that realizes a collective motion on a common circle according to Fig. 5.4(b), i.e., all agents move on a common circle with given radius ρ_0.

Theorem 5.8 (Collective Circular Motion on a Common Circle)**.** *Consider a group of $N > 1$ unicycles* (5.9) *with non-identical velocities $v_k > 0$, $k \in \mathcal{N}$, and consider the control law*

$$u_k = \omega_k \left(1 + \gamma \langle \tilde{r}_k, \dot{r}_k \rangle\right) + \gamma v_k^2 \frac{\partial U_{p_\theta}}{\partial \theta_k}, \tag{5.25}$$

with $\gamma > 0$, $\rho_0 > 0$, and $\omega_k = v_k/\rho_0$. Then, all solutions converge to a circular formation in which all agents move on a common circle of radius ρ_0 with angular frequencies ω_k.

Proof. See Appendix B.24. □

Note that (5.25) involves only the velocity v_k of agent k itself, and not the velocities of other agents in the group.

Example 5.5. A simulation is performed for a group of $N = 3$ unicycles (5.8) with velocities $v_1 = 1$, $v_2 = 1.5$, $v_3 = 2$, and control law (5.25) with $\rho_0 = 3$ and $\gamma = 0.6$. The motion of the agents in the plane is shown in Fig. 5.10. As expected, the agents move on a common circle with radius ρ_0 but with different angular frequencies $\omega_k = v_k/\rho_0$. △

Limited Communication and Formation Shape Control

The foregoing results have fostered further research on coordination control laws for groups on unicycle-type vehicles with non-identical and fixed velocities. Of practical interest are a relaxation of the all-to-all communication and more advanced formation control objectives. For vehicles with unit velocity, distributed control laws with reduced communication requirements were developed by Sepulchre et al. (2008). For non-identical velocities, most noteworthy are the Diploma thesis of Kianifar (2015) under the supervision of Seyboth and the work of Sun et al. (2015a,b). The focus of Kianifar (2015) is on communication constraints: the requirement of all-to-all communication is relaxed to limited communication according to an underlying graph $\mathcal{G}$. Moreover, the applicability of event-triggered control methods is explored. The focus of Sun et al. (2015a) is on formation shape control where the circle center points of all vehicles shall not coincide but form a desired predefined shape. The following development unifies and generalizes some of these results.

Suppose that $u_k = \omega_0$. Then, vehicle k moves on a circle of radius $\rho_k = v_k/|\omega_0|$ and the center of this circle is $c_k = r_k + \mathbf{i}\frac{v_k}{\omega_0}e^{\mathbf{i}\theta_k}$. As shown in the proof of Theorem 5.4, the control law (5.18) can equivalently be written as

$$u_k = \omega_0 + \gamma\omega_0 \left\langle P_k\mathbf{c}, v_k e^{\mathbf{i}\theta_k}\right\rangle, \tag{5.26}$$

where P_k is the k-th row of the projection matrix $P = I_N - \frac{1}{N}\mathbf{1}\mathbf{1}^\mathsf{T}$ and $\mathbf{c}$ is the stack vector of the circle center points c_k, $k \in \mathcal{N}$. With this control law, all circle center points converge to a common point in the plane. In order to achieve a desired formation shape, it is suggested by Sun et al. (2015a) to define a desired offset $\hat{c}_k \in \mathbb{C}$ from the common formation center point for each vehicle. The formation shape control objective is achieved when $P(\mathbf{c} - \hat{\mathbf{c}}) = \mathbf{0}$, where $\hat{\mathbf{c}}$ is the stack of $\hat{c}_k$, or equivalently, when $c_k - c_j = \hat{c}_k - \hat{c}_j$ for all pairs $j, k \in \mathcal{N}$. With this insight, we are ready to present the following result.

Theorem 5.9 (Circular Formation Stabilization: a Generalization)**.** *Consider a group of $N > 1$ unicycles (5.9) with non-identical velocities $v_k > 0$, $k \in \mathcal{N}$. Define $\tilde{\mathbf{c}} = \mathbf{c} - \hat{\mathbf{c}}$ for some $\hat{\mathbf{c}} \in \mathbb{C}^N$ and let $Q \in \mathbb{R}^{N\times N}$ be some symmetric positive semi-definite matrix, $Q = Q^\mathsf{T} \geq 0$. Consider the control law*

$$u_k = \omega_0 + \gamma\omega_0 \left\langle Q_k\tilde{\mathbf{c}}, v_k e^{\mathbf{i}\theta_k}\right\rangle \tag{5.27}$$

with $\gamma > 0$, $\omega_0 \neq 0$, and Q_k is the k-th row of Q. Then, all solutions converge to the null space of Q, i.e., $\tilde{\mathbf{c}}(t) \to \ker(Q)$ as $t \to \infty$.

Proof. There exists a matrix $H \in \mathbb{R}^{M\times N}$ of full row rank for some $M \leq N$ such that $Q = H^\mathsf{T}H$. We define the Lyapunov function $V = \frac{1}{2}\langle Q\tilde{\mathbf{c}}, \tilde{\mathbf{c}}\rangle = \|H\tilde{\mathbf{c}}\|^2$. Since H has full row rank, $\ker(H) = \ker(Q)$. The remainder of the proof employs LaSalle's Invariance Principle (Theorem A.5), analogously to the proof of Theorem 5.4. All solutions converge to a configuration where $Q\tilde{\mathbf{c}} = \mathbf{0}$. □

Theorem 5.9 specializes to the following particular control laws of interest:

- Suppose that $\hat{\mathbf{c}} = \mathbf{0}$ and $Q = P$. Then, (5.27) specializes to the control law (5.26).
- Suppose that $\hat{\mathbf{c}} = \mathbf{0}$ and $Q = L_\mathcal{G}$, where $L_\mathcal{G}$ is the Laplacian matrix of an undirected connected graph. Then, (5.27) specializes to the control law

  $$u_k = \omega_0 + \gamma\omega_0\langle L_{\mathcal{G},k}\mathbf{c}, v_k e^{\mathbf{i}\theta_k}\rangle,$$

  which is reported by Kianifar (2015). The benefit of this control law lies in the fact that it is distributed. It does not require all-to-all communication but only information from neighboring agents according to the graph $\mathcal{G}$. This becomes apparent by rewriting $u_k = \omega_0 + \gamma\omega_0 \sum_{j=1}^N a_{kj}\langle c_k - c_j, v_k e^{\mathbf{i}\theta_k}\rangle$. This control law achieves the same objective as (5.26) since $\ker(P) = \ker(L_\mathcal{G}) = \mathrm{im}(\mathbf{1})$.

- Suppose that $\hat{\mathbf{c}} \neq \mathbf{0}$ and $Q = I_N$. Then, (5.27) specializes to the control law

$$u_k = \omega_0 + \gamma\omega_0\langle c_k - \hat{c}_k, v_k e^{\mathrm{i}\theta_k}\rangle$$

  which is reported by Sun et al. (2015a). This control law is decentralized and does not require any information exchange among the vehicles. Each vehicle k is steered to a circular motion around its designated center point $\hat{c}_k$ since $(\mathbf{c}(t) - \hat{\mathbf{c}}) \to \ker(I_N) = \mathbf{0}$ as $t \to \infty$.

- Suppose that $\hat{\mathbf{c}} \neq \mathbf{0}$ and $Q = P$. Then, (5.27) specializes to the control law

$$u_k = \omega_0 + \gamma\omega_0\langle P_k(\mathbf{c} - \hat{\mathbf{c}}), v_k e^{\mathrm{i}\theta_k}\rangle$$

  which is reported by Sun et al. (2015a). This control law achieves the formation shape control objective $P(\mathbf{c} - \hat{\mathbf{c}}) = \mathbf{0}$, or equivalently, $c_k - c_j = \hat{c}_k - \hat{c}_j$ for all pairs $j, k \in \mathcal{N}$, asymptotically.

- Suppose that $\hat{\mathbf{c}} \neq \mathbf{0}$ and $Q = L_\mathcal{G}$, where $L_\mathcal{G}$ is the Laplacian matrix of an undirected connected graph. Then, (5.27) specializes to the control law

$$u_k = \omega_0 + \gamma\omega_0\langle L_{\mathcal{G},k}(\mathbf{c} - \hat{\mathbf{c}}), v_k e^{\mathrm{i}\theta_k}\rangle$$

  which is reported by Sun et al. (2015a). This control law achieves the same formation shape control objective as the control law above and it is distributed. It does not require all-to-all communication but only information from neighboring agents according to the graph $\mathcal{G}$. This becomes apparent by rewriting $u_k = \omega_0 + \gamma\omega_0 \sum_{j=1}^{N} a_{kj}\langle (c_k - c_j) - (\hat{c}_k - \hat{c}_j), v_k e^{\mathrm{i}\theta_k}\rangle$.

5.3 Summary and Discussion

Chapter 5 is dedicated to motion coordination problems for groups of nonholonomic mobile vehicles. Autonomous vehicles oftentimes need to be modeled as unicycle-type vehicles when single or double-integrator dynamics are oversimplified and the nonholonomic constraints need to be taken into account. Hence, specific distributed control algorithms need to be developed.

In Section 5.1, groups of mobile robots moving in the plane with a longitudinal force and a torque as control inputs are considered. Based on the distributed output regulation framework, we develop a novel distributed motion coordination algorithm. This algorithm achieves that the mobile robots form a dynamic formation, circling around the formation center point, while the formation as a whole tracks a desired path in the plane. The nonlinear dynamics are handled by feedback linearization and nonlinear observers are designed in order to estimate local external disturbances. The novel distributed control algorithm integrates various control theoretic concepts. Compared to prior results, the main benefits of this algorithm are that local disturbances acting on individual agents are asymptotically rejected, dynamic formation shapes can be realized via time-dependent local reference signals, and the deviation from the desired formation shape in transient phases is minimized by couplings on the agent level.

In Section 5.2, collective circular motion coordination problems for heterogeneous groups of nonholonomic mobile agents modeled as unicycle-type vehicles with non-identical and fixed forward velocities are addressed. In such groups, circular formations can only be realized in the following sense: (i) a circular motion on circles around a common center point with synchronized or balanced phase angles is only possible if non-identical radii of the circles are admitted and (ii) a circular motion of all vehicles on a common circle in the plane is only possible if non-identical angular frequencies are admitted. We generalize prior results for groups of unit speed vehicles to heterogeneous groups of vehicles with non-identical velocities. For both types of circular formations, we derive control laws which guarantee global convergence of the group to the desired formations, whereby the velocities of the individual vehicles in the group are explicitly taken into account. As further extensions, we show that the vehicles do not need to be aware of their neighbor velocities if a circular formation without a particular phase arrangement is sought, we show that distributed control laws can be obtained if the underlying graph is undirected, and we highlight the possibility to realize further formation shapes with predefined offsets between the vehicles' circle center points. These results show that the fundamental assumption of identical agents in prior work can be relaxed and hence contribute to the development of formation control methods tailored to heterogeneous groups of mobile vehicles.

It should be noted that the issue of collision avoidance is not taken into account in this chapter. There are various well-established collision avoidance strategies, cf., Dimarogonas et al. (2006); Qu (2009); Hernández-Martínez and Aranda-Bricaire (2011). The combination of these strategies with the present results is an interesting subject of future research.

Chapter 6

Conclusions

In this chapter, we offer a summary and discussion of the main results of this thesis and indicate possible directions for future research as an outlook.

6.1 Summary and Discussion

The research presented in this thesis focuses on the synthesis of cooperative behavior in networks of dynamical systems via distributed control. The growing size and complexity of modern engineering systems, along with the persistent desire for increasing autonomy, necessitate the development of a suitable systems theoretic framework and control methods.

In classical control applications, a single dynamical system (plant) is typically controlled by a single processor running a control algorithm in an isolated control loop. Already today, we are faced with control problems involving numerous dynamical systems instead of one isolated plant. Beyond stabilization, the control objective is typically the generation of a cooperative behavior of these systems. Distributed control algorithms are preferable to centralized algorithms, mainly due to their flexibility and scalability. A centralized control algorithm may even be infeasible in practical applications. The synthesis of a cooperative behavior among dynamical agents is desirable in numerous man-made systems including vehicle platooning, formation flight, mobile vehicle coordination, robot cooperation in production lines, or frequency synchronization in electrical grids. These developments gave rise to the novel and highly active research field of distributed and cooperative control within the field of systems theory and automatic control. A detailed account for its emergence is given in Chapter 2. The present thesis makes important contributions to this field through the development of distributed control laws and associated design methods for various classes of multi-agent systems and cooperative control tasks. The main contributions of this thesis are structured into the three Chapters 3–5.

Chapter 3 is focused on the output synchronization problem for groups of dynamical systems since the ability to reach an agreement on a quantity of interest, such as a meeting point, heading direction, or velocity is a basic requirement for cooperative behavior. The distributed consensus algorithm is a celebrated solution to this problem for networks consisting of agents with single-integrator dynamics and the objective to agree on a common constant state value. The need for more complex agent dynamics and for the ability to agree on time-dependent state and output trajectories, instead of a constant, motivates the developments in Chapter 3. In Section 3.1, state synchronization in networks of identical LTI systems is studied. After a review of necessary and sufficient existence conditions for distributed control laws in the form of static diffusive couplings, a variety of corresponding design methods is presented. A novel design method involving pole placement constraints is proposed which allows to conveniently impose a desired decay rate and damping ratio on the synchronization error as performance criteria of practical importance. In Section 3.2, heterogeneous networks consisting of non-identical linear dynamical systems with output synchronization objective are considered. Such networks have to fulfill the structural requirement known as the internal model principle for synchronization in order to be synchronizable. The corresponding result is revisited and discussed in detail for heterogeneous linear networks with static diffusive couplings. In order to develop a deeper understanding of the effects of heterogeneous agent dynamics on the behavior of the diffusively coupled multi-agent system, the robustness of synchronization under parameter perturbations is analyzed for two network types: double-integrators and harmonic oscillators. For each network, the role of the network topology is analyzed and the dynamic behavior is characterized. In this context, it is shown that networks of coupled harmonic oscillators are rendered asymptotically stable by an arbitrarily small frequency mismatch within the iSCC of the underlying graph. This surprising fact may harm motion coordination strategies built on top of such algorithms. The network of perturbed double-integrator agents proves to be less sensitive to the parameter perturbations and clock synchronization is discussed as a related application. In Section 3.3, state and output synchronization problems for networks consisting of agents with LPV dynamics are formulated and solved. This extension of the class of agent models compared to existing work is of practical interest and allows to describe homogeneous and heterogeneous groups intuitively through identical (global) and non-identical (local) time-varying parameters. After analyzing structural requirements, novel static and dynamic gain-scheduled distributed control laws are proposed for such networks in combination with suitable LMI-based design methods.

Concluding, our developments in Chapter 3 start out at the core of cooperative control and provide an improved distributed control design method taking performance specifications of practical importance into account. The subsequent developments in Chapter 3 contribute to the field by deepening the understanding of heterogeneous networks and pushing the boundary further into the direction of networks with increased agent complexity, with a focus on control design.

In Chapter 4, exogenous inputs to the multi-agent systems are added to the picture since, from a practical perspective, the ability to take external disturbances into account and to construct controllers coping with these disturbances is essential. In addition, reference signals need to be included in order to command a desired behavior to the group through an outside supervisor. The tasks of cooperative disturbance rejection and reference tracking are in the focus of Chapter 4. In Section 4.1, homogeneous linear networks under constant disturbances are considered. Distributed control laws with integral action are proposed as a solution to the robust output synchronization problem under disturbances. Like in classical single-loop regulation problems, the key idea is a suitable integrator augmentation of the agent dynamics. Existence conditions as well as synthesis methods for the corresponding control gains are provided and observer-based implementations of the distributed control law were provided, both for absolute and for relative output measurements. Moreover, a cooperative reference tracking problem is formulated and solved by means of the developed methods, generalizing the distributed dynamic input averaging problem. In Section 4.2, the distributed output regulation framework is revisited, refined, and extended. The geometric approach to linear systems theory, and output regulation theory in particular, proved useful in the context of cooperative control for the analysis of the consensus algorithm and for the formulation of the internal model principle for synchronization. The distributed output regulation framework strengthens this link. It allows to systematically formulate and solve a wide variety of multi-agent coordination tasks and takes disturbances acting on individual agents as well as global and local reference signals for all and for individual agents in the group into account. We present a novel, refined distributed regulator taking the structure of the exosystem explicitly into account. Moreover, for agents with identical or at least similar dynamics, we introduce a novel coupling on the agent level in order to achieve transient synchronization thus allowing for a cooperative reaction of the group on local disturbances. As a generalization of the problem formulation, we furthermore address coupled distributed output regulation problems, where couplings may occur among the agent states, the measurement outputs, or the regulation errors. Of particular importance are the last two cases. Coupled measurement outputs lead to a distributed estimation problem, an active research topic in its own right. A basic solution approach is presented for such scenarios. In case of coupled regulation errors, it is in general hard to find a distributed regulator due to a loss of structure in the regulator equations. Nevertheless, a class of problems admitting a distributed solution is identified which includes the platooning problem for strings of autonomous vehicles as a compelling application example.

Concluding, our results in Chapter 4 contribute to the development of a comprehensive cooperative control theory for linear multi-agent systems subject to exogenous inputs, which is an increase of the group objective complexity. The elaboration of distributed integral action and the distributed output regulation framework yields control design methods for a wide range of realistic distributed coordination and control tasks.

Chapter 5 is dedicated to motion coordination problems for groups of nonholonomic mobile vehicles such as mobile robots or UAVs. There is a demand for suitable control algorithms which take the nonlinear unicycle-type dynamics of such vehicles into account. Algorithms for two different motion coordination scenarios are developed. In Section 5.1, a dynamic formation control problem for a group of mobile robots in the plane is addressed. The exemplary task is to reach a formation in which the robots move equally spaced on a circle around the formation center point while the formation center follows a given linear path in the plane. The construction of a distributed motion coordination algorithm for this task demonstrates the integration of the distributed output regulation framework with the feedback linearization technique and nonlinear disturbance estimation. In Section 5.2, the focus is on the kinematic equations. Motivated by UAV coordination applications, each agent is assumed to travel with a constant longitudinal velocity which cannot be altered by the controller. Only the orientation can be controlled. Due to the nonlinearity of the dynamics, a different methodology as before is required for this problem. The challenge for the control design stems from the fact that the vehicle velocities are in general not identical, i.e., we are faced with a heterogeneous group of nonlinear dynamical agents. A variety of desirable circular motion patterns is defined for such groups and corresponding stabilizing nonlinear control laws are derived, thus generalizing existent results for homogeneous groups of unit speed vehicles.

Concluding, our results in Chapter 5 lie in one of the major application domains of distributed control and contribute to the development of flexible and powerful distributed motion coordination strategies which can cope with heterogeneous vehicle dynamics and disturbances.

It is an ongoing endeavor in the scientific community to translate well-established control methods from classical control scenarios to control of network systems. The present thesis contributes to this process. In particular, the concepts of state-feedback for LTI systems, state-feedback for LTI systems with pole placement constraints, state-feedback for LPV systems, control with integral action for LTI systems, and output regulation for LTI systems are brought forward to networks of dynamical systems, with a shift of the focus from stabilization to synchronization. Major challenges in this process are coping with complex network topologies, with heterogeneous network dynamics, and achieving scalability of the control algorithms and their synthesis. A key for these developments is the inclusion of graph theory into the methodological groundwork.

6.2 Outlook

The endeavor of developing a comprehensive systems and control theory for networks of dynamical systems is far from being completed. Even though significant advances have been made during the past decade, there is still a great amount of work lying ahead. As noted by Åström and Kumar (2014), feedback control is a young field. Yet, the spectrum of feedback control methods developed over the past half century is enormous. The selection and adaptation of existing successful methods to distributed and cooperative control problems and their evaluation requires further basic research as well as new challenging applications in the future.

Two alternative expedient approaches to the development of distributed control algorithms are the following. Oftentimes, cooperative control problems for multi-agent systems can be formulated as centralized single-loop control problems which lend themselves to classical control methods. The challenge is then to exploit the problem structure in order to adapt the control method and arrive at a distributed and scalable control strategy. An alternative approach is the decomposition of the multi-agent coordination problem into several subproblems, for which well-established methods are available. The remaining challenge is then to interconnect the components and combine the control algorithms into an integrated distributed control scheme for the network system. Besides the adaptation of existing control methods, it is of course essential to explore new directions and solution approaches, specifically for distributed control problems.

The work reported in the present thesis leads to the following list of some open questions and more concrete suggestions and ideas for future research.

- There is a variety of sophisticated controller synthesis methods for LPV systems, taking into account robustness and performance criteria for the closed-loop system, cf., Veenman and Scherer (2014); Briat (2015). As a continuation of the developments in Section 3.3, it is suggested to explore the applicability of advanced controller synthesis methods to networks of LPV systems. As an example, taking bounded parameter variation rates into account can be expected to reduce the conservatism significantly. Performance criteria may be incorporated analogously as was done for LTI networks, cf., Remark 3.3.

- In presence of actuator saturation, a well-known issue for control loops with integral action is the windup phenomenon. This issue can be overcome by an anti-windup mechanism, cf., Galeani et al. (2009). It is suggested to develop a distributed anti-windup strategy as an extension to the distributed control laws with integral action presented in Section 4.1.

- The distributed output regulation framework in Section 4.2 was developed for multi-agent systems consisting of LTI dynamical agents. Based on the results of Köroğlu and Scherer (2011) on output regulation for LPV systems and the results of Section 3.3, this framework may be generalized to networks of LPV systems.

- Distributed output regulation problems with coupled measurement outputs lead to distributed estimation problems, as explained in Section 4.2.5. A promising approach to solve such problems is the integration of distributed estimation methods, as developed by Wu et al. (2014), into the distributed regulator.

- A functionality of great practical importance, which was disregarded in Chapter 5, is the avoidance of collisions among the mobile vehicles. The design and implementation of distributed collision avoidance strategies is a research area in its own right, cf., Dimarogonas et al. (2006); Qu (2009); Hernández-Martínez and Aranda-Bricaire (2011). It is suggested to explore the compatibility of collision avoidance strategies, e.g., based on artificial potential functions, with the motion coordination strategies developed in Chapter 5.

- Another important research direction is distributed control over realistic communication networks. The imperfection of packet-based communication channels poses a challenge. Time delays, packet losses, quantization, and data rate constraints complicate the implementation and execution of time-critical control algorithms and may put the system stability at risk. The interplay of control algorithms and communication networks is the subject of the research area *networked control systems*, cf., Blind (2014). A concrete research goal in this direction is the robustness analysis of distributed control algorithms with respect to communication time delays. Another research goal are event-triggered realizations of the distributed control algorithms, which may facilitate the implementation on digital platforms, cf., Dimarogonas et al. (2011); Seyboth et al. (2013); De Persis and Frasca (2013).
- Last but not least, it is essential to test and evaluate distributed control algorithms thoroughly in experiments and practical applications. Not only is this necessary in order to prove the practicality and reliability of novel algorithms, but also have challenging applications always been a driving force for the theoretic progress.

Appendix A

Technical Background

This chapter provides a summary of the system theoretic concepts and methods, which form the basis for the developments in this thesis. Section A.1 is dedicated to linear systems and control theory including the relevant geometric concepts. Section A.2 is an introduction to graph theory which will be instrumental for the modeling and analysis of networks of dynamical systems.

A.1 Linear Systems and Control Theory

The purpose of this section is to provide a brief introduction to the relevant concepts and results from systems and control theory for dynamical systems, in particular linear dynamical systems, and the geometric approach to linear systems theory.

A.1.1 Preliminaries

Definition A.1 (Positive Definiteness (Horn and Johnson (1985))). A Hermitian matrix $P \in \mathbb{C}^{n\times n}$ is called *positive definite*, if

$$z^{\mathsf{H}} P z > 0 \quad \text{for all} \quad z \in \mathbb{C}^n,\ z \neq 0. \tag{A.1}$$

A symmetric matrix $P \in \mathbb{R}^{n\times n}$ is called *positive definite*, if

$$x^{\mathsf{T}} P x > 0 \quad \text{for all} \quad x \in \mathbb{R}^n,\ x \neq 0. \tag{A.2}$$

P is called *positive semi-definite*, *negative definite*, *negative semi-definite*, if (A.1) (or (A.2)) holds with $\geq, <, \leq$, respectively. As shorthand notation, we write $P \succ 0$, $P \succeq 0$, $P \prec 0$, $P \preceq 0$.

Lemma A.1 (Helmersson (1995)). *Suppose that $P \in \mathbb{C}^{n\times n}$ and $P = P^{\mathsf{H}}$. Denote $Q = \mathbf{Re}(P) \in \mathbb{R}^{n\times n}$ and $S = \mathbf{Im}(P) \in \mathbb{R}^{n\times n}$. Then, $Q = Q^{\mathsf{T}}$ and $S = -S^{\mathsf{T}}$. Moreover,*

$$P \succ 0 \qquad \Leftrightarrow \qquad \begin{bmatrix} Q & S \\ -S & Q \end{bmatrix} \succ 0.$$

The latter relation allows to convert complex linear matrix inequalities (LMIs) in real LMIs. The next lemma by Finsler (1937) is reported, e.g., in (de Oliveira and Skelton, 2001, Lem. 2).

Lemma A.2 (Finsler's Lemma). *Let $x \in \mathbb{R}^n$, $Q = Q^{\mathsf{T}} \in \mathbb{R}^{n\times n}$, and $B \in \mathbb{R}^{m\times n}$ such that* $\operatorname{rank}(B) < n$. *Then, the following statements are equivalent:*

i) $x^{\mathsf{T}} Q x < 0$, *for all* $x \in \ker(B)\backslash\{\mathbf{0}\}$.

ii) $(B^{\perp})^{\mathsf{T}}QB^{\perp} \prec 0$, *where* $B^{\perp}$ *is a basis matrix of* $\ker(B)$.

iii) $\exists\tau \in \mathbb{R}\ :\ Q - \tau B^{\mathsf{T}}B \prec 0$.

iv) $\exists X \in \mathbb{R}^{n\times m}\ :\ Q + XB + B^{\mathsf{T}}X^{\mathsf{T}} \prec 0$.

A.1.2 Stability

The fundamental stability definitions and results are first reviewed for nonlinear systems and then specialized to linear systems. The presentation in this section is mostly based on Khalil (2002). Consider the nonlinear system

$$\dot{x} = f(x) \tag{A.3}$$

with the state $x(t) \in D$, a vector field $f : D \to \mathbb{R}^n$, and an initial condition $x(0) = x_0 \in D$, where $D \subset \mathbb{R}^n$ is open and connected. In order to guarantee the existence and uniqueness of solutions of (A.3) for all initial conditions x_0, it is assumed that f is locally Lipschitz. Moreover, it is assumed that $f(\mathbf{0}) = \mathbf{0}$, i.e., (A.3) has an equilibrium point at the origin. The stability properties of this equilibrium point $x = \mathbf{0}$, or the zero solution $x(t) = \mathbf{0}$ for all $t \geq 0$, are characterized as follows.

Definition A.2 (Stability, (Khalil, 2002, Def. 4.1, 4.5))**.** The equilibrium point $x = \mathbf{0}$ of (A.3) is

- *stable* if for each $\epsilon > 0$ there is $\delta = \delta(\epsilon) > 0$ such that $\|x(0)\| < \delta \Rightarrow \|x(t)\| < \epsilon, \forall t \geq 0$.
- *unstable* if it is not stable.
- *asymptotically stable* if it is stable and δ can be chosen such that $\|x(0)\| < \delta$ implies that $\lim_{t\to\infty} x(t) = \mathbf{0}$.
- *globally asymptotically stable* if it is stable and $\lim_{t\to\infty} x(t) = \mathbf{0}$ for all initial states $x(0)$.
- *exponentially stable* if there exist constants $c, k, \gamma > 0$ such that $\|x(t)\| \leq k\|x(0)\|e^{-\gamma t}$, $\forall t \geq 0$, $\forall \|x(0)\| < c$. The constant γ is called the decay rate.
- *globally exponentially stable* if this holds for all initial states $x(0)$.

The stability properties can be assessed with the well-known stability theorems by Lyapunov.

Theorem A.3 (Lyapunov's 1st Theorem, (Khalil, 2002, Thm. 4.7))**.** *Let* $x = \mathbf{0}$ *be an equilibrium point of* (A.3), *where* f *is continuously differentiable and* D *is a neighborhood of the origin. Let*

$$A = \frac{\partial f}{\partial x}(x)\Big|_{x=\mathbf{0}}$$

be the Jacobian matrix of f *evaluated at* $x = \mathbf{0}$. *Then,*

- *the origin is asymptotically stable if* $\mathbf{Re}(\lambda_i) < 0$ *for all eigenvalues of* A.
- *the origin is unstable if* $\mathbf{Re}(\lambda_i) > 0$ *for one or more of the eigenvalues of* A.

Theorem A.4 (Lyapunov's 2nd Theorem & Barbashin-Krasovskii Theorem, (Khalil, 2002, Thm. 4.1, 4.2))**.** *Let* $x = \mathbf{0}$ *be an equilibrium point for* (A.3) *and* $D \subset \mathbb{R}^n$ *be a domain containing* $x = \mathbf{0}$. *Let* $V : D \to \mathbb{R}$ *be a continuously differentiable positive definite function, i.e.,* $V(\mathbf{0}) = 0$ *and* $V(x) > 0$ *in* $D\backslash\{\mathbf{0}\}$. *Then,*

- $x = \mathbf{0}$ *is stable if the derivative of V along the trajectories of* (A.3) *is negative semi-definite, i.e.,* $\dot{V}(x) \leq 0$ *in D.*
- $x = \mathbf{0}$ *is asymptotically stable if the derivative of V along the trajectories of* (A.3) *is negative definite, i.e., additionally* $\dot{V}(x) < 0$ *in* $D \backslash \{\mathbf{0}\}$.
- $x = \mathbf{0}$ *is globally asymptotically stable if* $D = \mathbb{R}^n$, *V is radially unbounded, i.e.,* $\|x\| \to \infty$ *implies* $V(x) \to \infty$, *and the derivative of V along the trajectories of* (A.3) *is negative definite.*

Another very useful stability result is the famous invariance principle by LaSalle (1976), which we present in the formulation of Khalil (2002).

Theorem A.5 (LaSalle's Invariance Principle, (Khalil, 2002, Thm. 3.4))**.** *Let* $\Omega \subset D$ *be a compact set that is positively invariant with respect to* (A.3). *Let* $V : D \to \mathbb{R}$ *be a continuously differentiable function such that* $\dot{V}(x) \leq 0$ *in* Ω. *Let* $E = \{x \in \Omega : \dot{V}(x) = 0\}$. *Let M be the largest invariant set in E. Then, every solution starting in* Ω *approaches M as* $t \to \infty$.

Corollary A.6 ((Khalil, 2002, Cor. 3.1, 3.2))**.** *Let* $x = \mathbf{0}$ *be an equilibrium point for* (A.3). *Let* $V : D \to \mathbb{R}$ *be a continuously differentiable positive definite function on a domain D containing the origin* $x = \mathbf{0}$, *such that* $\dot{V}(x) \leq 0$ *in D. Let* $S = \{x \in D : \dot{V}(x) = 0\}$ *and suppose that no solution can stay identically in S, other than the trivial solution. Then, the origin is asymptotically stable.*

Moreover, suppose that $D = \mathbb{R}^n$ *and, additionally, V is radially unbounded. Then, the origin is globally asymptotically stable.*

In the following, we consider the linear system

$$\dot{x} = Ax \tag{A.4}$$

with the state $x(t) \in \mathbb{R}^n$, system matrix $A \in \mathbb{R}^{n \times n}$, and an initial condition $x(0) = x_0 \in \mathbb{R}^n$.

Theorem A.7 (Stability of linear systems, (Khalil, 2002, Thm. 4.5))**.** *The equilibrium point* $x = \mathbf{0}$ *of* (A.4) *is stable if and only if all eigenvalues of A satisfy* $\mathbf{Re}(\lambda_i) \leq 0$ *and for every eigenvalue with* $\mathbf{Re}(\lambda_i) = 0$ *and algebraic multiplicity* $q_i \geq 2$, $\operatorname{rank}(A - \lambda_i I_n) = n - q_i$. *The equilibrium point* $x = \mathbf{0}$ *is (globally) asymptotically stable if and only if all eigenvalues of A satisfy* $\mathbf{Re}(\lambda_i) < 0$.

Theorem A.8 (Lyapunov equation, (Khalil, 2002, Thm. 4.6))**.** *A matrix A is Hurwitz; that is,* $\mathbf{Re}(\lambda_i) < 0$ *for all eigenvalues of A, if and only if for any given positive definite symmetric matrix Q there exists a positive definite symmetric matrix P that satisfies the Lyapunov equation*

$$PA + A^{\mathsf{T}}P = -Q. \tag{A.5}$$

Moreover, if A is Hurwitz, then P is the unique solution of (A.5).

Note that we also call the system (A.4) asymptotically stable if A is Hurwitz.

Theorem A.9 (Exponential stability of linear systems, (Bernstein, 2009, Fact 11.18.8))**.** *Suppose that A is Hurwitz and let* $\nu = \max\{\mathbf{Re}(\lambda) : \lambda \in \sigma(A)\}$. *Then,* $\nu < 0$ *and for every* $\gamma \in]0, -\nu[$, *there exists a constant k such that* $\|e^{At}\| \leq ke^{-\gamma t}$.

Since the solution of (A.4) is $x(t) = e^{At}x(0)$, it holds that $\|x(t)\| \leq k\|x(0)\|e^{-\gamma t}$. Hence, the equilibrium $x = \mathbf{0}$ is exponentially stable if and only if it is asymptotically stable and the decay rate is γ.

Lemma A.10 (Linear systems with exponentially decaying inputs (Wieland, 2010, Lem B.1))**.** *Let $x(t)$ be any solution of the linear system $\dot{x}(t) = Ax(t) + u(t)$ with input $u(\cdot) : \mathbb{R} \to \mathbb{R}^n$. Suppose that $u(t)$ is exponentially decaying, i.e., there exist constants $k_1, \gamma_1 > 0$ such that $\|u(t)\| \leq k_1 e^{-\gamma_1 t}$. Then, $x(t)$ converges exponentially to a solution of the unforced system $\dot{x} = Ax$, i.e., there exist some $\tilde{x} \in \mathbb{R}^n$ and constants $k_2, \gamma_2 > 0$ such that $\|x(t) - e^{At}\tilde{x}\| \leq k_2 e^{-\gamma_2 t}$ for all $t \geq 0$.*

A.1.3 Invariant Subspaces

Next, the concept of invariant and controlled invariant subspaces of linear systems is introduced. The presentation is based on Knobloch and Kwakernaak (1985); Trentelman et al. (2001), unless stated otherwise.

Definition A.3 (Invariant subspace)**.** The subspace $\mathfrak{U} \subseteq \mathbb{R}^n$ is called *invariant* with respect to $\dot{x} = Ax$, or shortly A-invariant, if $x(0) \in \mathfrak{U}$ implies $x(t) \in \mathfrak{U}$ for all $t \geq 0$.

Theorem A.11 (Invariant subspace)**.** *Consider the subspace $\mathfrak{U} \subseteq \mathbb{R}^n$. The following statements are equivalent:*

i) $\mathfrak{U}$ is A-invariant.

ii) $A\mathfrak{U} \subseteq \mathfrak{U}$, i.e., $Ax \in \mathfrak{U}$ for all $x \in \mathfrak{U}$.

iii) $e^{At}\mathfrak{U} \subseteq \mathfrak{U}$ for all t.

Theorem A.12 (Matrix criteria for invariant subspaces)**.** *Consider the subspace $\mathfrak{U} \subset \mathbb{R}^n$.*

i) Suppose that $\mathfrak{U} = \mathrm{im}(U)$ for some matrix $U \in \mathbb{R}^{n\times m}$. Then, $\mathfrak{U}$ is A-invariant if and only if there exists a matrix $S \in \mathbb{R}^{m\times m}$ such that $AU = US$. Moreover, if U has full column rank m, then every eigenvalue of S is an eigenvalue of A, and the associated eigenvector is in $\mathfrak{U}$.

ii) Suppose that $\mathfrak{U} = \ker(D)$ for some matrix $D \in \mathbb{R}^{m\times n}$. Then, $\mathfrak{U}$ is A-invariant if and only if there exists a matrix $P \in \mathbb{R}^{m\times m}$ such that $DA = PD$.

Proof. i): The subspace $\mathfrak{U}$ is A-invariant if and only if $Ax \in \mathfrak{U}$ for all $x \in \mathfrak{U}$. Since $\mathfrak{U} = \{x \in \mathbb{R}^n : x = Uy, y \in \mathbb{R}^m\}$, an equivalent condition is that for any $y \in \mathbb{R}^m$, there exists $y' \in \mathbb{R}^m$ such that $AUy = Uy'$. *(If:)* Suppose that $\mathfrak{U}$ is A-invariant. Then, for every canonical basis vector $e_i \in \mathbb{R}^m$, there exists $e_i' \in \mathbb{R}^m$ such that $AUe_i = Ue_i'$. Let $S = [e_1' \ \cdots \ e_m']$. Then, the latter equations can be combined into matrix form, which yields $AU = US$. *(Only if:)* Suppose that there exists a matrix $S \in \mathbb{R}^{m\times m}$ such that $AU = US$. Then, $AUy = USy$. Hence, for every $y \in \mathbb{R}^m$, a vector $y' \in \mathbb{R}^m$ such that $AUy = Uy'$ is given by $y' = Sy$.

Suppose now that $AU = US$ and $\mathrm{rank}(U) = m$. Let $\lambda \in \sigma(S)$ and v be the corresponding eigenvector, i.e., $Sv = \lambda v$. Then, $AUv = USv = U(\lambda v) = \lambda(Uv)$. Since $\mathrm{rank}(U) = m$, its columns are linearly independent, i.e., $Uv = \mathbf{0}$ only if $v = \mathbf{0}$. Consequently, λ is an eigenvalue of A and the corresponding eigenvector is $v' = Uv$. Obviously, $v' \in \mathfrak{U}$.

ii): See (Knobloch and Kwakernaak, 1985, Satz 5.1). □

Theorem A.13 (Rank-nullity Theorem (Horn and Johnson (1985))). *Consider the linear map from $\mathbb{R}^n$ to $\mathbb{R}^m$ defined by a matrix $A \in \mathbb{R}^{m\times n}$. Then, it holds that*

$$n = \dim(\ker(A)) + \dim(\mathrm{im}(A)). \tag{A.6}$$

Theorem A.14 (Kernel and image representations of a subspace (Bernstein, 2009, Thm. 2.4.3)). *Consider the linear map from $\mathbb{R}^n$ to $\mathbb{R}^m$ defined by a matrix $A \in \mathbb{R}^{m\times n}$. Then, it holds that*

$$\mathrm{im}(A)^\perp = \ker(A^\mathsf{T}).$$

Theorem A.15 ((Basile and Marro, 1992, Thm. 3.2.1)). *Let $\mathcal{U} \subseteq \mathbb{R}^n$ be an A-invariant subspace and $m = \dim(\mathcal{U})$. Then, there exists a transformation matrix $T = [U\ W] \in \mathbb{R}^{n\times n}$ such that*

$$A' = T^{-1}AT = \begin{bmatrix} A'_{11} & A'_{12} \\ 0 & A'_{22} \end{bmatrix}, \tag{A.7}$$

where $A'_{11} \in \mathbb{R}^{m\times m}$ and $\mathcal{U} = \mathrm{im}(U)$, $U \in \mathbb{R}^{n\times m}$.

Consider the change of variables $x = Tz$ with T as in Theorem A.15. Then, the transformed system $\dot{z} = A'z$ is decomposed into $\dot{z}_1 = A'_{11}z_1 + A'_{12}z_2$ and $\dot{z}_2 = A'_{22}z_2$, where $z = [z_1^\mathsf{T}\ z_2^\mathsf{T}]^\mathsf{T}$, $z_1 \in \mathbb{R}^m$, $z_2 \in \mathbb{R}^{n-m}$. In these coordinates, $z \in \mathcal{U}$ if and only if $z_2 = \mathbf{0}$. The motion on $\mathcal{U}$ is described by A'_{11}, while A'_{22} describes the motion on the complement $\mathcal{U}^\perp$. This observation leads to the following characterization of an asymptotically stable subspace. We write $x(t) \to \mathcal{U}$ for $t \to \infty$ as shorthand notation for $\forall\epsilon > 0\ \exists\tau > 0\ \forall t \geq \tau\ :\ \mathrm{dist}(x(t),\mathcal{U}) < \epsilon$, where $\mathrm{dist}(x(t),\mathcal{U}) = \inf_{\xi\in\mathcal{U}} \|x(t) - \xi\|$.

Definition A.4 (Asymptotically stable subspace). An A-invariant subspace $\mathcal{U}$ is called *asymptotically stable*, if all solutions of (A.4) satisfy $x(t) \to \mathcal{U}$ as $t \to \infty$.

Theorem A.16 (Asymptotically stable subspace). *An A-invariant subspace $\mathcal{U}$ is asymptotically stable, if and only if A'_{22} in* (A.7) *is Hurwitz.*

Note that in Basile and Marro (1992), an asymptotically stable subspace is referred to as an externally stable subspace, while an A-invariant subspace with the property that A'_{11} is Hurwitz is referred to as an internally stable subspace.

In the following, we consider the linear system with inputs and outputs, given by

$$\dot{x} = Ax + Bu \tag{A.8a}$$

$$y = Cx + Du \tag{A.8b}$$

with state $x(t) \in \mathbb{R}^{n_x}$, input $u(t) \in \mathbb{R}^{n_u}$, output $y(t) \in \mathbb{R}^{n_y}$, and matrices A, B, C, D of appropriate dimension.

Definition A.5 (Controlled invariant subspace). The subspace $\mathcal{U} \subseteq \mathbb{R}^{n_x}$ is a *controlled invariant* subspace of $\dot{x} = Ax + Bu$, or shortly (A,B)-invariant, if for every $x(0) \in \mathcal{U}$ there exists a control input $u(\cdot)$ such that the solution $x(t)$ satisfies $x(t) \in \mathcal{U}$ for all $t \geq 0$.

Theorem A.17 (Controlled invariant subspace). *Consider the subspace $\mathcal{U} \subseteq \mathbb{R}^{n_x}$. The following statements are equivalent:*

i) $\mathcal{U}$ is (A,B)-invariant.

ii) $A\mathcal{U} \subseteq \mathcal{U} + \mathrm{im}(B)$.

iii) For all $x \in \mathcal{U}$ exists $u \in \mathbb{R}^{n_u}$ such that $Ax + Bu \in \mathcal{U}$.

iv) There exists a feedback $u = Kx$ such that $\mathcal{U}$ is invariant with respect to $\dot{x} = (A + BK)x$.

A.1.4 The Linear Quadratic Regulation Problem

The classical linear quadratic regulation (LQR) problem plays an important role in this thesis.

Theorem A.18 (LQR (Anderson and Moore (1990))). *Consider $A, Q \in \mathbb{R}^{n_x \times n_x}$, $B \in \mathbb{R}^{n_x \times n_u}$, $R \in \mathbb{R}^{n_u \times n_u}$. Suppose that $Q = M^\mathsf{T} M \geq 0$ and $R = R^\mathsf{T} > 0$ and (A, B) is stabilizable and (A, M) is detectable. Then, among the solutions of the Algebraic Riccati Equation (ARE),*

$$PA + A^\mathsf{T} P - PBR^{-1}B^\mathsf{T} P + Q = 0 \tag{A.9}$$

there is a unique $P \in \mathbb{R}^{n_x \times n_x}$ with the property that $P = P^\mathsf{T} \geq 0$. Moreover, the matrix

$$A - BR^{-1}B^\mathsf{T} P \tag{A.10}$$

is Hurwitz. Furthermore, the input $u(t) = -R^{-1}B^\mathsf{T} P x(t)$ solves the Linear Quadratic Regulator (LQR) Problem

$$\begin{aligned} J^*(x_0) = \min_u \int_0^\infty & x(t)^\mathsf{T} Q x(t) + u(t)^\mathsf{T} R u(t) \mathrm{d}t \\ \text{s.t. } & \dot{x}(t) = Ax(t) + Bu(t) \\ & x(0) = x_0 \in \mathbb{R}^{n_x} \\ & u(t) \in \mathbb{R}^{n_u},\ t \in [0, \infty[, \end{aligned}$$

and $J^*(x_0) = x_0^\mathsf{T} P x_0$. If the pair (A, M) is observable, then $P > 0$.

Corollary A.19 (LQR Gain Margin). *Under the conditions of Theorem A.18 and with $Q > 0$, the matrix $A - \rho BR^{-1}B^\mathsf{T} P$ is Hurwitz for all $\rho \in \mathbb{C}$ with $\mathbf{Re}(\rho) \geq 0.5$.*

Proof. To see this, we define $\epsilon = \mathbf{Re}(\rho) - 1 \geq -0.5$ and compute

$$\begin{aligned} &(A - \rho BR^{-1}B^\mathsf{T} P)^\mathsf{H} P + P(A - \rho BR^{-1}B^\mathsf{T} P) \\ &\quad = (A - (\mathbf{Re}(\rho) - \mathbf{i}\,\mathbf{Im}(\rho))BR^{-1}B^\mathsf{T} P)^\mathsf{T} P + P(A - (\mathbf{Re}(\rho) + \mathbf{i}\,\mathbf{Im}(\rho))BR^{-1}B^\mathsf{T} P) \\ &\quad = (A - \mathbf{Re}(\rho)BR^{-1}B^\mathsf{T} P)^\mathsf{T} P + P(A - \mathbf{Re}(\rho)BR^{-1}B^\mathsf{T} P) \\ &\quad = (A - (\epsilon + 1)BR^{-1}B^\mathsf{T} P)^\mathsf{T} P + P(A - (\epsilon + 1)BR^{-1}B^\mathsf{T} P) \\ &\quad = (A - BR^{-1}B^\mathsf{T} P)^\mathsf{T} P + P(A - BR^{-1}B^\mathsf{T} P) - 2\epsilon PBR^{-1}B^\mathsf{T} P \\ &\quad = -Q - (1 + 2\epsilon)PBR^{-1}B^\mathsf{T} P < -Q. \end{aligned}$$

The statement follows since this is a complex Lyapunov inequality. □

A.1.5 The Output Regulation Problem

The classical output regulation theory plays an important role in the present thesis. Hence, the essential result is reviewed in the following. The theory was developed in the 1970s by Wonham (1973); Francis and Wonham (1975). Beyond set-point stabilization, this general framework allows to formulate both reference tracking and disturbance rejection objectives. The setup consists of a so-called *exosystem*, an autonomous system that generates all external signals (references as well as disturbances) acting on the plant, and a description of the plant. The signal generated by the exosystem is referred to as the *generalized disturbance*. The tracking and regulation requirements are formulated in terms of a *regulation error* depending on the plant state and the external signals. The objective is to find a control law, also called *regulator*, which ensures internal stability of the *generalized plant* and asymptotic convergence to zero of the regulation error for all initial conditions. For further details, the reader is referred to the monographs by Knobloch et al. (1993); Saberi et al. (2000); Trentelman et al. (2001), and Huang (2004).

We consider the generalized plant

$$\dot{x} = Ax + Bu + B_d d \tag{A.11a}$$

$$y = Cx + Du + D_d d \tag{A.11b}$$

$$e = C_e x + D_e u + D_{ed} d \tag{A.11c}$$

where $d(t) \in \mathbb{R}^{n_d}$ is a generalized disturbance, $y(t) \in \mathbb{R}^{n_y}$ is the measurement output and $e(t) \in \mathbb{R}^{n_e}$ is the regulation error. The generalized disturbance d is generated by an exosystem

$$\dot{d} = Sd, \tag{A.12}$$

where $\sigma(S) \subset \mathbb{C}^0 \cup \mathbb{C}^+$. The output regulation problem for (A.11), (A.12) is defined as follows.

Problem A.1 (Output regulation problem)**.** Find a dynamic measurement output feedback regulator of the form

$$\dot{z} = A^c z + B^c y \tag{A.13a}$$

$$u = C^c z \tag{A.13b}$$

with $z(t) \in \mathbb{R}^{n_z}$, such that the following two conditions are satisfied:

P1) Internal stability: Suppose that $d(0) = \mathbf{0}$. Then, for all initial conditions $x(0) = x_0 \in \mathbb{R}^{n_x}$ and $z(0) = z_0 \in \mathbb{R}^{n_z}$, it holds that $\lim_{t\to\infty} x(t) = \mathbf{0}$ and $\lim_{t\to\infty} z(t) = \mathbf{0}$.

P2) Regulation: For all initial conditions $d(0) = d_0 \in \mathbb{R}^{n_d}$, it holds that $\lim_{t\to\infty} e(t) = \mathbf{0}$.

The generalized disturbance d may consist of disturbances as well as reference signals acting on the plant (A.11). The reason for the great interest in Problem A.1 is its generality. It captures a wide range of control problems including disturbance rejection and reference tracking. The solution to the output regulation problem is due to the efforts of several researchers including Davison, Francis, and Wonham. The following formulation is found in (Huang, 2004, Thm. 1.14).

Theorem A.20 (Solvability of the output regulation problem)**.** *Let the pair* (A, B) *be stabilizable and the pair*

$$\left(\begin{bmatrix} A & B_d \\ 0 & S \end{bmatrix}, \begin{bmatrix} C & D_d \end{bmatrix}\right) \tag{A.14}$$

be detectable and suppose that $\sigma(S) \subset \mathbb{C}^0 \cup \mathbb{C}^+$. *Then, Problem A.1 has a solution* (A.13), *if and only if the regulator equation*

$$\begin{bmatrix} A & B \\ C_e & D_e \end{bmatrix}\begin{bmatrix} \Pi \\ \Gamma \end{bmatrix} - \begin{bmatrix} \Pi \\ 0 \end{bmatrix} S + \begin{bmatrix} B_d \\ D_{ed} \end{bmatrix} = 0 \tag{A.15}$$

is solvable with a solution $\Pi \in \mathbb{R}^{n_x \times n_d}$, $\Gamma \in \mathbb{R}^{n_u \times n_d}$.

In case all conditions in Theorem A.20 are fulfilled, a dynamic measurement output feedback regulator can be constructed as follows. Choose $F \in \mathbb{R}^{n_u \times n_x}$ and $L \in \mathbb{R}^{(n_x+n_d)\times n_y}$ such that

$$A - BF \qquad \text{and} \qquad \begin{bmatrix} A & B_d \\ 0 & S \end{bmatrix} - L \begin{bmatrix} C & D_d \end{bmatrix}$$

are Hurwitz. Define $G = \Gamma + F\Pi$ where Π, Γ solve (A.15). Then, the regulator (A.13) is given by

$$\begin{bmatrix} \dot{\hat{x}} \\ \dot{\hat{d}} \end{bmatrix} = \begin{bmatrix} A & B_d \\ 0 & S \end{bmatrix} \begin{bmatrix} \hat{x} \\ \hat{d} \end{bmatrix} + \begin{bmatrix} B \\ 0 \end{bmatrix} u + L\left(y - \hat{y}\right) \tag{A.16a}$$

$$u = -F\hat{x} + G\hat{d}, \tag{A.16b}$$

where $\hat{y} = C\hat{x} + Du + D_d\hat{d}$. The control law $u = -Fx + Gd$ is referred to as the *full information regulator*. Since x and d are not directly accessible, the observer (A.16a) is constructed to obtain estimates $\hat{x}$ and $\hat{d}$. The control law (A.16) is called *output feedback regulator*. In its original form, the regulator equation (A.15) was presented by Francis (1977).

In the following, we provide a geometric interpretation of Theorem A.20, referring to Knobloch et al. (1993). For ease of presentation, we consider the full information regulator $u = -Fx + Gd$. The plant dynamics (A.11a) and the exosystem (A.12) can be combined into the dynamical system

$$\begin{bmatrix} \dot{x} \\ \dot{d} \end{bmatrix} = \begin{bmatrix} A & B_d \\ 0 & S \end{bmatrix} \begin{bmatrix} x \\ d \end{bmatrix} + \begin{bmatrix} B \\ 0 \end{bmatrix} u. \tag{A.17}$$

We define a new variable $\epsilon = x - \Pi d$. Computing the time derivative of ϵ yields

$$\begin{aligned} \dot{\epsilon} &= \dot{x} - \Pi\dot{d} \\ &= (A - BF)x + (B_d + BG)d - \Pi S d \\ &= (A - BF)(\epsilon + \Pi d) + (B_d + B(\Gamma + F\Pi))d - \Pi S d \\ &= (A - BF)\epsilon + (B_d + B\Gamma + A\Pi - \Pi S)d. \end{aligned}$$

Note that the regulator equation (A.15) can equivalently be written as

$$A\Pi + B\Gamma - \Pi S + B_d = 0 \tag{A.18a}$$

$$C_e\Pi + D_e\Gamma + D_{ed} = 0. \tag{A.18b}$$

With (A.18a), it follows that

$$\dot{\epsilon} = (A - BF)\epsilon.$$

Obviously, $\epsilon = \mathbf{0}$ is asymptotically stable. Moreover, $\epsilon = \mathbf{0}$ is equivalent to $x = \Pi d$. The latter equation defines the subspace $\mathcal{V}^+ = \{(x, d) : x = \Pi d\} \subset \mathbb{R}^{n_x+n_d}$. This subspace can also be written as $\mathcal{V}^+ = \ker([I_{n_x} \; -\Pi])$. According to Theorem A.17, $\mathcal{V}^+$ is controlled invariant with respect to (A.17), if and only if there exists a control law $u = -Fx + Gd$ such that $\mathcal{V}^+$ is invariant with respect to

$$\begin{bmatrix} \dot{x} \\ \dot{d} \end{bmatrix} = \begin{bmatrix} A - BF & B_d + BG \\ 0 & S \end{bmatrix} \begin{bmatrix} x \\ d \end{bmatrix}. \tag{A.19}$$

With Theorem A.12, *ii)*, this is equivalent to the existence of a matrix $P \in \mathbb{R}^{n_x \times n_x}$ such that $A - BF = P$ and $B_d + BG - \Pi S = -P\Pi$, which yields the condition $B_d + B(G - F\Pi) + A\Pi - \Pi S = 0$.

With $\Gamma = G - F\Pi$, this is exactly the first regulator equation (A.18a). Hence, the condition $A\Pi + B\Gamma - \Pi S + B_d = 0$ guarantees that $\mathcal{V}^+$ is a controlled invariant subspace of (A.17). Moreover, $\mathcal{V}^+$ is rendered asymptotically stable by the feedback matrix F, or more particularly by $u = -Fx + Gd$. Since $\epsilon = \mathbf{0}$ is asymptotically stable, the variable ϵ is called the *transient state component* of $x = \Pi d + \epsilon$, whereas Πd is called the *persistent state component* and does not converge to zero since $\sigma(S) \subset \mathbb{C}^0 \cup \mathbb{C}^+$. The subspace $\mathcal{V}^+$ is referred to as the *persistent subspace*. Under the given control law, the regulation error e can be computed as

$$\begin{aligned} e &= C_e x + D_e u + D_{ed} d \\ &= (C_e - D_e F)x + (D_{ed} + D_e G)d \\ &= (C_e - D_e F)(\epsilon + \Pi d) + (D_{ed} + D_e(\Gamma + F\Pi))d \\ &= (C_e - D_e F)\epsilon + (C_e \Pi + D_{ed} + D_e \Gamma)d. \end{aligned}$$

With (A.18b), it follows that

$$e = (C_e - D_e F)\epsilon.$$

Obviously, $\epsilon(t) \to \mathbf{0}$ implies $e(t) \to \mathbf{0}$ as $t \to \infty$. More particularly, the second regulator equation (A.18b) guarantees that the regulation error e vanishes on the persistent subspace $\mathcal{V}^+$. Summarizing, the regulator equation (A.15) guarantees the existence of a controlled invariant subspace $\mathcal{V}^+$ for the system (A.17), on which the regulation error e vanishes. This subspace can be rendered asymptotically stable due to the stabilizability of (A, B), which leads to a solution of the output regulation problem.

The dynamic measurement output feedback regulator (A.16) contains a copy of the exosystem $\dot{d} = Sd$ generating the generalized disturbance d. In case $y = e$, this is a necessary structural criterion referred to as the *internal model principle* of control theory, which was discovered by Francis and Wonham (1975, 1976). An in-depth study can be found in the aforementioned books by Wonham (1985); Knobloch and Kwakernaak (1985); Basile and Marro (1992); Saberi et al. (2000); Trentelman et al. (2001). The output regulation theory was extended to nonlinear systems by Isidori (1995), see also Huang (2004).

A.1.6 Linear Parameter-Varying Systems

In the following, we consider the linear parameter-varying (LPV) system

$$\dot{x} = A(\rho(t))x + Bu \tag{A.20a}$$

$$y = Cx \tag{A.20b}$$

with state $x(t) \in \mathbb{R}^{n_x}$, input $u(t) \in \mathbb{R}^{n_u}$, and output $y(t) \in \mathbb{R}^{n_y}$. The matrix $A(\cdot)$ depends continuously on the time-varying parameter vector $\rho(t) \in \mathbb{R}^{n_\rho}$. In particular, ρ is an element of the parameter variation set $\mathcal{F}_\mathcal{P} = \{\rho : \mathbb{R}_0^+ \to \mathcal{P} \subset \mathbb{R}^{n_\rho}, \rho \text{ piece-wise continuous}\}$, where $\mathcal{P}$ is a compact set and $A : \mathbb{R}^{n_\rho} \to \mathbb{R}^{n_x \times n_x}$ is a continuous bounded map. A standard stability notion for LPV systems is quadratic stability.

Definition A.6 (Quadratic stability (Becker et al. (1993))). The LPV system (A.20) is said to be *quadratically stable* if there exists a matrix $P \in \mathbb{R}^{n_x \times n_x}$ such that $P = P^\mathsf{T} > 0$ and

$$A(\rho)^\mathsf{T} P + PA(\rho) < 0 \quad \text{for all } \rho \in \mathcal{P}. \tag{A.21}$$

For a quadratically stable LPV system, $V(x) = x^\mathsf{T} P x$ serves as Lyapunov function. In fact, quadratic stability implies global uniform exponential stability of $x = \mathbf{0}$, i.e., there exist constants $k, \gamma > 0$ such that all solutions of (A.20) satisfy $\|x(t)\| \leq k\|x(t_0)\|e^{-\gamma(t-t_0)}$ for all $t \geq t_0$. This follows from the observation that $\dot{V} \leq -\mu V$ for some $\mu > 0$ and the comparison lemma.

Definition A.7 (Polytopic LPV system (Apkarian et al. (1995))). The LPV system (A.20) is called *polytopic* if it has the following two properties.

i) $A(\cdot)$ depends affinely on $\rho = [\rho^{(1)} \ \cdots \ \rho^{(n_\rho)}]^\mathsf{T}$, i.e., $A(\rho) = A_0 + \rho^{(1)} A_1 + \cdots + \rho^{(n_\rho)} A_{n_\rho}$.

ii) $\mathcal{P}$ is a convex polytope defined by some vertices $\omega_i \in \mathbb{R}^{n_\rho}$, $i = 1, ..., p$, i.e.,

$$\mathcal{P} = \operatorname{conv}\{\omega_1, ..., \omega_p\} := \left\{ \textstyle\sum_{i=1}^p \alpha_i \omega_i : \alpha_i \geq 0, \sum_{i=1}^p \alpha_i = 1 \right\}.$$

For a polytopic LPV system, each parameter vector $\rho(t)$ is a convex combination of the vertices ω_i, $i = 1, ..., p$. The vector $\alpha = [\alpha_1 \ \cdots \ \alpha_p]^\mathsf{T}$ is an element of the unit p-simplex $\Delta^p = \{\alpha \in \mathbb{R}^p : \sum_{i=1}^p \alpha_i = 1, \alpha_i \geq 0 \text{ for all } i\}$. The system matrix can be written as convex combination of corresponding vertex matrices as well,

$$A(\rho(t)) = A\left(\textstyle\sum_{i=1}^p \alpha_i(t)\omega_i\right) = \sum_{i=1}^p \alpha_i(t) A(\omega_i) \in \operatorname{conv}\{A(\omega_1), ..., A(\omega_p)\},$$

where $\alpha(t) \in \Delta^p$.

Lemma A.21 (Vertex property (Apkarian et al. (1995))). *Suppose that the LPV system* (A.20) *is polytopic and* $\mathcal{P} = \operatorname{conv}\{\omega_1, ..., \omega_p\}$. *Then,* (A.21) *is equivalent to*

$$A(\omega_i)^\mathsf{T} P + P A(\omega_i) \prec 0 \quad \textit{for all } i \in \{1, ..., p\}. \tag{A.22}$$

Lemma A.21 reduces the infinite number of LMI conditions (A.21) to a finite number in (A.22), which makes the feasibility problem numerically tractable. Inspired by Apkarian et al. (1995), quadratic stabilizability and detectability are defined as follows.

Definition A.8 (Quadratic stabilizability). A polytopic LPV system (A.20) is said to be *quadratically stabilizable* if there exists an affine $K : \mathbb{R}^{n_\rho} \to \mathbb{R}^{n_u \times n_x}$ and $P \in \mathbb{R}^{n_x \times n_x}$ such that $P = P^\mathsf{T} \succ 0$ and

$$(A(\rho) - BK(\rho))^\mathsf{T} P + P(A(\rho) - BK(\rho)) \prec 0 \quad \text{for all } \rho \in \mathcal{P}. \tag{A.23}$$

In short, $(A(\rho), B)$ is called quadratically stabilizable.

Definition A.9 (Quadratic detectability). A polytopic LPV system (A.20) is said to be *quadratically detectable* if there exists an affine $L : \mathbb{R}^{n_\rho} \to \mathbb{R}^{n_x \times n_y}$ and $P \in \mathbb{R}^{n_x \times n_x}$ such that $P = P^\mathsf{T} \succ 0$ and

$$(A(\rho) - L(\rho)C)^\mathsf{T} P + P(A(\rho) - L(\rho)C) \prec 0 \quad \text{for all } \rho \in \mathcal{P}. \tag{A.24}$$

In short, $(A(\rho), C)$ is called quadratically detectable.

Lemma A.22 (Gain-scheduled state feedback synthesis via LMIs). *Consider the polytopic LPV system* (A.20) *and suppose that* $(A(\rho), B)$ *is quadratically stabilizable. Then, there exists a matrix* $X = X^\mathsf{T} \succ 0$ *and real scalars* $\tau_i > 0$ *such that*

$$X A(\omega_i)^\mathsf{T} + A(\omega_i) X - \tau_i B B^\mathsf{T} \prec 0 \tag{A.25}$$

for $i = 1, ..., p$, *and the gain-scheduled state feedback gain* $K(\rho)$ *defined by* $K(\omega_i) = \frac{\tau_i}{2} B^\mathsf{T} X^{-1}$ *is such that* $\dot{x} = (A(\rho(t)) - BK(\rho(t)))x$ *is quadratically stable.*

Moreover, $\dot{x} = (A(\rho(t)) - \mu BK(\rho(t)))x$ *is quadratically stable for all* $\mu \in \mathbb{C}$ *with* $\mathbf{Re}(\mu) \geq 1$.

Proof. Due to the polytopic structure of $A(\rho)$ and $K(\rho)$, quadratic stabilizability ensures that there exists a $P = P^\mathsf{T} \succ 0$ such that $(A(\omega_i) - BK(\omega_i))^\mathsf{T} P + P(A(\omega_i) - BK(\omega_i)) \prec 0$ for all $i \in 1,...,p$. Analogously to the proof of Lemma 3.3, the change of variables $X = P^{-1}$, $Y_i = K(\omega_i)X$ and Finsler's Lemma A.2 show that this is equivalent to the existence of scalars $\tau_i > 0$, $i = 1,...,p$, such that (A.25) is satisfied. The remainder of the proof is analogous to proof of Lemma 3.3 while exploiting the vertex property of the polytopic LPV system: Let $X \succ 0$, $\tau_i > 0$ be such that (A.25) holds and set $K(\omega_i) = \frac{\tau_i}{2} B^\mathsf{T} X^{-1}$. Then, $P = X^{-1} \succ 0$ and

$$\begin{aligned}(A(\omega_i) - \tfrac{\tau_i}{2} BB^\mathsf{T} P)^\mathsf{T} P + P(A(\omega_i) - \tfrac{\tau_i}{2} BB^\mathsf{T} P) &= A(\omega_i)^\mathsf{T} P + PA(\omega_i) - \tau_i PBB^\mathsf{T} P \\ &= P(XA(\omega_i)^\mathsf{T} + A(\omega_i)X - \tau_i BB^\mathsf{T})P \prec 0,\end{aligned}$$

i.e., $\dot{x} = (A(\rho(t)) - BK(\rho(t)))x$ is quadratically stable. Let $\mu \in \mathbb{C}$ with $\mathbf{Re}(\mu) \geq 1$ and define $\epsilon = \mathbf{Re}(\mu) - 1 \geq 0$. Then,

$$\begin{aligned}&(A(\omega_i) - \rho\tfrac{\tau_i}{2} BB^\mathsf{T} P)^\mathsf{H} P + P(A(\omega_i) - \rho\tfrac{\tau_i}{2} BB^\mathsf{T} P) \\ &\quad = (A(\omega_i) - (\mathbf{Re}(\rho) - \mathbf{Im}(\rho)\mathbf{i})\tfrac{\tau_i}{2} BB^\mathsf{T} P)^\mathsf{T} P + P(A(\omega_i) - (\mathbf{Re}(\rho) + \mathbf{Im}(\rho)\mathbf{i})\tfrac{\tau_i}{2} BB^\mathsf{T} P) \\ &\quad = (A(\omega_i) - \mathbf{Re}(\rho)\tfrac{\tau_i}{2} BB^\mathsf{T} P)^\mathsf{T} P + P(A(\omega_i) - \mathbf{Re}(\rho)\tfrac{\tau_i}{2} BB^\mathsf{T} P) \\ &\quad = (A(\omega_i) - (1+\epsilon)\tfrac{\tau_i}{2} BB^\mathsf{T} P)^\mathsf{T} P + P(A(\omega_i) - (1+\epsilon)\tfrac{\tau_i}{2} BB^\mathsf{T} P) \\ &\quad = A(\omega_i)^\mathsf{T} P + PA(\omega_i) - (1+\epsilon)\tau_i PBB^\mathsf{T} P \\ &\quad = P(XA(\omega_i)^\mathsf{T} + A(\omega_i)X - (1+\epsilon)\tau_i BB^\mathsf{T})P \prec 0.\end{aligned}$$

Consequently, $\dot{x} = (A(\rho(t)) - \mu BK(\rho(t)))x$ is quadratically stable for all such μ. □

It is worth noting that quadratic stabilizability is also sufficient for the existence of a constant state feedback gain matrix K which robustly stabilizes the system for all $\rho \in \mathcal{F}_\mathcal{P}$.

Lemma A.23 (Robust state feedback synthesis via LMIs). *Consider the polytopic LPV system* (A.20) *and suppose that* $(A(\rho), B)$ *is quadratically stabilizable. Then, there exists a matrix* $X = X^\mathsf{T} \succ 0$ *and a real scalar* $\tau > 0$ *such that*

$$XA(\omega_i)^\mathsf{T} + A(\omega_i)X - \tau BB^\mathsf{T} \prec 0 \tag{A.26}$$

for $i = 1,...,p$, *and the state feedback gain* $K = \frac{\tau}{2} B^\mathsf{T} X^{-1}$ *is such that* $\dot{x} = (A(\rho(t)) - BK)x$ *is quadratically stable.*

Moreover, $\dot{x} = (A(\rho(t)) - \mu BK)x$ *is quadratically stable for all* $\mu \in \mathbb{C}$ *with* $\mathbf{Re}(\mu) \geq 1$.

Proof. Suppose that the conditions of Lemma A.22 are satisfied, i.e., there exists a matrix $X = X^\mathsf{T} \succ 0$ and real scalars $\tau_i > 0$ such that the LMIs (A.25) hold for $i = 1,...,p$. Then, (A.25) hold as well for $\tau = \max\{\tau_1,...,\tau_p\}$ since $BB^\mathsf{T} \succeq 0$. □

The following result shows that there is no loss of generality in the assumption that the matrices $A_1, A_2, ..., A_{n_\rho}$ of a polytopic LPV system (A.20) are linearly independent.

Lemma A.24. *Let* $\mathcal{P}$ *be a convex polytope defined by* $\mathcal{P} = \operatorname{conv}\{\omega_1,...,\omega_p\} \subset \mathbb{R}^{n_\rho}$. *Let* $A : \mathbb{R}^{n_\rho} \to \mathbb{R}^{n_x \times n_x}$ *be an affine map defined by* $A(\rho) = A_0 + \sum_{i=1}^{n_\rho} \rho^{(i)} A_i$, *where* $A_i \in \mathbb{R}^{n_x \times n_x}$ *for* $i = 0,1,...,n_\rho$. *Suppose that the matrices* $\{A_1,...,A_{n_\rho}\}$ *are linearly dependent. Then, there exists a polytope* $\tilde{\mathcal{P}} = \operatorname{conv}\{\tilde{\omega}_1,...,\tilde{\omega}_{\tilde{p}}\} \subset \mathbb{R}^{n_{\tilde{\rho}}}$ *and a linearly independent subset* $\{A_{i_1},...,A_{i_{n_{\tilde{\rho}}}}\}$ *with* $i_j \in \{1,...,n_\rho\}$, $j = 1,...,n_{\tilde{\rho}}$, $n_{\tilde{\rho}} < n_\rho$, *such that for all* $\rho \in \mathcal{P}$ *there exists a* $\tilde{\rho} \in \tilde{\mathcal{P}}$ *such that* $\tilde{A}(\tilde{\rho}) = A(\rho)$, *where* $\tilde{A}(\tilde{\rho}) = A_0 + \sum_{j=1}^{n_{\tilde{\rho}}} \tilde{\rho}^{(j)} A_{i_j}$. *Moreover, for all* $\tilde{\rho} \in \tilde{\mathcal{P}}$ *there exists a* $\rho \in \mathcal{P}$ *such that* $\tilde{A}(\tilde{\rho}) = A(\rho)$. *The reduced map* $\tilde{A}$ *is injective.*

Proof. Suppose, without loss of generality, that there exist coefficients $\mu^{(i)}$ such that $A_{n_\rho} = \sum_{i=1}^{n_\rho-1} \mu^{(i)} A_i$. Then, it follows that

$$\begin{aligned} A(\rho) &= A_0 + \textstyle\sum_{i=1}^{n_\rho} \rho^{(i)} A_i \\ &= A_0 + \textstyle\sum_{i=1}^{n_\rho-1} \rho^{(i)} A_i + \rho^{(n_\rho)} (\sum_{i=1}^{n_\rho-1} \mu^{(i)} A_i) \\ &= A_0 + \textstyle\sum_{i=1}^{n_\rho-1} \underbrace{(\rho^{(i)} + \mu^{(i)} \rho^{(n_\rho)})}_{=: \tilde{\rho}^{(i)}} A_i =: \tilde{A}(\tilde{\rho}). \end{aligned}$$

Hence, the parameter vector ρ is reduced to $\tilde{\rho}$ according to $\tilde{\rho} = H\rho$ with $H = [I_{n_\rho-1}\ \mu] \in \mathbb{R}^{(n_\rho-1)\times n_\rho}$. This yields $\tilde{\mathcal{P}} = H\mathcal{P} = \text{conv}\{H\omega_1, ..., H\omega_p\}$. If the set $\{A_1, ..., A_{n_\rho-1}\}$ is still linearly dependent, the procedure is repeated until a linearly independent subset $\{A_{i_1}, ..., A_{i_{n_{\tilde{\rho}}}}\}$ is reached. The resulting map $\tilde{A}(\tilde{\rho})$ is injective due to the linear independence of $\{A_{i_1}, ..., A_{i_{n_{\tilde{\rho}}}}\}$. □

A.2 Graph Theory

Graph theory plays a fundamental role in modeling, analysis, and design of networks of dynamical systems. Graphs consisting of nodes and links among the nodes are well suited to describe the information structure and interactions in groups of dynamical agents. In the present section, we review selected definitions and results which are instrumental throughout this thesis. The basis are selected works on algebraic graph theory by Fiedler (1973); Brualdi and Ryser (1991); Godsil and Royle (2001); as well as more recent works on networks of dynamical systems by Wu (2007); Ren and Beard (2008); Wieland (2010).

A graph $\mathcal{G}$ is defined as triple $\mathcal{G} = (\mathcal{V}, \mathcal{E}, A_\mathcal{G})$, where $\mathcal{V}$ is a vertex set, $\mathcal{E}$ is an edge set, and $A_\mathcal{G}$ is a weighted adjacency matrix. Let $\mathcal{N}$ be the index set $\mathcal{N} = \{1, ..., N\}$. Each vertex (or node) v_k in the set $\mathcal{V} = \{v_1, ..., v_N\}$ corresponds to a dynamical subsystem (agent) $k \in \mathcal{N}$ in the network. The edge (or link) set $\mathcal{E} \subseteq \mathcal{V}\times\mathcal{V}$ consists of directed edges (v_j, v_k), $j, k \in \mathcal{N}$. In particular, $(v_j, v_k) \in \mathcal{E}$ if and only if there is a directed edge (or link) from vertex v_j to v_k. In the context of networks of dynamical systems, $(v_j, v_k) \in \mathcal{E}$ means that system k is influenced by (receives information from) system j. The vertices v_k, v_j are called head and tail of the edge (v_j, v_k), respectively. Since the edges in $\mathcal{E}$ are in general directed, we call $\mathcal{G}$ a directed graph. A consecutive sequence of directed edges is called a directed path. The adjacency matrix $A_\mathcal{G} \in \mathbb{R}^{N\times N}$ describes the graph structure and edge weights, i.e., $a_{kj} > 0 \Leftrightarrow (v_j, v_k) \in \mathcal{E}$ and $a_{kj} = 0$ otherwise. If the edge weights take only values in $\{0, 1\}$, the graph $\mathcal{G}$ is called unweighted. Unweighted graphs are completely described by $\mathcal{G} = (\mathcal{V}, \mathcal{E})$. A graph $\mathcal{G}$ is called undirected if $(v_j, v_k) \in \mathcal{E} \Leftrightarrow (v_k, v_j) \in \mathcal{E}$ and, more particularly, $a_{kj} = a_{jk}$ for all $j, k \in \mathcal{N}$. The neighbor set of a vertex v_k is the index set $\mathcal{N}_k = \{j \in \mathcal{N} : (v_k, v_j) \in \mathcal{E}\}$ of all vertices which have an outgoing edge to v_k. The degree of a vertex v_k is defined as the sum of the weights of all its incoming edges, i.e., $d_k = \sum_{j=1}^{N} a_{kj}$. The vector consisting of the degrees of all vertices can be computed as $d = A_\mathcal{G}\mathbf{1}$. The diagonal matrix $D_\mathcal{G} = \text{diag}(A_\mathcal{G}\mathbf{1})$ is called the degree matrix of $\mathcal{G}$. The Laplacian matrix $L_\mathcal{G} \in \mathbb{R}^{N\times N}$ of $\mathcal{G}$ is defined as

$$L_\mathcal{G} = D_\mathcal{G} - A_\mathcal{G}. \tag{A.27}$$

By construction, $L_\mathcal{G}$ is a Metzler matrix, i.e., it has nonnegative off-diagonal elements, and it has zero row sums, i.e., $L_\mathcal{G}\mathbf{1} = \mathbf{0}$. The vector of ones $\mathbf{1}$ is the eigenvector corresponding to the zero eigenvalue $\lambda_1(L_\mathcal{G}) = 0$. The normalized Laplacian matrix of a graph $\mathcal{G}$ is defined as $\tilde{L}_\mathcal{G} = D_\mathcal{G}^{-1} L_\mathcal{G}$.

Definition A.10 (Connected graph). The graph $\mathcal{G}$ is called *connected* if it contains a directed spanning tree, i.e., if there exists a vertex v_k such that there is a directed path from v_k to every other vertex $v_j \in \mathcal{V}$. In this case, v_k is called *centroid.*

Note that graphs satisfying Definition A.10 are also referred to as quasi-strongly connected graphs in the literature, cf., Wu (2007). Our simpler terminology is in accordance with Scardovi and Sepulchre (2009); Wieland (2010).

Definition A.11 (Strongly connected graph). The graph $\mathcal{G}$ is called *strongly connected* if there exists a directed path from any vertex to any other vertex in $\mathcal{V}$. In this case, every vertex is a centroid.

The following connection between the connectivity of $\mathcal{G}$ and the spectrum of $L_{\mathcal{G}}$ was obtained independently by Ren et al. (2005b) and Lin et al. (2005).

Theorem A.25 (Connectedness and the spectrum of $L_{\mathcal{G}}$). *All eigenvalues of $L_{\mathcal{G}}$ are contained in the closed right-half plane, i.e., $\lambda_k(L_{\mathcal{G}}) \in \mathbb{C}^0 \cup \mathbb{C}^+$ for all $k \in \mathcal{N}$. Moreover, the zero eigenvalue $\lambda_1(L_{\mathcal{G}}) = 0$ is simple and all other eigenvalues have positive real parts $\mathbf{Re}(\lambda_k(L_{\mathcal{G}})) > 0$ for all $k \in \mathcal{N}\backslash\{1\}$, if and only if $\mathcal{G}$ is connected.*

As a convention, the indices of the eigenvalues of $L_{\mathcal{G}}$ are ordered according to the magnitude of the real parts such that $0 = \lambda_1(L_{\mathcal{G}}) \leq \mathbf{Re}(\lambda_2(L_{\mathcal{G}})) \leq \cdots \leq \mathbf{Re}(\lambda_N(L_{\mathcal{G}}))$. In the sequel, the argument $L_{\mathcal{G}}$ will be omitted if there is no ambiguity. If the graph $\mathcal{G}$ is undirected, the Laplacian matrix $L_{\mathcal{G}}$ is symmetric and has only real eigenvalues. In this case, the second smallest eigenvalue $\lambda_2(L_{\mathcal{G}}) \geq 0$ is called the Fiedler eigenvalue or algebraic connectivity of the graph, cf., Fiedler (1973); Godsil and Royle (2001).

Definition A.12 (Induced subgraph). An *induced subgraph* of $\mathcal{G} = (\mathcal{V}, \mathcal{E})$ is a graph $\tilde{\mathcal{G}} = (\tilde{\mathcal{V}}, \tilde{\mathcal{E}})$ with $\tilde{\mathcal{V}} \subseteq \mathcal{V}$ and $\tilde{\mathcal{E}} = \{(v, w) \in \mathcal{E} : v, w \in \tilde{\mathcal{V}}\}$. The graph $\tilde{\mathcal{G}}$ is said to be *induced* by $\tilde{\mathcal{V}}$.

In the following, the concept of independent strongly connected components for directed graphs is reviewed, which was developed in Wieland (2010); Wieland et al. (2011a). It generalizes the notion of components of undirected graphs to directed graphs.

Definition A.13 (Strongly connected component (SCC)). A *strongly connected component (SCC)* of a directed graph $\mathcal{G} = (\mathcal{V}, \mathcal{E})$ is an induced subgraph $\tilde{\mathcal{G}} = (\tilde{\mathcal{V}}, \tilde{\mathcal{E}})$ which is maximal, subject to being strongly connected.

In other words, a SCC is an induced subgraph $\tilde{\mathcal{G}} = (\tilde{\mathcal{V}}, \tilde{\mathcal{E}})$ with the properties that (i) it is strongly connected and (ii) the subgraph induced by any set $\hat{\mathcal{V}}$ with $\tilde{\mathcal{V}} \subseteq \hat{\mathcal{V}} \subseteq \mathcal{V}$ is strongly connected if and only if $\tilde{\mathcal{V}} = \hat{\mathcal{V}}$. Note that the graph $\mathcal{G} = (\{v_1\}, \emptyset)$ consisting of a single node is strongly connected.

Definition A.14 (Independent strongly connected component (iSCC)). An *independent strongly connected component (iSCC)* of a directed graph $\mathcal{G} = (\mathcal{V}, \mathcal{E})$ is a SCC $\tilde{\mathcal{G}} = (\tilde{\mathcal{V}}, \tilde{\mathcal{E}})$, with the property that $(v, \tilde{v}) \notin \mathcal{E}$ for any $v \in \mathcal{V}\backslash\tilde{\mathcal{V}}$ and $\tilde{v} \in \tilde{\mathcal{V}}$.

In words, an iSCC $\tilde{\mathcal{G}} = (\tilde{\mathcal{V}}, \tilde{\mathcal{E}})$ is a SCC of $\mathcal{G} = (\mathcal{V}, \mathcal{E})$ with the additional property that $\tilde{\mathcal{V}}$ has no incoming edges from $\mathcal{V}\backslash\tilde{\mathcal{V}}$, i.e., there is no edge in $\mathcal{E}$ with tail outside $\tilde{\mathcal{V}}$ and head inside $\tilde{\mathcal{V}}$. Note that iSCCs are also called isolated strongly connected components, cf., Bauer (2012).

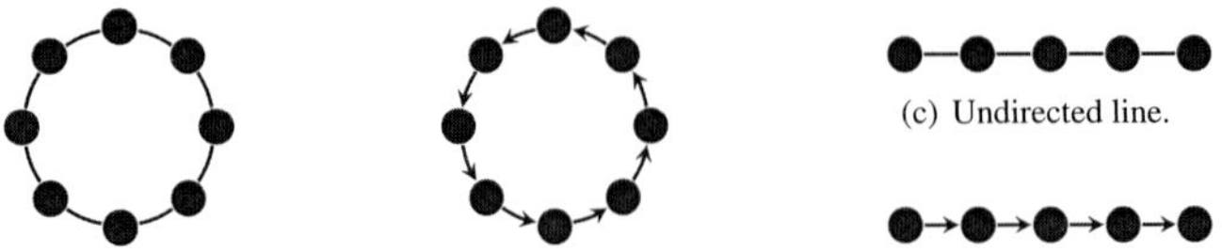

(a) Undirected cycle. (b) Directed cycle. (c) Undirected line. (d) Directed line.

Figure A.1: Four standard undirected and directed graphs.

Lemma A.26 (Existence of iSCCs (Bauer, 2012, Lem. 4.2)). *Every graph $\mathcal{G}$ contains at least one iSCC. Furthermore, if $\mathcal{G}$ is connected, then $\mathcal{G}$ has exactly one iSCC.*

Theorem A.27 ((Wieland, 2010, Thm. 2.13)). *Suppose that $\mathcal{G} = (\mathcal{V}, \mathcal{E})$ has $r \geq 1$ distinct iSCCs induced by $\tilde{\mathcal{V}}_i \subseteq \mathcal{V}$, $i = 1, \ldots, r$. Then, $\dim(\ker(L_{\mathcal{G}}^{\mathsf{T}})) = r$, i.e., $\operatorname{rank}(L_{\mathcal{G}}) = N - r$, and there exist unique (modulo vertex permutations) vectors $p^i \in \mathbb{R}^N$, $i = 1, \ldots, r$, which satisfy*

i) element-wise $p_j^i > 0$ if $v_j \in \tilde{\mathcal{V}}_i$ and $p_j^i = 0$ if $v_j \notin \tilde{\mathcal{V}}_i$, for $j \in \mathcal{N}$.

ii) $(p^i)^{\mathsf{T}} \mathbf{1} = 1$,

iii) $\ker(L_{\mathcal{G}}^{\mathsf{T}}) = \operatorname{span}\{p^1, \ldots, p^r\}$.

Corollary A.28. *If $\mathcal{G}$ is connected, then $\ker(L_{\mathcal{G}}) = \operatorname{im}(\mathbf{1})$ and the left eigenvector p of $L_{\mathcal{G}}$ corresponding to eigenvalue zero with $p^{\mathsf{T}}\mathbf{1} = 1$ is non-negative, i.e., $p^{\mathsf{T}} L = \mathbf{0}^{\mathsf{T}}$ and $p \geq \mathbf{0}$ element-wise. Moreover, if $\mathcal{G}$ is strongly connected, then p is positive, i.e., $p > \mathbf{0}$ element-wise.*

Lemma A.29 (Zhang et al. (2012)). *Suppose that $\mathcal{G}$ is strongly connected and $P = \operatorname{diag}(p)$, where $p^{\mathsf{T}} L_{\mathcal{G}} = \mathbf{0}^{\mathsf{T}}$ and $p^{\mathsf{T}}\mathbf{1} = 1$. Then, $P > 0$ and*

i) $PL_{\mathcal{G}} + L_{\mathcal{G}}^{\mathsf{T}} P \geq 0$,

ii) $\ker(PL_{\mathcal{G}} + L_{\mathcal{G}}^{\mathsf{T}} P) = \operatorname{im}(\mathbf{1})$.

Definition A.15 (Reducible and irreducible matrices (Brualdi and Ryser (1991))). A square matrix $M \in \mathbb{R}^{N \times N}$ is called *reducible* if there exists a permutation matrix T such that

$$TMT^{\mathsf{T}} = \begin{bmatrix} M_{11} & 0 \\ M_{21} & M_{22} \end{bmatrix},$$

where M_{11} and M_{22} are square matrices of order at least one. Otherwise, it is called *irreducible*.

Theorem A.30 (Strong connectedness and irreducibility (Brualdi and Ryser, 1991, Thm. 3.2.1)). *The matrix $L_{\mathcal{G}}$ (as well as $A_{\mathcal{G}}$) is irreducible, if and only if $\mathcal{G}$ is strongly connected.*

Oftentimes, it is desirable to write the Laplacian matrix of a connected graph in a special form. This canonical form is reported, e.g., by Wu (2007); Qu (2009); Lewis et al. (2014).

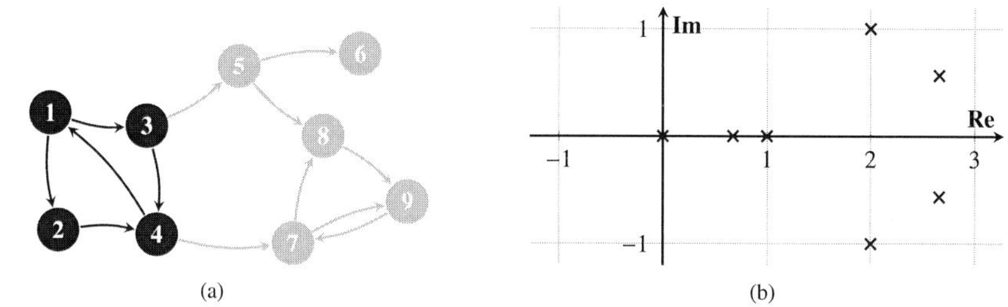

Figure A.2: Directed graph $\mathcal{G}$ from Example A.1 (a) and the spectrum $\sigma(L_{\mathcal{G}})$ of its Laplacian matrix (b).

Theorem A.31 (Frobenius normal form of $L_{\mathcal{G}}$)**.** *If the graph $\mathcal{G}$ is connected, then there exists a vertex permutation such that $L_{\mathcal{G}}$ reduces to the lower block-triangular form*

$$L_{\mathcal{G}} = \begin{bmatrix} L_{11} & & & 0 \\ L_{21} & L_{22} & & \\ \vdots & \vdots & \ddots & \\ L_{m1} & L_{m2} & \cdots & L_{mm} \end{bmatrix}, \tag{A.28}$$

where L_{ii}, $i = 1, ..., m$, are irreducible square matrices and each L_{ii}, $i = 2, ..., m$, has at least one row with positive row sum. If L_{11} is scalar, then $L_{11} = 0$.

Note that the Frobenius normal form of a matrix commonly has an upper triangular structure in the literature (Brualdi and Ryser, 1991, Thm. 3.2.4). A lower triangular structure as in (A.28), however, is easily obtained by a similarity transformation with a suitable permutation matrix. A lower triangular structure is also presented in Brualdi and Shader (2004). The form (A.28) admits a nice interpretation in terms of the graph $\mathcal{G}$. The block matrices on the diagonal correspond to the strongly connected components of $\mathcal{G}$. The block L_{11}, in particular, is a Laplacian matrix of the unique iSCC of $\mathcal{G}$.

In order to illustrate some of the definitions and results above, we discuss a particular graph in the following example.

Example A.1. We consider the unweighted graph $\mathcal{G} = (\mathcal{V}, \mathcal{E})$ shown in Fig. A.2(a). This graph $\mathcal{G}$ is directed and connected. It consists of the nine vertices $\mathcal{V} = \{v_1, ..., v_9\}$ and 13 directed edges $\mathcal{E} = \{(v_1, v_2), (v_1, v_3), (v_2, v_4), (v_3, v_4), (v_3, v_5), (v_4, v_1), (v_4, v_7), (v_5, v_6), (v_5, v_8), (v_7, v_8), (v_7, v_9), (v_8, v_9), (v_9, v_7)\}$. The corresponding adjacency matrix $A_{\mathcal{G}}$, degree matrix $D_{\mathcal{G}}$, and Laplacian matrix $L_{\mathcal{G}}$ are given by

$$A_{\mathcal{G}} = \begin{bmatrix} 0&0&0&1&0&0&0&0&0 \\ 1&0&0&0&0&0&0&0&0 \\ 1&0&0&0&0&0&0&0&0 \\ 0&1&1&0&0&0&0&0&0 \\ 0&0&1&0&0&0&0&0&0 \\ 0&0&0&0&1&0&0&0&0 \\ 0&0&0&1&0&0&0&0&1 \\ 0&0&0&0&1&0&1&0&0 \\ 0&0&0&0&0&0&1&1&0 \end{bmatrix}, \quad D_{\mathcal{G}} = \begin{bmatrix} 1&0&0&0&0&0&0&0&0&0 \\ 0&1&0&0&0&0&0&0&0 \\ 0&0&1&0&0&0&0&0&0 \\ 0&0&0&2&0&0&0&0&0 \\ 0&0&0&0&1&0&0&0&0 \\ 0&0&0&0&0&1&0&0&0 \\ 0&0&0&0&0&0&2&0&0 \\ 0&0&0&0&0&0&0&2&0 \\ 0&0&0&0&0&0&0&0&2 \end{bmatrix}, \quad L_{\mathcal{G}} = \begin{bmatrix} 1&0&0&-1&0&0&0&0&0 \\ -1&1&0&0&0&0&0&0&0 \\ -1&0&1&0&0&0&0&0&0 \\ 0&-1&-1&2&0&0&0&0&0 \\ 0&0&-1&0&1&0&0&0&0 \\ 0&0&0&0&-1&1&0&0&0 \\ 0&0&0&-1&0&0&2&0&-1 \\ 0&0&0&0&-1&0&-1&2&0 \\ 0&0&0&0&0&0&-1&-1&2 \end{bmatrix}.$$

The spectrum $\sigma(L_{\mathcal{G}})$ is plotted in Fig. A.2(b). It has a simple eigenvalue at zero and the eigenvalue at $+1$ has algebraic multiplicity 3. It is easy to see that $\mathcal{G}$ is connected since each of the vertices v_1, v_2, v_3, v_4, is a centroid. Since $\mathbf{Re}(\lambda_2(L_{\mathcal{G}})) = 0.6753 > 0$, this is in accordance with Theorem A.25.

The graph $\mathcal{G}$ is not strongly connected since the nodes $v_5, \ldots, v_9$ are no roots of a spanning tree. In particular, v_6 has no outgoing edges. The unique iSCC of $\mathcal{G}$ consists of vertices $\tilde{\mathcal{V}} = \{v_1, v_2, v_3, v_4\}$ and the edges among them, colored as •→• in Fig. A.2(a). The normalized left eigenvector p corresponding to the zero eigenvalue is given by

$$p^{\mathsf{T}} = \begin{bmatrix} 0.4 & 0.2 & 0.2 & 0.2 & 0 & 0 & 0 & 0 & 0 \end{bmatrix}.$$

As expected, $p \geq \mathbf{0}$ and the elements $p_k > 0$ correspond to the nodes which form the iSCC of $\mathcal{G}$, cf., Theorem A.27. The matrix $L_{\mathcal{G}}$ is already given in canonical form (A.28), where the top left 4×4 block L_{11} corresponds to the iSCC, $L_{22} = 1$, $L_{33} = 1$, and $L_{44} \in \mathbb{R}^{3\times 3}$. △

Appendix B

Technical Proofs

B.1 On the proof of Theorem 3.1

Construction of the transformation matrix T in the proof of Theorem 3.1, p. 18.

i) Choose any iSCC of $\mathcal{G}$ induced by $\tilde{\mathcal{V}} \subseteq \mathcal{V}$ and let $p \in \mathbb{R}^N$ be the unique nonnegative vector satisfying $p_k > 0$ if $v_k \in \tilde{\mathcal{V}}$, $k \in \mathcal{N}$, $p^\mathsf{T} L_\mathcal{G} = \mathbf{0}^\mathsf{T}$ and $p^\mathsf{T}\mathbf{1} = 1$. Such p exists by Theorem A.27.

ii) Choose $U \in \mathbb{R}^{N\times(N-1)}$ such that the columns of U form an orthonormal basis of $\ker(p^\mathsf{T})$, i.e., $p^\mathsf{T} U = \mathbf{0}^\mathsf{T}$ and $\operatorname{rank}(U) = N-1$.

iii) Define the matrix $H_1 = [\mathbf{1}\ U]$. By construction, H_1 has full rank and is hence invertible. The inverse of H_1 has the form $H_1^{-1} = [p\ W^\mathsf{T}]^\mathsf{T}$, where $W \in \mathbb{R}^{(N-1)\times N}$ is the left inverse[1] of U. A similarity transformation of $L_\mathcal{G}$ with H_1 yields

$$H_1^{-1} L_\mathcal{G} H_1 = \left[\begin{array}{c|c} 0 & 0 \\ \hline 0 & W L_\mathcal{G} U \end{array}\right].$$

iv) Due to the real version of Schur's triangularization theorem (Horn and Johnson, 1985, Thm. 2.3.1), there exists a real orthogonal matrix $H_2 \in \mathbb{R}^{(N-1)\times(N-1)}$ such that $H_2^\mathsf{T} W L_\mathcal{G} U H_2 = \Delta_\mathrm{r}$. Using such H_2, we define

$$T = H_1 \left[\begin{array}{c|c} 1 & 0 \\ \hline 0 & H_2 \end{array}\right] = \begin{bmatrix} \mathbf{1} & U H_2 \end{bmatrix}.$$

With this transformation matrix T, we obtain (3.11). The inverse of T has the form

$$T^{-1} = \left[\begin{array}{c|c} 1 & 0 \\ \hline 0 & H_2^T \end{array}\right] H_1^{-1} = \begin{bmatrix} p^\mathsf{T} \\ H_2^T W \end{bmatrix}.$$

Note that H_2 in the last step may as well be chosen as unitary matrix $H_2 \in \mathbb{C}^{(N-1)\times(N-1)}$ such that $H_2^\mathsf{H} W L_\mathcal{G} U H_2$ is an upper triangular matrix with diagonal elements λ_k, $k = 2, \dots, N$. Such H_2 exists by Schur's triangularization theorem. In this case, the transformation matrix T defined analogously yields $\Lambda = T^{-1} L_\mathcal{G} T$ upper triangular with $\lambda_1, \dots, \lambda_N$ on the diagonal. □

[1] The left inverse of a matrix $A \in \mathbb{R}^{n\times m}$ of full column rank, $n > m$, is given by $A_\mathrm{left}^{-1} = (A^\mathsf{T} A)^{-1} A^\mathsf{T}$ and has the property that $A_\mathrm{left}^{-1} A = I_m$.

B.2 Proof of Lemma 3.3

Proof of Lemma 3.3, p. 20. Since (A, B) is stabilizable, there exist matrices $P > 0$ and K such that $(A - BK)^\mathsf{T} P + P(A - BK) < 0$. With the change of variables $X = P^{-1}$ and $Y = KX$, we obtain $XA^\mathsf{T} + AX - Y^\mathsf{T} B^\mathsf{T} - BY < 0$. By Finsler's Lemma A.2, such a matrix Y exists if and only if there exists a scalar $\tau > 0$ such that the LMI (3.17) is satisfied. Note that τ may be chosen positive since $BB^\mathsf{T} \geq 0$, and we may set $\tau = 2$ by redefining X as $\frac{2}{\tau}X$ since X and τ are both decision variables. Let $X > 0$, $\tau > 0$ be a solution of (3.17) and set $K = \frac{\tau}{2}B^\mathsf{T} X^{-1}$. Then, $P = X^{-1} > 0$ and

$$(A - \tfrac{\tau}{2}BB^\mathsf{T} P)^\mathsf{T} P + P(A - \tfrac{\tau}{2}BB^\mathsf{T} P) = A^\mathsf{T} P + PA - \tau PBB^\mathsf{T} P = P(XA^\mathsf{T} + AX - \tau BB^\mathsf{T})P < 0,$$

i.e., $A - BK$ is Hurwitz. Let $\rho \in \mathbb{C}$ with $\mathbf{Re}(\rho) \geq 1$ and define $\epsilon = \mathbf{Re}(\rho) - 1 \geq 0$. Then,

$$\begin{aligned}
&(A - \rho\tfrac{\tau}{2}BB^\mathsf{T} P)^\mathsf{H} P + P(A - \rho\tfrac{\tau}{2}BB^\mathsf{T} P) \\
&\quad = (A - (\mathbf{Re}(\rho) - \mathbf{Im}(\rho)\mathbf{i})\tfrac{\tau}{2}BB^\mathsf{T} P)^\mathsf{T} P + P(A - (\mathbf{Re}(\rho) + \mathbf{Im}(\rho)\mathbf{i})\tfrac{\tau}{2}BB^\mathsf{T} P) \\
&\quad = (A - \mathbf{Re}(\rho)\tfrac{\tau}{2}BB^\mathsf{T} P)^\mathsf{T} P + P(A - \mathbf{Re}(\rho)\tfrac{\tau}{2}BB^\mathsf{T} P) \\
&\quad = (A - (1+\epsilon)\tfrac{\tau}{2}BB^\mathsf{T} P)^\mathsf{T} P + P(A - (1+\epsilon)\tfrac{\tau}{2}BB^\mathsf{T} P) \\
&\quad = A^\mathsf{T} P + PA - (1+\epsilon)\tau PBB^\mathsf{T} P \\
&\quad = P(XA^\mathsf{T} + AX - (1+\epsilon)\tau BB^\mathsf{T})P < 0.
\end{aligned}$$

Consequently, $A - \rho BK$ is Hurwitz for all such ρ. □

B.3 Proof of Theorem 3.12

Proof of Theorem 3.12, p. 29. Suppose that the coupled agents (3.27), (3.33) reach output synchronization, i.e., $x(t) \to \mathcal{S}_y$ as $t \to \infty$, where $x = [x_1^\mathsf{T} \ \cdots \ x_N^\mathsf{T}]^\mathsf{T} \in \mathbb{R}^{\hat{n}_x}$, and the synchronous subspace $\mathcal{S}_y$ is defined in (3.29). Obviously, $u_k = \mathbf{0}$ for all $k \in \mathcal{N}$ if $x \in \mathcal{S}_y$. Hence, the dynamics on $\mathcal{S}_y$ are

$$\dot{x} = \operatorname{diag}(A_1, ..., A_N)x, \quad y = \operatorname{diag}(C_1, ..., C_N)x.$$

By assumption, the agents synchronize to a non-trivial trajectory in $\mathcal{S}_y$ for all initial conditions. Therefore, there exists an invariant asymptotically stable subspace $\mathcal{S}_y^* \subseteq \mathcal{S}_y$ such that (i) $\mathcal{S}_y^*$ contains no exponentially stable modes, (ii) $\mathcal{S}_y^*$ contains only modes that are observable at y, and (iii) $\mathcal{S}_y^*$ is non-trivial with dimension m. Thus, by Theorem A.12, there exists a matrix $\Psi \in \mathbb{R}^{\hat{n}_x \times m}$ with full column rank and $S \in \mathbb{R}^{m \times m}$ such that $\mathcal{S}_y^* = \operatorname{im}(\Psi)$ and

$$\operatorname{diag}(A_1, ..., A_N)\Psi = \Psi S.$$

Let Ψ be partitioned as $\Psi = [\Pi_1^\mathsf{T} \ \cdots \ \Pi_N^\mathsf{T}]^\mathsf{T}$ with $\Pi_k \in \mathbb{R}^{n_k^x \times m}$. Then, we obtain the matrix equations

$$A_k \Pi_k = \Pi_k S$$

for all $k \in \mathcal{N}$. Consequently, $\mathcal{S}_{y,k}^* = \operatorname{im}(\Pi_k) \subset \mathbb{R}^{n_k^x}$ is an A_k-invariant subspace and contains no exponentially stable modes. On $\mathcal{S}_y^*$, it holds that $y_i = y_j$ for all pairs $i, j \in \mathcal{N}$, and hence $C_i x_i = C_j x_j$. Note that $x \in \mathcal{S}_y^* \Leftrightarrow \exists w \in \mathbb{R}^m : x = \Psi w$. In this case, $x_k = \Pi_k w$ and hence $C_i \Pi_i w = C_j \Pi_j w$. This has to hold for all $w \in \mathbb{R}^m$, which shows that there exists a matrix $R \in \mathbb{R}^{n_y \times m}$ such that

$$C_k \Pi_k = R.$$

The matrices Π_k, $k \in \mathcal{N}$, have full column rank m. This follows from the following contradiction. For any $w_0 \in \mathbb{R}^m : w_0 \neq \mathbf{0}$, there exists at least one $i \in \mathcal{N}$ such that $\Pi_i w_0 \neq \mathbf{0}$ since $\text{rank}(\Psi) = m$. Let $x_i(0) = \Pi_i w_0$. Since (A_i, C_i) is detectable and $x_i(0)$ lies in the unstable subspace of A_i, $y_i(t) = C_i e^{A_i t} x_i(0)$ is not identically zero for all $t \geq 0$. Suppose that there exists $j \in \mathcal{N}$ such that $x_j(0) = \Pi_j w_0 = \mathbf{0}$. Then, $x_j(t) = \mathbf{0}$ and $y_j(t) = \mathbf{0}$ for all $t \geq 0$, which is a contradiction to $y_i(t) = y_j(t)$. Consequently, such Π_j with rank $< m$ cannot exist.

The outputs converge to a common trajectory in $\mathcal{S}_y^*$ generated by the internal model $\dot{w} = Sw$, i.e., there exists a $w_0 \in \mathbb{R}^m$ such that $\lim_{t\to\infty} \|y_k(t) - Re^{St} w_0\| = \mathbf{0}$. □

B.4 Proof of Theorem 3.16

Proof of Theorem 3.16, p. 33. Let $s_k, v_k \in \mathbb{R}$ be the components of $x_k = [s_k \; v_k]^\mathsf{T}$ and define the stack vectors $s = [s_1 \; \cdots \; s_N]^\mathsf{T}$ and $v = [v_1 \; \cdots \; v_N]^\mathsf{T}$. Then, (3.35), (3.36) can be compactly written as $\dot{s} = -L_\mathcal{G} s + Dv$ and $\dot{v} = -\alpha L_\mathcal{G} s$. We perform the change of variables $\tilde{s} = Hs$, where $H = I_N - \frac{1}{N}\mathbf{1}\mathbf{1}^\mathsf{T}$ defines an orthogonal projection on $\text{im}(\mathbf{1})^\perp$, i.e., on the subspace orthogonal to $\text{im}(\mathbf{1})$. Since $L_\mathcal{G}\mathbf{1} = \mathbf{0}$ and $\mathbf{1}^\mathsf{T} L_\mathcal{G} = \mathbf{0}^\mathsf{T}$, it holds that $L_\mathcal{G} H = H L_\mathcal{G} = L_\mathcal{G}$ and hence $L_\mathcal{G}\tilde{s} = L_\mathcal{G} H s = L_\mathcal{G} s$. The change of variable yields

$$\begin{bmatrix} \dot{\tilde{s}} \\ \dot{v} \end{bmatrix} = \begin{bmatrix} -L_\mathcal{G} & HD \\ -\alpha L_\mathcal{G} & 0 \end{bmatrix} \begin{bmatrix} \tilde{s} \\ v \end{bmatrix}. \tag{B.1}$$

Note that the state $\tilde{s}$ is restricted to the subspace $\text{im}(\mathbf{1})^\perp$. Since the system is linear, the solution $s(t)$, $v(t)$ exists for all times $t \geq 0$. Hence, we can analyze the behavior of the state component $\tilde{s}(t)$ in $\text{im}(\mathbf{1})^\perp$ using the Lyapunov function $V = \alpha\tilde{s}^\mathsf{T} L_\mathcal{G}\tilde{s} + v^\mathsf{T} D v$. V is positive definite since $D > 0$ and $\tilde{s}^\mathsf{T} L_\mathcal{G}\tilde{s} > 0$ for all $\tilde{s} \in \text{im}(\mathbf{1})^\perp$, $s \neq \mathbf{0}$. In particular, $V \geq 0$ for all $\tilde{s}$, v and $V = 0 \Leftrightarrow \tilde{s} = v = \mathbf{0}$. The Lie-derivative $\dot{V}$ can be computed as $\dot{V} = -2\alpha\tilde{s}^\mathsf{T} L_\mathcal{G} L_\mathcal{G}\tilde{s}$. Hence, $\dot{V}$ is negative semi-definite and the set $\mathcal{S}$ on which $\dot{V}$ vanishes is given by $\mathcal{S} = \{\tilde{s}, v : \tilde{s} = \mathbf{0}\}$. Since V is positive definite and $\dot{V}$ is negative semi-definite, we can conclude that the trajectories $\tilde{s}(t)$, $v(t)$ are bounded for all $t \geq 0$. Hence, LaSalle's Invariance Principle (Theorem A.5) is applicable. It follows that $\tilde{s}(t)$, $v(t)$ converge to the largest invariant set $\mathcal{J}$ contained in $\mathcal{S}$. The trajectories contained in $\mathcal{S}$ have to satisfy $\tilde{s}(t) = \mathbf{0}$ and $\dot{\tilde{s}}(t) = \mathbf{0}$ for all $t \geq 0$. The dynamics (B.1) restricted to $\mathcal{S}$ are $\dot{\tilde{s}} = HDv$ and $\dot{v} = \mathbf{0}$, and therefore $HDv(t) = \mathbf{0}$, or equivalently, $Dv(t) \in \text{im}(\mathbf{1})$. Thus, the largest invariant set $\mathcal{J} \subset \mathcal{S}$ is $\mathcal{J} = \{\tilde{s}, v : \tilde{s} = \mathbf{0}, Dv \in \text{im}(\mathbf{1})\} \subset \mathcal{S}$. By LaSalle's Invariance Principle, the solutions of (B.1) converge to $\mathcal{J}$ asymptotically. Therefore, in original coordinates, we can conclude that, for some $c \in \mathbb{R}$, it holds that $\dot{v}(t) \to \mathbf{0}$, $Dv(t) \to c\mathbf{1}$, $\dot{s}(t) \to c\mathbf{1}$, and $s(t) \to \text{im}(\mathbf{1})$ as $t \to \infty$. □

B.5 Proof of Theorem 3.17

Proof of Theorem 3.17, p. 35. With stack vectors s, v, and matrix $\Delta = \text{diag}(\delta)$, the dynamics (3.38), (3.39) become

$$\dot{s} = -L_\mathcal{G} s + (I_N + \Delta)v, \tag{B.2a}$$

$$\dot{v} = -L_\mathcal{G} v. \tag{B.2b}$$

The network (B.2b) converges to consensus, i.e., for $t \to \infty$, $v(t) \to \mathbf{1}p^\mathsf{T} v_0$, where $v(0) = v_0$, cf., Section 2.1. Suppose that $s, v \in \text{im}(\mathbf{1})$. Then, $\dot{s} = (I_N + \Delta)v \notin \text{im}(\mathbf{1})$ since $\delta \notin \text{im}(\mathbf{1})$ by assumption. Thus, $\mathrm{im}(\mathbf{1})$ is not invariant for (B.2a) and the states $s(t)$ do not synchronize. Let $\xi = \dot{s}$. Then, with (B.2a) and (B.2b), $\dot{\xi} = -L_{\mathcal{G}}\xi - (I_N + \Delta)L_{\mathcal{G}}v$. It holds that $L_{\mathcal{G}}v(t) \to L_{\mathcal{G}}\mathbf{1}p^{\mathsf{T}}v_0 = \mathbf{0}$ as $t \to \infty$, and $\xi(t)$ converges exponentially to a solution of the unforced system $\dot{\xi} = -L_{\mathcal{G}}\xi$. Hence, for $t \to \infty$, $\dot{s}(t) = \xi(t) \to \mathrm{im}(\mathbf{1})$. Asymptotically, the states $s(t)$ grow with constant and identical velocity. With (B.2a) and $v(t) \to \mathbf{1}p^{\mathsf{T}}v_0$, it follows that

$$-L_{\mathcal{G}}s(t) + \delta p^{\mathsf{T}}v_0 \to \mathrm{im}(\mathbf{1}) \tag{B.3}$$

as $t \to \infty$. The state s can be decomposed into a sum of two components, one component in the subspace $\mathrm{im}(\mathbf{1})$ and the other, denoted by $s_\perp$, in the orthogonal complement $\mathrm{im}(\mathbf{1})^\perp$. We are interested in the component $s_\perp$ since it determines the distance of s from $\mathrm{im}(\mathbf{1})$. Since $L_{\mathcal{G}}\mathbf{1} = \mathbf{0}$, it holds that $L_{\mathcal{G}}s = L_{\mathcal{G}}s_\perp$ and therefore with (B.3), $-L_{\mathcal{G}}s_\perp + \delta p^{\mathsf{T}}v_0 \in \mathrm{im}(\mathbf{1})$. This can be rewritten as $L_{\mathcal{G}}s_\perp + c\mathbf{1} = \delta p^{\mathsf{T}}v_0$ for some $c \in \mathbb{R}$, or equivalently,

$$\begin{bmatrix} L_{\mathcal{G}} & \mathbf{1} \\ \mathbf{1}^{\mathsf{T}} & 0 \end{bmatrix} \begin{bmatrix} s_\perp \\ c \end{bmatrix} = \begin{bmatrix} \delta p^{\mathsf{T}}v_0 \\ 0 \end{bmatrix} \tag{B.4}$$

It holds that $\mathrm{im}(L_{\mathcal{G}})^\perp = \ker(L_{\mathcal{G}}^{\mathsf{T}}) = \mathrm{im}(p)$, where $p^{\mathsf{T}}L_{\mathcal{G}} = \mathbf{0}^{\mathsf{T}}$, $p^{\mathsf{T}}\mathbf{1} = 1$. Since $p^{\mathsf{T}}\mathbf{1} \neq 0$, it follows that $\mathrm{im}([L_{\mathcal{G}}\ \mathbf{1}]) = \mathbb{R}^N$, i.e., the rank of the matrix $[L_{\mathcal{G}}\ \mathbf{1}] \in \mathbb{R}^{N\times(N+1)}$ is N. It holds that $[L_{\mathcal{G}}\ \mathbf{1}][\mathbf{1}^{\mathsf{T}}\ 0]^{\mathsf{T}} = \mathbf{0}$ and $[\mathbf{1}^{\mathsf{T}}\ 0][\mathbf{1}^{\mathsf{T}}\ 0]^{\mathsf{T}} \neq 0$. Therefore, the coefficient matrix in (B.4) has full rank $N + 1$ and (B.4) has the unique solution (3.40). With (B.2a), we can finally conclude that $\dot{s}(t) \to \mathbf{1}(p^{\mathsf{T}}v_0 + c)$ as $t \to \infty$, i.e., the constant c is the deviation of the agents' velocity from the nominal case, in which $\dot{s}(t) \to \mathbf{1}p^{\mathsf{T}}v_0$ as $t \to \infty$. □

B.6 Proof of Theorem 3.18

Proof of Theorem 3.18, p. 36. Define $D = \mathrm{diag}((\omega + \delta_1)^2, \ldots, (\omega + \delta_N)^2)$ and denote $x_k = [s_k\ v_k]^{\mathsf{T}} \in \mathbb{R}^2$. Then,

$$\begin{bmatrix} \dot{s} \\ \dot{v} \end{bmatrix} = \begin{bmatrix} 0 & I_N \\ -D & -L_{\mathcal{G}} \end{bmatrix} \begin{bmatrix} s \\ v \end{bmatrix}.$$

At first, we assume that the graph $\mathcal{G}$ is strongly connected. Afterwards, we will relax this assumption and prove stability for general connected graphs. It is shown by Ren (2008) that the network reaches (non-trivial) state synchronization if the frequencies of all oscillators are identical. Hence, it remains to show that the network is rendered asymptotically stable by frequency perturbations within the iSCC of $\mathcal{G}$. We consider the Lyapunov function $V = s^{\mathsf{T}}PDs + v^{\mathsf{T}}Pv$, where $P = \mathrm{diag}(p)$ and p is the left eigenvector of $L_{\mathcal{G}}$ corresponding to zero. Since $\mathcal{G}$ is strongly connected, $\mathcal{V}_{\mathrm{iSCC}} = \mathcal{V}$ and $P > 0$. The Lie-derivative of V is $\dot{V} = -v^{\mathsf{T}}(PL_{\mathcal{G}} + L_{\mathcal{G}}^{\mathsf{T}}P)v$, which is negative semi-definite by Lemma A.29. The set on which $\dot{V} = 0$ is given by $\mathcal{J} = \{s, v : v \in \mathrm{im}(\mathbf{1})\}$. Since $L_{\mathcal{G}}\mathbf{1} = \mathbf{0}$, the dynamics on $\mathcal{J}$ are given by $\dot{s} = v$ and $\dot{v} = -Ds$. Every solution in $\mathcal{J}$ has to satisfy $\dot{v}(t) = -Ds(t) \in \mathrm{im}(\mathbf{1})$ and $D\dot{s}(t) = Dv(t) \in \mathrm{im}(\mathbf{1})$. This can only be true if $s = v = \mathbf{0}$ since $D \neq I_N$. By Theorem A.6, it follows that the origin $s = v = \mathbf{0}$ is asymptotically stable.

Suppose now that $\mathcal{G}$ is connected but not necessarily strongly connected. Then, there exists a vertex permutation such that $L_{\mathcal{G}}$ reduces to the Frobenius normal form as stated in (A.28). We have seen that $x_k(t) \to \mathbf{0}$ as $t \to \infty$ for all $k : v_k \in \mathcal{V}_{\mathrm{iSCC}}$. It remains to show that this implies $x_j(t) \to \mathbf{0}$ as $t \to \infty$ for $j \in \mathcal{V}\backslash\mathcal{V}_{\mathrm{iSCC}}$. We partition the vectors s, v according to the size of the blocks on the diagonal of $L_{\mathcal{G}}$, i.e., $s = [s_1^{\mathsf{T}}\ \cdots\ s_m^{\mathsf{T}}]^{\mathsf{T}}$, $v = [v_1^{\mathsf{T}}\ \cdots\ v_m^{\mathsf{T}}]^{\mathsf{T}}$, and

$D = \mathrm{diag}(D_{11}, \ldots, D_{mm})$. This yields $\dot{s}_i = v_i$ and $\dot{v}_i = -D_{ii}s_i - \sum_{l=1}^{i-1} L_{il}v_l - L_{ii}v_i$, for $i = 2, \ldots, m$. Each L_{ii} has at least one row with positive row sum. Therefore it is possible to decompose $L_{ii} = \tilde{L}_{ii} + M_{ii}$, such that $\tilde{L}_{ii}$ is the Laplacian matrix corresponding to a strongly connected graph $\tilde{\mathcal{G}}_{ii}$ and M_{ii} is a non-negative diagonal matrix with at least one positive element. We prove asymptotic stability block-wise by induction. For block $i = 1$, exponential stability follows since L_{11} corresponds to a strongly connected graph. For any block $i > 1$, it can be shown that $s_i(t), v_i(t) \to \mathbf{0}$ as $t \to \infty$ if $s_l(t), v_l(t) \to \mathbf{0}$ as $t \to \infty$ for $l = 1, \ldots, i-1$ by the following argument. If $s_l(t), v_l(t) \to \mathbf{0}$ as $t \to \infty$ for $l = 1, \ldots, i-1$, then the dynamics of s_i, v_i are asymptotically described by $\dot{s}_i = v_i$, $\dot{v}_i = -D_{ii}s_i - (\tilde{L}_{ii} + M_{ii})v_i$. Consider the Lyapunov function $V_i = s_i^\mathsf{T} P_{ii} D_{ii} s_i + v_i^\mathsf{T} P_{ii} v_i$, where $P_{ii} = \mathrm{diag}(\tilde{p}_i)$ and $\tilde{p}_i$ is the left eigenvector of $\tilde{L}_{ii}$ corresponding to zero. Since $\tilde{\mathcal{G}}_{ii}$ is strongly connected, $P_{ii} > 0$ and hence V_i is positive definite. Furthermore, we obtain $\dot{V}_i = -v_i^\mathsf{T}(P_{ii}\tilde{L}_{ii} + \tilde{L}_{ii}^\mathsf{T} P_{ii})v_i - 2v_i^\mathsf{T} P_{ii} M_{ii} v_i$. It holds that $(P_{ii}\tilde{L}_{ii} + \tilde{L}_{ii}^\mathsf{T} P_{ii}) \geq 0$ and $\ker(P_{ii}\tilde{L}_{ii} + \tilde{L}_{ii}^\mathsf{T} P_{ii}) = \mathrm{im}(\mathbf{1})$. Since $P_{ii}M_{ii}$ is a non-negative diagonal matrix with at least one positive element, $\mathbf{1}^\mathsf{T} P_{ii} M_{ii} \mathbf{1} > 0$ and therefore $\dot{V}_i < 0$. This proves that $s_i(t), v_i(t) \to \mathbf{0}$ as $t \to \infty$. By induction, we conclude that $s(t), v(t) \to \mathbf{0}$ as $t \to \infty$. □

B.7 Proof of Theorem 3.19

Proof of Theorem 3.19, p. 43. The closed-loop system consisting of agents (3.46) and dynamic controllers (3.50) can be written compactly as $\dot{x}^* = A^*(\rho(t))x^* + B^*(\rho(t))v$, $y = C^*x^*$, $\zeta = D^*x^*$, $v = -(L_\mathcal{G} \otimes I_{n_\zeta})\zeta$, where $x^* \in \mathbb{R}^{N(n_x+n_z)}$ is the stack vector $x^* = [x_1^\mathsf{T}\ z_1^\mathsf{T}\ \cdots\ x_N^\mathsf{T}\ z_N^\mathsf{T}]^\mathsf{T}$ and A^*, B^*, C^*, D^* are the resulting block diagonal matrices. By $\rho(t)$ we denote the vector of all global and local parameters $\rho(t) = [\rho^g(t)^\mathsf{T}\ \rho_1^\ell(t)^\mathsf{T}\ \cdots\ \rho_N^\ell(t)^\mathsf{T}]^\mathsf{T}$. The diagonal blocks of $A^*(\rho(t))$ are

$$\begin{bmatrix} A(\rho_k(t)) + BM(\rho_k(t))C & BG(\rho_k(t)) \\ E(\rho_k(t))C & D(\rho_k(t)) \end{bmatrix},$$

and the diagonal blocks of C^* are $[C\ \ 0]$. By assumption, all solutions $x^*(t) \to \mathcal{S}^*$ as $t \to \infty$ uniformly exponentially. Since the couplings v vanish on $\mathcal{S}^*$, the dynamics of the closed-loop system on $\mathcal{S}^*$ are $\dot{x}^* = A^*(\rho(t))x^*$ and the invariance condition reads

$$A^*(\rho)\mathcal{S}^* \subseteq \mathcal{S}^*.$$

Let $\mathcal{S}^* = \mathrm{im}(\Psi)$ for some $\Psi \in \mathbb{R}^{N(n_x+n_z)\times m}$ with full column rank. For any fixed ρ there exists a matrix $S(\rho)$ such that

$$A^*(\rho)\Psi = \Psi S(\rho),$$

cf., Theorem A.12. Let Ψ be partitioned as $\Psi = [\Pi_1^\mathsf{T}\ \Sigma_1^\mathsf{T}\ \cdots\ \Pi_N^\mathsf{T}\ \Sigma_N^\mathsf{T}]^\mathsf{T}$ according to the partition of x^* into agent and controller states, i.e., $\Pi_k \in \mathbb{R}^{n_x \times m}$ and $\Sigma_k \in \mathbb{R}^{n_z \times m}$ for all $k \in \mathcal{N}$. Due to the block diagonal structure of $A^*(\rho)$, this is equivalent to

$$\begin{bmatrix} A(\rho_k) + BM(\rho_k)C & BG(\rho_k) \\ E(\rho_k)C & D(\rho_k) \end{bmatrix} \begin{bmatrix} \Pi_k \\ \Sigma_k \end{bmatrix} = \begin{bmatrix} \Pi_k \\ \Sigma_k \end{bmatrix} S(\rho).$$

Each pair Π_k, Σ_k solves the equation for all $\rho^g \in \mathcal{P}^g$, $\rho_k^\ell \in \mathcal{P}^\ell$. Since all local parameters ρ_k^ℓ take values in the same set $\mathcal{P}^\ell$, we solve in fact the same equation N times, i.e., there exist Π and Σ such that

$$\begin{bmatrix} A(\rho_k) + BM(\rho_k)C & BG(\rho_k) \\ E(\rho_k)C & D(\rho_k) \end{bmatrix} \begin{bmatrix} \Pi \\ \Sigma \end{bmatrix} = \begin{bmatrix} \Pi \\ \Sigma \end{bmatrix} S(\rho) \tag{B.5}$$

for all $\rho^g \in \mathcal{P}^g$, $\rho_k^\ell \in \mathcal{P}^\ell$, $k = 1, ..., N$. Note that $[\Pi^\mathsf{T}\ \Sigma^\mathsf{T}]^\mathsf{T}$ has full column rank. Hence, we can multiply (B.5) with its left inverse and obtain

$$\begin{bmatrix}\Pi\\ \Sigma\end{bmatrix}_{\text{left}}^{-1} \begin{bmatrix} A(\rho_k) + BM(\rho_k)C & BG(\rho_k) \\ E(\rho_k)C & D(\rho_k)\end{bmatrix} \begin{bmatrix}\Pi\\ \Sigma\end{bmatrix} = S(\rho) \tag{B.6}$$

Since equation (B.6) is satisfied for all $\rho^g \in \mathcal{P}^g$, $\rho_k^\ell \in \mathcal{P}^\ell$, $k \in \mathcal{N}$, the parameter-dependent matrix $S(\rho)$ must be independent of the generally non-identical local parameters ρ_k^ℓ, i.e., it is solely dependent on the global parameters, $S(\rho^g)$, $S : \mathcal{P}^g \to \mathbb{R}^{m\times m}$. Equation (B.6) also reveals that affine parameter dependence of the system and controller matrices carries over to S. Finally, by choosing $\Gamma(\rho_k) = M(\rho_k)C\Pi + H(\rho_k)\Sigma$, we conclude that there exist matrices Π, $\Gamma(\rho_k)$, and $S(\rho^g)$, such that

$$A(\rho_k)\Pi + B\Gamma(\rho_k) = \Pi S(\rho^g)$$

for all $\rho^g \in \mathcal{P}^g$, $\rho_k^\ell \in \mathcal{P}^\ell$. The affine parameter dependence carries over to Γ. □

B.8 Proof of Lemma 3.22

Proof of Lemma 3.22, p. 45. For the unforced system $\dot{x} = A(\rho(t))x$ with skew-symmetric $A(\cdot)$, it holds that $\frac{\mathrm{d}}{\mathrm{d}t}\|x(t)\|^2 = x(t)^\mathsf{T}(A(\rho(t)) + A(\rho(t))^\mathsf{T})x(t) = 0$ and hence $\|x(t)\| = \|x(0)\|$ for all $t \geq 0$. Consequently, the state transition matrix satisfies $\|\Phi(t,\tau)\| = 1$ for all $t, \tau \geq 0$ and $\rho \in \mathcal{F}_\mathcal{P}$, cf., Köroğlu and Scherer (2011). The solution $x(t)$ of (3.57) is given by

$$x(t) = \Phi(t,0)\left(x_0 + \int_0^t \Phi(0,\tau)\delta(\tau)\mathrm{d}\tau\right).$$

With the given bounds, we obtain $c_1 e^{-\mu\tau}$ as norm bound for the integrand. Hence there exists a vector θ such that $\int_0^\infty \Phi(0,\tau)\delta(\tau)\mathrm{d}\tau = \theta$. Consequently,

$$\left\|\int_0^t \Phi(0,\tau)\delta(\tau)\mathrm{d}\tau - \theta\right\| = \left\|\int_t^\infty \Phi(0,\tau)\delta(\tau)\mathrm{d}\tau\right\| \leq c_1 \int_t^\infty e^{-\mu\tau}\mathrm{d}\tau = \frac{c_1}{\mu}e^{-\mu t}.$$

Finally, we obtain

$$\|x(t) - \Phi(t,0)(x_0 + \theta)\| = \left\|\Phi(t,0)\left(\int_0^t \Phi(0,\tau)\delta(\tau)\mathrm{d}\tau - \theta\right)\right\| \leq \frac{c_1}{\mu}e^{-\mu t}.$$

The statement follows with $\tilde{x}_0 = x_0 + \theta$, $c_2 = \frac{c_1}{\mu}$. □

B.9 Proof of Theorem 3.26

Proof of Theorem 3.26, p. 47. From Theorem 3.24 and Theorem 3.25, we know that the systems (3.66a) achieve state synchronization and $\zeta_k(t) - w(t) \to \mathbf{0}$ uniformly exponentially as $t \to \infty$ for all $k \in \mathcal{N}$, where $w(t)$ is the solution of $\dot{w} = S(\rho^g(t))w$ with $w(0) = (p^\mathsf{T} \otimes I_{n_x})\zeta(0)$. Consider the new state variables $\epsilon_k = x_k - \Pi\zeta_k$ and $e_k = x_k - \hat{x}_k$. From (3.46), (3.66), we obtain

$\dot{e}_k = (A(\rho_k(t)) - L(\rho_k(t))C)\, e_k$, which is quadratically stable. Furthermore, with (3.55), we have

$$\begin{aligned}\dot{\epsilon}_k &= A(\rho_k(t))x_k + B\left(-F(\rho_k(t))(\hat{x}_k - \Pi\zeta_k) + \Gamma(\rho_k(t))\zeta_k\right)\\ &\quad - \Pi(S(\rho^g(t))\zeta_k + K(\rho^g(t)) \textstyle\sum_{j=1}^N a_{kj}(\zeta_j - \zeta_k))\\ &= (A(\rho_k(t)) - BF(\rho_k(t)))\, \epsilon_k + BF(\rho_k(t))e_k - \Pi K(\rho^g(t)) \textstyle\sum_{j=1}^N a_{kj}(\zeta_j - \zeta_k)\\ &\quad + (A(\rho_k(t))\Pi + B\Gamma(\rho_k(t)) - \Pi S(\rho^g(t)))\zeta_k.\\ &= (A(\rho_k(t)) - BF(\rho_k(t)))\, \epsilon_k + BF(\rho_k(t))e_k - \Pi K(\rho^g(t)) \textstyle\sum_{j=1}^N a_{kj}(\zeta_j - \zeta_k).\end{aligned}$$

The two additive terms vanish due to the uniform exponential stability of $\zeta_j - \zeta_k$ and e_k. Since $\dot{\epsilon}_k = (A(\rho_k(t)) - BF(\rho_k(t)))\, \epsilon_k$ is quadratically stable, $\epsilon_k \to \mathbf{0}$ uniformly exponentially as $t \to \infty$. Consequently, $x_k(t) - \Pi\zeta_k(t) \to \mathbf{0}$ uniformly exponentially as $t \to \infty$, i.e., the Output Synchronization Problem 3.4 is solved. In fact, the agents achieve state synchronization, i.e., $x_k(t) - x_j(t) \to \mathbf{0}$ as $t \to \infty$, which implies synchronization of the outputs. The solution $w(t)$ of $\dot{w} = S(\rho^g(t))w$ with $w(0) = (p^\mathsf{T} \otimes I_{n_x})\zeta(0)$ describes the synchronous state trajectory $\Pi w(t)$ and $s(t) = C\Pi w(t)$ is a synchronous output trajectory of the network (3.46), (3.66). □

B.10 Proof of Theorem 4.1

Proof of Theorem 4.1, p. 54. The control laws (4.3b) can be written in stacked form as $u = -(L_\mathcal{G} \otimes K_\mathrm{P})x + (I_N \otimes K_\mathrm{I})\zeta$. The agent dynamics (4.2) in stacked form are

$$\dot{z} = \left(I_N \otimes \begin{bmatrix} 0 & 0 \\ 0 & A \end{bmatrix} - L_\mathcal{G} \otimes \begin{bmatrix} 0 & C \\ 0 & 0 \end{bmatrix}\right) z + (I_N \otimes B_\mathrm{e})\, u + (I_N \otimes P_\mathrm{e})\, d.$$

With $x = (I_N \otimes [0\ I_{n_x}])z$ and $\zeta = (I_N \otimes [I_{n_y}\ 0])z$, the closed loop with (4.3b) can consequently be written as

$$\dot{z} = \left(I_N \otimes \begin{bmatrix} 0 & 0 \\ BK_\mathrm{I} & A \end{bmatrix} - L_\mathcal{G} \otimes \begin{bmatrix} 0 & C \\ 0 & BK_\mathrm{P} \end{bmatrix}\right) z + (I_N \otimes P_\mathrm{e})\, d.$$

Analogously to Theorem 3.1, we apply the state transformation $\tilde{z} = (T^{-1} \otimes I_{n_x+n_y})z$ with T chosen such that $\Lambda = T^{-1}L_\mathcal{G}T$ is upper triangular with $0, \lambda_2, ..., \lambda_N$ on the diagonal, the first column of T is the vector of ones $\mathbf{1}$, and the first row of T^{-1} is p^T, where $p^\mathsf{T}L_\mathcal{G} = \mathbf{0}^\mathsf{T}$ and $p^\mathsf{T}\mathbf{1} = 1$. This yields

$$\dot{\tilde{z}} = \left(I_N \otimes \begin{bmatrix} 0 & 0 \\ BK_\mathrm{I} & A \end{bmatrix} - \Lambda \otimes \begin{bmatrix} 0 & C \\ 0 & BK_\mathrm{P} \end{bmatrix}\right) \tilde{z} + \left(T^{-1} \otimes P_\mathrm{e}\right) d.$$

The transformed system has the structure

$$\dot{\tilde{z}} = \left[\begin{array}{c|ccc} M_1 & & 0 & \\ \hline & M_2 & \cdots & \star \\ 0 & & \ddots & \vdots \\ & 0 & & M_N \end{array}\right] \tilde{z} + \left(T^{-1} \otimes P_\mathrm{e}\right) d$$

with block matrices

$$M_k = \begin{bmatrix} 0 & -\lambda_k C \\ BK_\mathrm{I} & A - \lambda_k BK_\mathrm{P} \end{bmatrix}. \tag{B.7}$$

Let the transformed state be partitioned as $\tilde{z}^\mathsf{T} = [\tilde{z}_1^\mathsf{T} \cdots \tilde{z}_N^\mathsf{T}]$ such that $\tilde{z}_k \in \mathbb{R}^{n_x+n_y}$, $k \in \mathcal{N}$. The stacked outputs are $y = (I_N \otimes C_\mathrm{e})z = (T \otimes C_\mathrm{e})\tilde{z}$. Since the first column of T is the vector of ones $\mathbf{1}$, the outputs synchronize if and only if $C_\mathrm{e}\tilde{z}_k(t) \to \mathbf{0}$ as $t \to \infty$ for all $k \in \mathcal{N}\backslash\{1\}$. In this case, $y(t) \to (\mathbf{1} \otimes C_\mathrm{e})\tilde{z}_1(t)$. Due to the block triangular structure of the transformed system matrix, a sufficient condition is that the matrices (B.7) are Hurwitz for all $k \in \mathcal{N}\backslash\{1\}$. In particular, if all these blocks are Hurwitz, a constant disturbance d leads to constant steady-states $\tilde{z}_k(t) \to \tilde{z}_k^\circ$ as $t \to \infty$ for all $k \in \mathcal{N}\backslash\{1\}$. Let $\tilde{z}_k$ be partitioned as $\tilde{z}_k^\mathsf{T} = [\tilde{\zeta}_k^\mathsf{T}\ \tilde{x}_k^\mathsf{T}]$ with $\tilde{\zeta}_k \in \mathbb{R}^{n_y}$. It holds that $\tilde{\zeta}_N(t) \to \tilde{\zeta}_N^\circ$ and hence $\frac{\mathrm{d}}{\mathrm{d}t}\tilde{\zeta}_N(t) = C\tilde{x}_N(t) \to \mathbf{0}$ as $t \to \infty$. Since $\frac{\mathrm{d}}{\mathrm{d}t}\tilde{\zeta}_k(t) \to \mathbf{0}$ for all $k \in \mathcal{N}\backslash\{1\}$, it follows that $C\tilde{x}_k(t) \to \mathbf{0}$ by induction. Consequently, $C_\mathrm{e}\tilde{z}_k(t) = C\tilde{x}_k(t) \to \mathbf{0}$ as $t \to \infty$ for all $k \in \mathcal{N}\backslash\{1\}$. The matrices (B.7) can only be stable for non-zero λ_k, i.e., $\mathcal{G}$ must be connected. A similarity transformation of (B.7) with $\mathrm{diag}(-\lambda_k I_{n_y}, I_{n_x})$ reveals that this is equivalent to the matrices (4.4) being stable.

It follows that $y_k(t) - C_\mathrm{e}\tilde{z}_1(t) \to \mathbf{0}$ as $t \to \infty$ for all $k \in \mathcal{N}$. The dynamics of $\tilde{z}_1$ are given by

$$\dot{\tilde{z}}_1 = \begin{bmatrix} 0 & 0 \\ BK_\mathrm{I} & A \end{bmatrix} \tilde{z}_1 + (p^\mathsf{T} \otimes P_\mathrm{e})d$$

with initial condition $\tilde{z}_1(0) = (p^\mathsf{T} \otimes I_{n_x+n_y})z(0)$. The synchronous output trajectory of the network is $s(t) = C_\mathrm{e}\tilde{z}_1(t)$. Since $\zeta_k(0) = \mathbf{0}$ for all $k \in \mathcal{N}$, it holds that $\tilde{\zeta}_1(t) = \mathbf{0}$ for all $t \geq 0$ and hence $s(t) = C\tilde{x}_1(t)$, where $\dot{\tilde{x}}_1 = A\tilde{x}_1 + (p^\mathsf{T} \otimes P)d$ and $\tilde{x}_1(0) = (p^\mathsf{T} \otimes I_{n_x})x(0)$.

The rank condition (4.6) and stabilizability of (A, B) imply stabilizability of $(A_\mathrm{e}, B_\mathrm{e})$, which can be shown via the Hautus test. Along with connectedness of $\mathcal{G}$, this guarantees the existence of K_e in view of Theorem 3.2. □

B.11 Proof of Theorem 4.2

Proof of Theorem 4.2, p. 56. The dynamics of the observation errors $e_k^x = \hat{x}_k - x_k$ and $e_k^d = \hat{d}_k - d_k$ can be computed as

$$\begin{bmatrix} \dot{e}_k^x \\ \dot{e}_k^d \end{bmatrix} = \left(\begin{bmatrix} A & P \\ 0 & 0 \end{bmatrix} - H \begin{bmatrix} C & 0 \end{bmatrix} \right) \begin{bmatrix} e_k^x \\ e_k^d \end{bmatrix}. \tag{B.8}$$

Since (4.9) is Hurwitz, $e_k^x(t) \to \mathbf{0}$ and $e_k^d(t) \to \mathbf{0}$ as $t \to \infty$. The couplings (4.10c) can be rewritten as $u_k = K_\mathrm{P} \sum_{j=1}^N a_{kj}(x_j - x_k) + K_\mathrm{P} \sum_{j=1}^N a_{kj}(e_j^x - e_k^x) + K_\mathrm{I}\zeta_k$. Let $e^x = [e_1^{x\mathsf{T}} \cdots e_N^{x\mathsf{T}}]^\mathsf{T}$. Then, analogously to the proof of Theorem 4.1, the dynamics of the coupled agents are

$$\dot{z} = \left(I_N \otimes \begin{bmatrix} 0 & 0 \\ BK_\mathrm{I} & A \end{bmatrix} - L_\mathcal{G} \otimes \begin{bmatrix} 0 & C \\ 0 & BK_\mathrm{P} \end{bmatrix} \right) z + (I_N \otimes P_\mathrm{e})\, d - (L_\mathcal{G} \otimes (B_\mathrm{e}K_\mathrm{P}))\, e^x, \tag{B.9}$$

where e^x acts now as an additional disturbance. In transformed coordinates $\tilde{z} = (T^{-1} \otimes I_{n_x+n_y})z$, this yields

$$\dot{\tilde{z}} = \left(I_N \otimes \begin{bmatrix} 0 & 0 \\ BK_\mathrm{I} & A \end{bmatrix} - \Lambda \otimes \begin{bmatrix} 0 & C \\ 0 & BK_\mathrm{P} \end{bmatrix} \right) \tilde{z} + \left(T^{-1} \otimes P_\mathrm{e}\right) d - (\Lambda \otimes (B_\mathrm{e}K_\mathrm{P}))\, \tilde{e}^x,$$

where $\tilde{e}^x = (T^{-1} \otimes I_{n_x})e^x$. Since the first row of Λ is zero, the error $\tilde{e}^x$ has no influence on the synchronous motion $\tilde{z}_1$ of the group. Moreover, since $\tilde{e}^x$ decays to zero exponentially, convergence of the state components $\tilde{z}_k$, $k = 2, \ldots, N$, is guaranteed. The rest of the statement follows by the same arguments as in the proof of Theorem 4.1. □

B.12 Proof of Theorem 4.3

Proof of Theorem 4.3, p. 57. We define the observer errors $e^x_{\mathcal{N}_k} = \hat{x}_{\mathcal{N}_k} - x_{\mathcal{N}_k}$ and $e^d_{\mathcal{N}_k} = \hat{d}_{\mathcal{N}_k} - d_{\mathcal{N}_k}$. Using the equation $\dot{x}_{\mathcal{N}_k} = Ax_{\mathcal{N}_k} + Bu_{\mathcal{N}_k} + Pd_{\mathcal{N}_k}$, we obtain the error dynamics

$$\begin{bmatrix} \dot{e}^x_{\mathcal{N}_k} \\ \dot{e}^d_{\mathcal{N}_k} \end{bmatrix} = \left(\begin{bmatrix} A & P \\ 0 & 0 \end{bmatrix} - H \begin{bmatrix} C & 0 \end{bmatrix} \right) \begin{bmatrix} e^x_{\mathcal{N}_k} \\ e^d_{\mathcal{N}_k} \end{bmatrix}.$$

Since (4.9) is Hurwitz, $e^x_{\mathcal{N}_k}(t) \to \mathbf{0}$ and $e^d_{\mathcal{N}_k}(t) \to \mathbf{0}$ as $t \to \infty$. The couplings (4.11c) can be rewritten as $u_k = K_{\mathrm{P}} \sum_{j=1}^N a_{kj}(x_j - x_k) + K_{\mathrm{P}} e^x_{\mathcal{N}_k} + K_{\mathrm{I}} \zeta_k$. Let $e^x_{\mathcal{N}} = [e^{x\,\mathsf{T}}_{\mathcal{N}_1} \cdots e^{x\,\mathsf{T}}_{\mathcal{N}_N}]^{\mathsf{T}}$. Then, analogously to the proof of Theorem 4.1, the dynamics of the coupled agents are

$$\dot{z} = \left(I_N \otimes \begin{bmatrix} 0 & 0 \\ BK_{\mathrm{I}} & A \end{bmatrix} - L_{\mathcal{G}} \otimes \begin{bmatrix} 0 & C \\ 0 & BK_{\mathrm{P}} \end{bmatrix} \right) z + (I_N \otimes P_{\mathrm{e}})\, d + (I_N \otimes B_{\mathrm{e}} K_{\mathrm{P}})\, e^x_{\mathcal{N}},$$

where $e^x_{\mathcal{N}}$ acts now as an additional disturbance. By the same arguments as in the proof of Theorem 4.2, the agents achieve robust output synchronization. The decisive difference to (B.9) is that, here, the additional disturbance does not necessarily lie in the range space of $L_{\mathcal{G}} \otimes I_{n_x}$. In other words, there does not necessarily exist a function $e(t)$ such that $e^x_{\mathcal{N}}(t) = -(L_{\mathcal{G}} \otimes I_{n_x}) e(t)$.

The synchronous output trajectory $s(t) = C\tilde{x}_1(t)$, however, is affected by the observer errors:

$$\dot{\tilde{x}}_1 = A\tilde{x}_1 + (p^{\mathsf{T}} \otimes P)d + (p^{\mathsf{T}} \otimes BK_{\mathrm{P}}) e^x_{\mathcal{N}}.$$

As we will show next, the influence of the observer errors on the synchronous trajectory can be avoided by a suitable choice of the initial conditions of the observers. Note that the corresponding requirement can be rewritten as

$$(p^{\mathsf{T}} \otimes I_{n_x}) e^x_{\mathcal{N}}(t) = \mathbf{0}$$

for all $t \geq 0$. We show that $(p^{\mathsf{T}} \otimes I_{n_x}) e^x_{\mathcal{N}}(0) = \mathbf{0}$ implies $(p^{\mathsf{T}} \otimes I_{n_x}) e^x_{\mathcal{N}}(t) = \mathbf{0}$ for all $t \geq 0$. More precisely, we show that the null space of $p^{\mathsf{T}} \otimes I_{n_x}$ is invariant with respect to the observer dynamics. Let $y_{\mathcal{N}}, x_{\mathcal{N}}, d_{\mathcal{N}}, u_{\mathcal{N}}, \hat{x}_{\mathcal{N}}, \hat{d}_{\mathcal{N}}$ be the stack vectors of $y_{\mathcal{N}_k}, x_{\mathcal{N}_k}, d_{\mathcal{N}_k}, u_{\mathcal{N}_k}, \hat{x}_{\mathcal{N}_k}, \hat{d}_{\mathcal{N}_k}$, $k \in \mathcal{N}$, respectively. Then, it holds that $y_{\mathcal{N}} = -(L_{\mathcal{G}} \otimes I_{n_y})y$, $x_{\mathcal{N}} = -(L_{\mathcal{G}} \otimes I_{n_x})x$, $d_{\mathcal{N}} = -(L_{\mathcal{G}} \otimes I_{n_d})d$, $u_{\mathcal{N}} = -(L_{\mathcal{G}} \otimes I_{n_u})u$. Note that such a decomposition does *not* necessarily exist for the observer states $\hat{x}_{\mathcal{N}}$, $\hat{d}_{\mathcal{N}}$. Let H be partitioned as $H = [H_1^{\mathsf{T}}\ H_2^{\mathsf{T}}]^{\mathsf{T}}$, $H_1 \in \mathbb{R}^{n_x \times n_y}$. The dynamics of all observers (4.11b) can be written compactly as

$$\begin{bmatrix} \dot{\hat{x}}_{\mathcal{N}} \\ \dot{\hat{d}}_{\mathcal{N}} \end{bmatrix} = \begin{bmatrix} I_N \otimes A & I_N \otimes P \\ 0 & 0 \end{bmatrix} \begin{bmatrix} \hat{x}_{\mathcal{N}} \\ \hat{d}_{\mathcal{N}} \end{bmatrix} - \begin{bmatrix} L_{\mathcal{G}} \otimes B \\ 0 \end{bmatrix} u - \begin{bmatrix} L_{\mathcal{G}} \otimes H_1 C \\ L_{\mathcal{G}} \otimes H_2 C \end{bmatrix} x - \begin{bmatrix} I_N \otimes H_1 C \\ I_N \otimes H_2 C \end{bmatrix} \hat{x}_{\mathcal{N}}.$$

Suppose that $(p^{\mathsf{T}} \otimes I_{n_x})\hat{x}_{\mathcal{N}} = \mathbf{0}$ and $(p^{\mathsf{T}} \otimes I_{n_d})\hat{d}_{\mathcal{N}} = \mathbf{0}$. Then, with $p^{\mathsf{T}} L_{\mathcal{G}} = \mathbf{0}^{\mathsf{T}}$, it follows that

$$\begin{aligned} \begin{bmatrix} (p^{\mathsf{T}} \otimes I_{n_x})\dot{\hat{x}}_{\mathcal{N}} \\ (p^{\mathsf{T}} \otimes I_{n_d})\dot{\hat{d}}_{\mathcal{N}} \end{bmatrix} &= \begin{bmatrix} p^{\mathsf{T}} \otimes A & p^{\mathsf{T}} \otimes P \\ 0 & 0 \end{bmatrix} \begin{bmatrix} \hat{x}_{\mathcal{N}} \\ \hat{d}_{\mathcal{N}} \end{bmatrix} - \begin{bmatrix} p^{\mathsf{T}} \otimes H_1 C \\ p^{\mathsf{T}} \otimes H_2 C \end{bmatrix} \hat{x}_{\mathcal{N}} \\ &= \begin{bmatrix} A & P \\ 0 & 0 \end{bmatrix} \begin{bmatrix} (p^{\mathsf{T}} \otimes I_{n_x})\hat{x}_{\mathcal{N}} \\ (p^{\mathsf{T}} \otimes I_{n_d})\hat{d}_{\mathcal{N}} \end{bmatrix} - \begin{bmatrix} H_1 C \\ H_2 C \end{bmatrix} (p^{\mathsf{T}} \otimes I_{n_x})\hat{x}_{\mathcal{N}} = \mathbf{0}. \end{aligned}$$

This shows that $(p^{\mathsf{T}} \otimes I_{n_x})\hat{x}_{\mathcal{N}} = \mathbf{0}$, $(p^{\mathsf{T}} \otimes I_{n_d})\hat{d}_{\mathcal{N}} = \mathbf{0}$ is invariant with respect to the observer dynamics. The observer errors satisfy this condition at the same time since

$$\begin{aligned} (p^{\mathsf{T}} \otimes I_{n_x}) e^x_{\mathcal{N}} &= (p^{\mathsf{T}} \otimes I_{n_x})(\hat{x}_{\mathcal{N}} + (L_{\mathcal{G}} \otimes I_{n_x})x) = (p^{\mathsf{T}} \otimes I_{n_x})\hat{x}_{\mathcal{N}} \\ (p^{\mathsf{T}} \otimes I_{n_d}) e^d_{\mathcal{N}} &= (p^{\mathsf{T}} \otimes I_{n_x})(\hat{d}_{\mathcal{N}} + (L_{\mathcal{G}} \otimes I_{n_d})d) = (p^{\mathsf{T}} \otimes I_{n_d})\hat{d}_{\mathcal{N}}. \end{aligned}$$

Consequently, if the observers are initialized such that $(p^\top \otimes I_{n_x})\hat{x}_{\mathcal{N}}(0) = \mathbf{0}$, then $(p^\top \otimes I_{n_x})e^x_{\mathcal{N}}(t) = \mathbf{0}$ for all $t \geq 0$ and $s(t)$ is not affected by the observer errors. A feasible initialization is simply $\hat{x}_{\mathcal{N}_k}(0) = \mathbf{0}$, $\hat{d}_{\mathcal{N}_k}(0) = \mathbf{0}$, for all $k \in \mathcal{N}$. □

B.13 Proof of Lemma 4.5

Proof of Lemma 4.5, p. 65. We prove the statement by showing that every solution Π, Γ of (4.26) yields a solution Π^g_k, Γ^g_k, Π^ℓ_k, Γ^ℓ_k, $k \in \mathcal{N}$ of (4.27), (4.28), and vice versa. Consider (4.26) and partition Π, Γ according to

$$\Pi = \begin{bmatrix} \Pi^g_1 & \Pi^\ell_{11} & \cdots & \Pi^\ell_{1N} \\ \vdots & \vdots & & \vdots \\ \Pi^g_N & \Pi^\ell_{N1} & \cdots & \Pi^\ell_{NN} \end{bmatrix}, \qquad \Gamma = \begin{bmatrix} \Gamma^g_1 & \Gamma^\ell_{11} & \cdots & \Gamma^\ell_{1N} \\ \vdots & \vdots & & \vdots \\ \Gamma^g_N & \Gamma^\ell_{N1} & \cdots & \Gamma^\ell_{NN} \end{bmatrix}. \tag{B.10}$$

Using (B.10) and the structure of $\mathcal{A}$, $\mathcal{B}$, $\mathcal{B}_d$, and $\mathcal{S}$, the first equation $\mathcal{A}\Pi + \mathcal{B}\Gamma - \Pi\mathcal{S} + \mathcal{B}_d = 0$ of (4.26) yields the set of matrix equations

$$A_k\Pi^g_k + B_k\Gamma^g_k - \Pi^g_k S^g + B^{d^g}_k = 0, \tag{B.11}$$
$$A_k\Pi^\ell_{kk} + B_k\Gamma^\ell_{kk} - \Pi^\ell_{kk} S^\ell_k + B^{d^\ell}_k = 0 \tag{B.12}$$

and for $j \neq k$,

$$A_k\Pi^\ell_{kj} + B_k\Gamma^\ell_{kj} - \Pi^\ell_{kj} S^\ell_j = 0. \tag{B.13}$$

Using (B.10) and the structure of $\mathcal{C}_e$, $\mathcal{D}_e$, $\mathcal{D}_{ed}$, the second equation $\mathcal{C}_e\Pi + \mathcal{D}_e\Gamma + \mathcal{D}_{ed} = 0$ of (4.26) yields the set of matrix equations

$$C^e_k\Pi^g_k + D^e_k\Gamma^g_k + D^{ed^g}_k = 0, \tag{B.14}$$
$$C^e_k\Pi^\ell_{kk} + D^e_k\Gamma^\ell_{kk} + D^{ed^\ell}_k = 0 \tag{B.15}$$

and for $j \neq k$,

$$C^e_k\Pi^\ell_{kj} + D^e_k\Gamma^\ell_{kj} = 0. \tag{B.16}$$

Only If: Suppose that (B.10) solve (4.26). Then, according to (B.11), (B.14) and (B.12), (B.15), the regulator equations (4.27) and (4.28) are solved by Π^g_k, Γ^g_k and $\Pi^\ell_k = \Pi^\ell_{kk}$, $\Gamma^\ell_k = \Gamma^\ell_{kk}$.

If: Suppose that Π^g_k, Γ^g_k and Π^ℓ_{kk}, Γ^ℓ_{kk} solve (4.27) and (4.28). It is easy to see that (B.13) and (B.16) can be satisfied by the choice $\Pi^\ell_{kj} = 0$ and $\Gamma^\ell_{kj} = 0$ for $j,k \in \mathcal{N}$ with $k \neq j$. Consequently, (B.10) with $\Pi^\ell_{kj} = 0$ and $\Gamma^\ell_{kj} = 0$ for $j,k \in \mathcal{N}$, $k \neq j$, solve the regulator equation (4.26). □

B.14 Proof of Lemma 4.7

Proof of Lemma 4.7, p. 72. As in (4.40), the dynamics of ϵ_k can be computed as

$$\dot{\epsilon}_k = (A - BF)\epsilon_k + BH \sum_{j\in\mathcal{N}_k} \left(\epsilon_j - \epsilon_k\right). \tag{B.17}$$

The novel term in the control law (4.44) couples the transient state components of the agents. Eq. (B.17) is a diffusively coupled network of N identical stable linear systems. We apply the transformation $\tilde{\epsilon} = (T^{-1} \otimes I_{n^x})\epsilon$ with T such that $\Lambda = T^{-1} L_{\mathcal{G}} T$ is upper-triangular, the first column of T is $\mathbf{1}$, and the first row of T^{-1} is p^T, with $p^\mathsf{T} L_{\mathcal{G}} = \mathbf{0}^\mathsf{T}$, $p^\mathsf{T}\mathbf{1} = 1$, cf., Theorem 3.1. This leads to $\dot{\tilde{\epsilon}} = (I_N \otimes (A - BF) - \Lambda \otimes BH)\tilde{\epsilon}$. Let $\tilde{\epsilon}$ be partitioned into $\tilde{\epsilon}'$ and $\tilde{\epsilon}''$. Then, $\dot{\tilde{\epsilon}}' = (A - BF)\tilde{\epsilon}'$ and

$$\dot{\tilde{\epsilon}}'' = \begin{bmatrix} A - BF - \lambda_2 BH & \cdots & \star \\ & \ddots & \vdots \\ 0 & & A - BF - \lambda_N BH \end{bmatrix} \tilde{\epsilon}''.$$

By (4.43), the maximal real part of the eigenvalues of the latter matrix is smaller than $-\gamma$. Such a gain H exists by Assumptions 4.8 and Theorem 3.2. Hence, there exists a constant $\tilde{c} > 0$ such that $\|\tilde{\epsilon}''(t)\| \leq \tilde{c}\|\tilde{\epsilon}''(0)\|e^{-\gamma t}$, for all $t \geq 0$, cf., Bernstein (2009). We define the projection matrix $P = I_N - \frac{1}{N}\mathbf{1}\mathbf{1}^\mathsf{T}$ and partition $T = [\mathbf{1} \;\; T'']$. Then, the stack vector of synchronization errors (4.42) is given by $\epsilon^s = (P \otimes I_{n^x})\epsilon = (PT \otimes I_{n^x})\tilde{\epsilon} = (PT'' \otimes I_{n^x})\tilde{\epsilon}''$ since $P\mathbf{1} = \mathbf{0}$. Hence, it follows that there exists a constant $c > 0$ such that $\|\epsilon^s(t)\| \leq c\|(P \otimes I_{n^x})\epsilon(0)\|e^{-\gamma t}$. Consequently, the synchronization error is exponentially stable with a decay rate of at least γ. The matrix $A - BF$ is Hurwitz by the choice of F, which guarantees that $\tilde{\epsilon}'(t) \to \mathbf{0}$ as $t \to \infty$. Therefore, it holds that $\epsilon(t) = (T \otimes I_{n^x})\tilde{\epsilon}(t) \to \mathbf{0}$ as $t \to \infty$. Since $e_k = (C^e - D^e F)\epsilon_k$, Problem 4.2 is solved. □

B.15 Proof of Theorem 4.8

Proof of Theorem 4.8, p. 72. The novel term in (4.45) has no influence on the dynamics of the observer errors ϵ_k^x, $\epsilon_k^{d^\ell}$, $\epsilon_k^{d^g}$. They obey the same dynamics as in the proof of Theorem 4.6. Moreover, by assumption, there exist constants $c_k > 0$ such that

$$\left\| \begin{bmatrix} \epsilon_k^x(t) \\ \epsilon_k^{d^g}(t) \\ \epsilon_k^{d^\ell}(t) \end{bmatrix} \right\| \leq c_k \left\| \begin{bmatrix} \epsilon_k^x(0) \\ \epsilon_k^{d^g}(0) \\ \epsilon_k^{d^\ell}(0) \end{bmatrix} \right\| e^{-\eta t} \tag{B.18}$$

for all $t \geq 0$. Note that

$$\begin{aligned} \hat{\epsilon}_k &= \hat{x}_k - \Pi_k^g \hat{d}_k^g - \Pi_k^\ell \hat{d}_k^\ell \\ &= x_k - \epsilon_k^x - \Pi_k^g (d^g - \epsilon_k^{d^g}) - \Pi_k^\ell (d_k^\ell - \epsilon_k^{d^\ell}) \\ &= \epsilon_k - \epsilon_k^x + \Pi_k^g \epsilon_k^{d^g} + \Pi_k^\ell \epsilon_k^{d^\ell}. \end{aligned}$$

Analogously to the proof of Theorem 4.6, the dynamics of ϵ_k can be computed as $\dot{\epsilon}_k = (A - BF)\epsilon_k + BH \sum_{j \in \mathcal{N}_k} (\epsilon_j - \epsilon_k) + \delta_k(t)$, where $\delta_k(t)$ captures the influence of the observer errors $\delta_k(t) = BF\epsilon_k^x - BG_k^g \epsilon_k^{d^g} - BG_k^\ell \epsilon_k^{d^\ell} - BH \sum_{j \in \mathcal{N}_k} (\epsilon_j^x - \Pi_j^g \epsilon_j^{d^g} - \Pi_j^\ell \epsilon_j^{d^\ell} - \epsilon_k^x + \Pi_k^g \epsilon_k^{d^g} + \Pi_k^\ell \epsilon_k^{d^\ell})$. We define the stack vector $\delta = [\delta_1^\mathsf{T} \; \cdots \; \delta_N^\mathsf{T}]^\mathsf{T}$. From (B.18) it follows that there exists a constant c_δ such that for all $t \geq 0$,

$$\|\delta(t)\| \leq c_\delta \|\delta(0)\| e^{-\eta t}. \tag{B.19}$$

Analogously to the proof of Lemma 4.7, we obtain $\dot{\tilde{\epsilon}} = (I_N \otimes (A - BF) - \Lambda \otimes BH)\tilde{\epsilon} + (T^{-1} \otimes I_{n^x})\delta$. With (B.19) and since $\eta > \gamma$, it follows that $\tilde{\epsilon}''(t) \to \mathbf{0}$ exponentially as $t \to \infty$ with decay rate γ. Moreover, we also have $\tilde{\epsilon}'(t) \to \mathbf{0}$ as $t \to \infty$. Since $\epsilon_k(t) \to \mathbf{0}$ as $t \to \infty$ for all $k \in \mathcal{N}$, the coupling term in (4.45) vanishes asymptotically. Consequently, analogously to the proof of Theorem 4.6, it holds that $e_k(t) \to \mathbf{0}$ as $t \to \infty$ and Problem 4.2 is solved. □

B.16 Proof of Lemma 4.9

Proof of Lemma 4.9, p. 76. A state-space description of (4.50) is given by

$$\dot{z}_k = \tilde{A} z_k + \tilde{B} v_k, \tag{B.20a}$$

$$\zeta_k = v_k, \tag{B.20b}$$

$$\epsilon_k = z_k + \omega_k. \tag{B.20c}$$

With couplings $v_k = H \sum_{j \in \mathcal{N}_k} (\epsilon_j - \epsilon_k)$ between the systems and the representation (B.20) for each agent, we obtain

$$\begin{aligned} T^{\omega\zeta}: \quad & \dot{z} = (I_N \otimes \tilde{A} - L_{\mathcal{G}} \otimes \tilde{B}H)z - (L_{\mathcal{G}} \otimes \tilde{B}H)\omega \\ & \zeta = -(L_{\mathcal{G}} \otimes H)z - (L_{\mathcal{G}} \otimes H)\omega \\ \Delta: \quad & \omega = \operatorname{diag}(\Delta_k)\zeta. \end{aligned}$$

Since $\mathcal{G}$ is undirected and connected, there exists an orthogonal matrix U such that $U^\top L_{\mathcal{G}} U = \Lambda = \operatorname{diag}(\lambda_k)$ and $\lambda_1 = 0$ is the top left diagonal element of Λ. With the coordinate transformation $\tilde{z} = (U^\top \otimes I_N)z$, $\tilde{\omega} = (U^\top \otimes I_N)\omega$, $\tilde{\zeta} = (U^\top \otimes I_N)\zeta$, similarly to Li et al. (2011); Trentelman et al. (2013), we obtain

$$\begin{aligned} T^{\tilde{\omega}\tilde{\zeta}}: \quad & \dot{\tilde{z}} = (I_N \otimes \tilde{A} - \Lambda \otimes \tilde{B}H)\tilde{z} - (\Lambda \otimes \tilde{B}H)\tilde{\omega} \\ & \tilde{\zeta} = -(\Lambda \otimes H)\tilde{z} - (\Lambda \otimes H)\tilde{\omega} \end{aligned}$$

and $\tilde{\omega} = (U^\top \otimes I) \operatorname{diag}(\Delta_k)(U \otimes I)\tilde{\zeta}$. The system $T^{\tilde{\omega}\tilde{\zeta}}$ consists of the decoupled systems

$$\begin{aligned} T_k^{\tilde{\omega}\tilde{\zeta}}: \quad & \dot{\tilde{z}}_k = (\tilde{A} - \lambda_k \tilde{B}H)\tilde{z}_k - \lambda_k \tilde{B}H\tilde{\omega}_k \\ & \tilde{\zeta}_k = -\lambda_k H\tilde{z}_k - \lambda_k H\tilde{\omega}_k. \end{aligned}$$

According to the Bounded Real Lemma (Zhou and Doyle (1998)), the system $T_k^{\tilde{\omega}\tilde{\zeta}}$ is asymptotically stable and has $\mathcal{H}_\infty$ norm less than or equal $\eta > 0$, if and only if there exists $X_k > 0$ such that (4.51) holds. Furthermore, if $\|T_k^{\tilde{\omega}\tilde{\zeta}}\|_\infty < \eta$ for all $k \in \mathcal{N}$, then $\|T^{\tilde{\omega}\tilde{\zeta}}\|_\infty < \eta$ due to the block diagonal structure. Moreover, since $\|U\|_\infty = \|U^\top\|_\infty = 1$ and the maximal singular value of a block diagonal matrix equals the maximum of the maximal singular values of each block, it follows that $\|(U^\top \otimes I) \operatorname{diag}(\Delta_k)(U \otimes I)\|_\infty = \max_{k \in \mathcal{N}} \|\Delta_k\|_\infty < 1/\eta$. Asymptotic stability of the perturbed closed loop $T^{\omega\zeta}$, Δ is a direct consequence of the Small Gain Theorem (Trentelman et al. (2001)). □

B.17 Proof of Theorem 4.10

Proof of Theorem 4.10, p. 78. In view of Theorem 4.6, it only remains to show that the distributed estimator (4.54) achieves $\hat{x}_k(t) - x_k(t) \to \mathbf{0}$ and $\hat{d}_k^\ell(t) - d_k^\ell(t) \to \mathbf{0}$ as $t \to \infty$ for all $k \in \mathcal{N}$. We consider the observer errors $\epsilon_k^x = x_k - \hat{x}_k$, $\epsilon_k^{d^\ell} = d_k^\ell - \hat{d}_k^\ell$, and $\epsilon_k^{d^g} = d^g - \hat{d}_k^g$ and obtain

$$\begin{bmatrix} \dot{\epsilon}_k^x \\ \dot{\epsilon}_k^{d^\ell} \end{bmatrix} = \left(\begin{bmatrix} A_k & B_k^{d^\ell} \\ 0 & S_k^\ell \end{bmatrix} - L_k \begin{bmatrix} C_k & D_k^{d^\ell} \end{bmatrix} \right) \begin{bmatrix} \epsilon_k^x \\ \epsilon_k^{d^\ell} \end{bmatrix} - L_k \sum_{j \in \mathcal{N}_k} \left(C_{kj}\epsilon_j^x + D_{kj}^{d^\ell}\epsilon_j^{d^\ell} \right) + \left(\begin{bmatrix} B_k^{d^g} \\ 0 \end{bmatrix} - L_k D_k^{d^g} \right) \epsilon_k^{d^g}.$$

By design of (4.34), it holds that $\epsilon_k^{d^g}(t) \to \mathbf{0}$ as $t \to \infty$ for all $k \in \mathcal{N}$, hence $\epsilon_k^{d^g}$ can be disregarded for the stability analysis. With the stack vector $\epsilon = [\epsilon_1^{x\mathsf{T}}\ \epsilon_1^{d^\ell\mathsf{T}}\ \cdots\ \epsilon_N^{x\ \mathsf{T}}\ \epsilon_N^{d^\ell\mathsf{T}}]^\mathsf{T}$, the remaining error dynamics are

$$\dot{\epsilon} = (\mathcal{A} - L\mathcal{C})\epsilon,$$

where $L = \text{diag}(L_1, ..., L_N)$. Noting that $L = P^{-1}Y$, the matrix $\mathcal{A} - L\mathcal{C}$ is guaranteed to be Hurwitz by (4.53). Consequently, $\epsilon_k^x(t) \to \mathbf{0}$ and $\epsilon_k^{d^\ell}(t) \to \mathbf{0}$ as $t \to \infty$ for all initial conditions. □

B.18 Proof of Theorem 4.11

Proof of Theorem 4.11, p. 82. It can be verified through substitution that Π, Γ solve (4.26) for the present problem. Intuitively, the persistent state component satisfies $x = \Pi d$. With $x_1 = \Pi_1 d_1$ and $C_e\Pi_1 + D_{ed} = 0$, it follows that $e_1 = \mathbf{0}$. Next, $e_2 = C_e(x_2 - x_1) + D_{ed}d_2 = \mathbf{0}$ is satisfied for $(x_2 - x_1) = \Pi_1 d_2$, leading to $x_2 = \Pi_1(d_1 + d_2)$. This shows that the second regulator equation is satisfied. The first regulator equation reduces to $(I_N \otimes A)\Pi - \Pi(I_N \otimes S)$ since $B\Gamma_1 + B_d = 0$. It is satisfied since $A\Pi_1 - \Pi_1 S = 0$.

It holds that $\sigma(\tilde{L}_\mathcal{G}) \subset \mathbb{C}^+$, i.e., all its eigenvalues are positive (Qu, 2009, Cor. 4.33). By the same arguments as in Theorem 3.1, the feedback $u = -(\tilde{L}_\mathcal{G} \otimes F)x$ stabilizes $\dot{x} = (I_N \otimes A)x + (I_N \otimes B)u$ and such F can be found by means of Theorem 3.2. Due to the tridiagonal structure of $\tilde{L}_\mathcal{G}$ and the structure of Π, the feed-forward gain matrix $\mathsf{G} = \Gamma + (\tilde{L}_\mathcal{G} \otimes F)\Pi$ is sparse and respects the information structure. In particular,

$$(\tilde{L}_\mathcal{G} \otimes F)\Pi = \left(\tilde{L}_\mathcal{G} \begin{bmatrix} 1 & & 0 \\ \vdots & \ddots & \\ 1 & \cdots & 1 \end{bmatrix}\right) \otimes (F\Pi_1) = \begin{bmatrix} \alpha & -1 & 0 & \cdots & & \\ 0 & 1 & -1 & 0 & \cdots & \\ & & \ddots & \ddots & & \end{bmatrix} \otimes (F\Pi_1)$$

Agent k needs to communicate only with agents $k-1$, $k+1$ in order to implement its output regulation control law u_k. □

B.19 Proof of Theorem 5.1

Proof of Theorem 5.1, p. 97. The control law (5.12) is a gradient control law for the potential U_{p_θ} in (5.13). Its gradient can be computed as

$$\frac{\partial U_{p_\theta}}{\partial \theta_k} = \frac{\partial}{\partial \theta_k}\frac{N}{2}\langle p_\theta, p_\theta\rangle = \frac{N}{2}\left(\left\langle \frac{\partial p_\theta}{\partial \theta_k}, p_\theta \right\rangle + \left\langle p_\theta, \frac{\partial p_\theta}{\partial \theta_k}\right\rangle\right) = N\left\langle \frac{\partial p_\theta}{\partial \theta_k}, p_\theta\right\rangle = \left\langle \mathbf{i}e^{\mathrm{i}\theta_k}, p_\theta \right\rangle, \quad \text{(B.22)}$$

which leads to the control law

$$u_k = \gamma \frac{\partial U_{p_\theta}}{\partial \theta_k} = \frac{\gamma}{N}\sum_{j=1}^{N} \sin(\theta_j - \theta_k).$$

The dynamics of the phase angles (5.9b) are decoupled from (5.9a) and hence identical to the homogeneous case in Sepulchre et al. (2007). The stability properties of the gradient dynamics $\dot{\theta} = \gamma\nabla U_{p_\theta}$ are characterized in (Sepulchre et al., 2007, Thm. 1), which concludes the proof for control law (5.12). □

B.20 Proof of Theorem 5.2

Proof of Theorem 5.2, p. 98. The control law (5.14) is a gradient control law for the potential $U_{\dot{R}}$ in (5.15). The gradient of $U_{\dot{R}}$ can be computed as

$$\frac{\partial U_{\dot{R}}}{\partial \theta_k} = \frac{\partial}{\partial \theta_k}\frac{N}{2}\left\langle \dot{R}, \dot{R}\right\rangle = N\left\langle \frac{\partial \dot{R}}{\partial \theta_k}, \dot{R}\right\rangle = \left\langle \mathbf{i}v_k e^{\mathbf{i}\theta_k}, \dot{R}\right\rangle. \tag{B.23}$$

Furthermore, evaluating the scalar product yields

$$\frac{\partial U_{\dot{R}}}{\partial \theta_k} = \mathbf{Re}\left(-\mathbf{i}v_k e^{-\mathbf{i}\theta_k}\left(\frac{1}{N}\sum_{j=1}^{N} v_j e^{\mathbf{i}\theta_j}\right)\right) = \frac{1}{N}\sum_{j=1}^{N} v_j v_k\, \mathbf{Re}\left(-\mathbf{i}e^{\mathbf{i}(\theta_j-\theta_k)}\right) = \frac{1}{N}\sum_{j=1}^{N} v_j v_k \sin\left(\theta_j - \theta_k\right). \tag{B.24}$$

Hence, the dynamics of the phase angles (5.9b) with control law (5.14) are

$$\dot{\boldsymbol{\theta}} = \gamma \nabla U_{\dot{R}}. \tag{B.25}$$

This is a gradient flow for the potential function $U_{\dot{R}}$. Since $U_{\dot{R}} \geq 0$, it can be used as Lyapunov function. The derivative along trajectories can be computed as

$$\dot{U}_{\dot{R}} = (\nabla U_{\dot{R}})^{\mathsf{T}}\dot{\boldsymbol{\theta}} = \gamma(\nabla U_{\dot{R}})^{\mathsf{T}}\nabla U_{\dot{R}} = \gamma\sum_{k=1}^{N}\left(\frac{\partial U_{\dot{R}}}{\partial \theta_k}\right)^2.$$

Obviously, $\dot{U}_{\dot{R}} \geq 0$ for $\gamma > 0$ and $\dot{U}_{\dot{R}} \leq 0$ for $\gamma < 0$. Furthermore, $\dot{U}_{\dot{R}} = 0$ if and only if $\nabla U_{\dot{R}} = \mathbf{0}$, i.e., on the critical set of $U_{\dot{R}}$. The critical set of $U_{\dot{R}}$ is the set of all $\boldsymbol{\theta}$ on the N-dimensional torus $\mathbb{T}^N$, for which $\nabla U_{\dot{R}} = \mathbf{0}$. Since $\boldsymbol{\theta} \in \mathbb{T}^N$ and $\mathbb{T}^N \subset \mathbb{R}^{N+1}$ is compact, it follows from LaSalle's Invariance Principle (Theorem A.5) that all solutions of (B.25) converge to the critical set of $U_{\dot{R}}$.

Analysis of the critical set: The stability properties of the critical points can be assessed by Lyapunov's indirect method, i.e., by linearization. Note that the Jacobian of the right-hand side of (B.25) is $\gamma H_{U_{\dot{R}}}$, where $H_{U_{\dot{R}}}$ is the Hessian matrix of $U_{\dot{R}}$. Let $\dot{R} = |\dot{R}|e^{\mathbf{i}\psi}$ with $\psi = \arg(\dot{R})$. Then,

$$\frac{\partial U_{\dot{R}}}{\partial \theta_k} = \left\langle \mathbf{i}v_k e^{\mathbf{i}\theta_k}, |\dot{R}|e^{\mathbf{i}\psi}\right\rangle = v_k|\dot{R}|\sin\left(\psi - \theta_k\right).$$

The critical points with $\dot{R} = 0$ are minima of $U_{\dot{R}}$ and correspond to a constant average linear momentum. Hence these are desirable equilibria. For $\gamma < 0$, the set on which $\dot{R} = 0$ is locally asymptotically stable.

Next, we focus on those critical points for which $|\dot{R}| > 0$, i.e., $\sin(\psi - \theta_k) = 0$ for $k = 1, ..., N$. This implies that $\theta_k \in \{\psi \mod 2\pi, \psi + \pi \mod 2\pi\}$ for all k. Let M be the number of agents with phase $\psi + \pi \mod 2\pi$. Then $0 \leq M \leq N - 1$, and $M = 0$ corresponds to synchronized phase angles and maximal $|\dot{R}|$. These are also desirable equilibria. For $\gamma > 0$, the synchronized state is locally asymptotically stable.

Without loss of generality, we renumber the agents such that $\theta_k = \psi + \pi \mod 2\pi$ for $k = 1, ..., M$ and $\theta_k = \psi \mod 2\pi$ for $k = M + 1, ..., N$. Assume that $M = N - 1$. Then, since $|\dot{R}| > 0$, it follows that $v_N > \sum_{j=1}^{N-1} v_j$. This is a contradiction to Assumption 5.1. Therefore, it remains to analyze the critical points with $1 \leq M \leq N - 2$. The second derivatives of $U_{\dot{R}}$ are given by

$$\frac{\partial^2 U_{\dot{R}}}{\partial \theta_k^2} = \frac{\partial}{\partial \theta_k}\left\langle \mathbf{i}v_k e^{\mathbf{i}\theta_k}, \dot{R}\right\rangle = \left\langle \mathbf{i}v_k e^{\mathbf{i}\theta_k}, \frac{\partial \dot{R}}{\partial \theta_k}\right\rangle - \left\langle v_k e^{\mathbf{i}\theta_k}, \dot{R}\right\rangle = \frac{1}{N}v_k^2 - v_k|\dot{R}|\cos(\psi - \theta_k).$$

$$\frac{\partial^2 U_{\dot{R}}}{\partial \theta_j \partial \theta_k} = \frac{\partial}{\partial \theta_j}\left\langle \mathbf{i}v_k e^{\mathbf{i}\theta_k}, \dot{R}\right\rangle = \left\langle \mathbf{i}v_k e^{\mathbf{i}\theta_k}, \frac{\partial \dot{R}}{\partial \theta_j}\right\rangle = \frac{1}{N}v_j v_k \cos(\theta_j - \theta_k).$$

Hence, it holds that

$$\frac{\partial^2 U_{\dot{R}}}{\partial \theta_k^2} = \begin{cases} (1/N)v_k^2 + |\dot{R}|v_k, & k \in \{1,...,M\} \\ (1/N)v_k^2 - |\dot{R}|v_k, & k \in \{M+1,...,N\}. \end{cases}$$

Since $(1/N)v_k^2 + |\dot{R}|v_k > 0$ the Hessian matrix $H_{U_{\dot{R}}}$ has at least one positive eigenvalue. In order to show that all critical points $1 \leq M \leq N-2$ are saddle points, we verify that the Hessian matrix $H_{U_{\dot{R}}}$ is indefinite by showing that it also has at least one negative eigenvalue. It holds that

$$\frac{\partial^2 U_{\dot{R}}}{\partial \theta_j \partial \theta_k} = \begin{cases} \frac{1}{N}v_j v_k, & j,k \in \{1,...,M\} \text{ or } j,k \in \{M+1,...,N\} \\ -\frac{1}{N}v_j v_k, & \text{else.} \end{cases}$$

Define $\tilde{\mathbf{v}} = [v_1,...,v_M,-v_{M+1},...,-v_N]^\mathsf{T}$. Then the Hessian matrix can be compactly written as

$$H_{U_{\dot{R}}} = \frac{1}{N}\tilde{\mathbf{v}}\tilde{\mathbf{v}}^\mathsf{T} + |\dot{R}|\text{diag}(\tilde{\mathbf{v}}).$$

Define the vector $\mathbf{q} \in \mathbb{R}^N$ with two non-zero elements $q_N = -v_{N-1}$ and $q_{N-1} = v_N$. By construction, $\tilde{\mathbf{v}}^\mathsf{T}\mathbf{q} = 0$ and it follows that

$$\mathbf{q}^\mathsf{T} H_{U_{\dot{R}}}\mathbf{q} = |\dot{R}|\mathbf{q}^\mathsf{T}\text{diag}(\tilde{\mathbf{v}})\mathbf{q} = |\dot{R}|\left(-v_N^2 v_{N-1} - v_N v_{N-1}^2\right) < 0.$$

Consequently, all critical points for which the phase angles are not synchronized and $\dot{R} \neq 0$ the Hessian $H_{U_{\dot{R}}}$ is indefinite. These critical points are hence saddle points and unstable, both for $\gamma < 0$ and $\gamma > 0$. □

B.21 Proof of Theorem 5.4

Proof of Theorem 5.4, p. 100. Suppose that agent k is driven by a constant control input $u_k = \omega_0$ with some angular frequency ω_0. Then, it moves on a circle with radius $\rho_k = v_k/|\omega_0|$. The center of this circle is

$$c_k = r_k + \mathbf{i}\frac{v_k}{\omega_0}e^{\mathrm{i}\theta_k}.$$

The agents are supposed to circle around a common point in the plane, with identical angular frequency ω_0. The circle center points c_k of all agents coincide if and only if

$$P\mathbf{c} = \mathbf{0},$$

where $\mathbf{c} = [c_1 \ \cdots \ c_N]^\mathsf{T}$ and $P = I_N - \frac{1}{N}\mathbf{1}\mathbf{1}^\mathsf{T}$ is a projection matrix. Therefore we define the Lyapunov function

$$V(\mathbf{r},\boldsymbol{\theta}) = \frac{1}{2}\|P\mathbf{c}\|^2$$

and construct a controller that minimizes V. The Lyapunov function V is positive semi-definite in $\mathbf{c}$ and positive definite in $P\mathbf{c}$. In particular, $V \geq 0$ for all $\mathbf{c} \in \mathbb{C}^N$ and $V = 0$ if and only if $P\mathbf{c} = 0$. Furthermore, $V \to \infty$ for $\|P\mathbf{c}\| \to \infty$. Hence, V is a suitable Lyapunov function in order to assess agreement on a common circle center point. Note that

$$\dot{c}_k = \frac{\mathrm{d}}{\mathrm{d}t}\left(r_k + \mathbf{i}\frac{v_k}{\omega_0}e^{\mathrm{i}\theta_k}\right) = v_k e^{\mathrm{i}\theta_k} - \frac{v_k}{\omega_0}e^{\mathrm{i}\theta_k}u_k = \frac{v_k}{\omega_0}e^{\mathrm{i}\theta_k}(\omega_0 - u_k).$$

The matrix P defines a projection, i.e., $PP = P$, and $V = \frac{1}{2}\langle P\mathbf{c}, P\mathbf{c}\rangle = \frac{1}{2}\langle P\mathbf{c}, \mathbf{c}\rangle$. Let P_k denote the k-th row of matrix P. Then, the Lie-derivative of V can be computed as

$$\dot{V} = \langle P\mathbf{c}, \dot{\mathbf{c}}\rangle = \sum_{k=1}^{N} \langle P_k\mathbf{c}, \dot{c}_k\rangle = \sum_{k=1}^{N} \left\langle P_k\mathbf{c}, \frac{v_k}{\omega_0} e^{\mathbf{i}\theta_k}(\omega_0 - u_k)\right\rangle.$$

We choose the control law

$$u_k = \omega_0 + \gamma\omega_0 \left\langle P_k\mathbf{c}, v_k e^{\mathbf{i}\theta_k}\right\rangle, \tag{B.26}$$

where $\gamma > 0$ is a positive constant. This yields

$$\dot{V} = -\gamma \sum_{k=1}^{N} \left\langle P_k\mathbf{c}, v_k e^{\mathbf{i}\theta_k}\right\rangle^2 \leq 0,$$

i.e., the Lie-derivative $\dot{V}$ is rendered negative semi-definite by the control law (B.26). Let $\mathcal{J}$ be the set of all $(\mathbf{r},\boldsymbol{\theta})$, for which $\dot{V} = 0$. By LaSalle's Invariance Principle (Theorem A.5), all solutions converge to the largest invariant set contained in $\mathcal{J}$. Furthermore, $\dot{V} = 0$ is equivalent to $\langle P_k\mathbf{c}, v_k e^{\mathbf{i}\theta_k}\rangle = 0$ for $k = 1,\dots,N$. Hence, on $\mathcal{J}$, $u_k = \omega_0$ and $\dot{c}_k = 0$ for $k = 1,\dots,N$, i.e., $P_k\mathbf{c}$ is constant on $\mathcal{J}$. Since $\langle P_k\mathbf{c}, v_k e^{\mathbf{i}\theta_k}\rangle = 0$ can hold for all $k = 1,\dots,N$ and $t \geq 0$ only if $P\mathbf{c} = \mathbf{0}$, the solutions converge to a configuration where $P\mathbf{c} = \mathbf{0}$, as desired. Note that

$$P_k\mathbf{c} = c_k - \frac{1}{N}\mathbf{1}^{\mathsf{T}}\mathbf{c} = r_k + \mathbf{i}\frac{v_k}{\omega_0}e^{\mathbf{i}\theta_k} - \frac{1}{N}\sum_{j=1}^{N}\left(r_j + \mathbf{i}\frac{v_j}{\omega_0}e^{\mathbf{i}\theta_j}\right) = r_k + \mathbf{i}\frac{v_k}{\omega_0}e^{\mathbf{i}\theta_k} - \left(R + \mathbf{i}\frac{1}{\omega_0}\dot{R}\right),$$

with $\dot{R}$ as defined in (5.11) and $R = 1/N\sum_{j=1}^{N} r_j$. Hence

$$\begin{aligned}\left\langle P_k\mathbf{c}, v_k e^{\mathbf{i}\theta_k}\right\rangle &= \mathbf{Re}\left(v_k e^{-\mathbf{i}\theta_k}\left(r_k + \mathbf{i}\frac{v_k}{\omega_0}e^{\mathbf{i}\theta_k} - \left(R + \mathbf{i}\frac{1}{\omega_0}\dot{R}\right)\right)\right)\\ &= \mathbf{Re}\left(\left(v_k e^{-\mathbf{i}\theta_k}(r_k - R)\right) - v_k e^{-\mathbf{i}\theta_k}\mathbf{i}\frac{1}{\omega_0}\dot{R}\right) = \langle \tilde{r}_k, \dot{r}_k\rangle + \frac{1}{\omega_0}\left\langle \dot{R}, \mathbf{i}v_k e^{\mathbf{i}\theta_k}\right\rangle,\end{aligned}$$

where $\tilde{r}_k = r_k - R$. With potential (5.15) and gradient (B.23), the control law (B.26) is given by

$$u_k = \omega_0 + \gamma\omega_0\left\langle P_k\mathbf{c}, v_k e^{\mathbf{i}\theta_k}\right\rangle = \omega_0 + \gamma\omega_0\langle \tilde{r}_k, \dot{r}_k\rangle + \gamma\left\langle \dot{R}, \mathbf{i}v_k e^{\mathbf{i}\theta_k}\right\rangle = \omega_0(1 + \gamma\langle \tilde{r}_k, \dot{r}_k\rangle) + \gamma\frac{\partial U_{\dot{R}}}{\partial\theta_k},$$

as stated in (5.18). In order to express (5.18) in real variables, we compute

$$\begin{aligned}\langle \tilde{r}_k, \dot{r}_k\rangle &= \mathbf{Re}\left(\left(\bar{r}_k - \frac{1}{N}\sum_{j=1}^{N}\bar{r}_j\right)v_k e^{\mathbf{i}\theta_k}\right) = \mathbf{Re}\left(\bar{r}_k v_k e^{\mathbf{i}\theta_k}\right) - \frac{1}{N}\sum_{j=1}^{N}\mathbf{Re}\left(\bar{r}_j v_k e^{\mathbf{i}\theta_k}\right)\\ &= |r_k| v_k \cos(\phi_k - \theta_k) - \frac{1}{N}\sum_{j=1}^{N}|r_j| v_k \cos\left(\phi_j - \theta_k\right).\end{aligned} \tag{B.27}$$

□

B.22 Proof of Theorem 5.5

Proof of Theorem 5.5, p. 101. We choose the Lyapunov function $W(\mathbf{r},\boldsymbol{\theta}) = \gamma\omega_0^2 V(\mathbf{r},\boldsymbol{\theta}) + U(\boldsymbol{\theta})$. With $\langle \nabla U, \mathbf{1}\rangle = 0$, we obtain

$$\dot{W} = \sum_{k=1}^{N}\left(\gamma\omega_0^2\left\langle P_k\mathbf{c}, \frac{v_k}{\omega_0}e^{\mathbf{i}\theta_k}\right\rangle(\omega_0 - u_k) + \frac{\partial U}{\partial\theta_k}u_k\right) = \sum_{k=1}^{N}\left(\gamma\omega_0\left\langle P_k\mathbf{c}, v_k e^{\mathbf{i}\theta_k}\right\rangle - \frac{\partial U}{\partial\theta_k}\right)(\omega_0 - u_k).$$

The control law (5.20) can equivalently be written as

$$u_k = \omega_0 + \gamma\omega_0\left\langle P_k\mathbf{c}, v_k e^{\mathbf{i}\theta_k}\right\rangle - \frac{\partial U}{\partial\theta_k},$$

which leads to $\dot{W} = -\sum_{k=1}^{N}(\omega_0 - u_k)^2 \leq 0$. The rest of the proof follows the lines of (Sepulchre et al., 2007, Thm. 3). □

B.23 Proof of Theorem 5.7

Proof of Theorem 5.7, p. 103. Define the potential function

$$U(\boldsymbol{\theta}) = \gamma\left(U_{\dot{R}}(\boldsymbol{\theta}) - U_{p_\theta}(\boldsymbol{\theta})\right).$$

With this potential function, control law (5.20) in Theorem 5.5 results in (5.24). The potentials U_{p_θ} and $U_{\dot{R}}$ satisfy $\langle \nabla U_{p_\theta}, \mathbf{1}\rangle = 0$ and $\langle \nabla U_{\dot{R}}, \mathbf{1}\rangle = 0$, respectively. Hence, the potential U satisfies $\langle \nabla U, \mathbf{1}\rangle = 0$. From Theorem 5.5, it follows that all solutions of (5.9), (5.24) converge to a circular formation where all agents k move on circles of radius $\rho_k = v_k/|\omega_0|$ with common center point and direction given by the sign of ω_0, which concludes the proof. □

B.24 Proof of Theorem 5.8

Proof of Theorem 5.8, p. 104. Similarly to Theorem 5.4, we consider the circle center point of agent k assuming a constant control input $u_k = \omega_k = v_k/\rho_0$, i.e.,

$$c_k = r_k + \mathbf{i}\rho_0 e^{\mathbf{i}\theta_k}.$$

The agents are supposed to circle around a common point in the plane, which is achieved if $P\mathbf{c} = \mathbf{0}$. We use again the Lyapunov function

$$V(\mathbf{r},\boldsymbol{\theta}) = \frac{1}{2}||P\mathbf{c}||^2$$

and construct a controller that minimizes V. Here,

$$\dot{c}_k = v_k e^{\mathbf{i}\theta_k}\left(1 - \frac{u_k}{\omega_k}\right).$$

The Lie-derivative of V can be computed as

$$\dot{V} = \langle P\mathbf{c}, \dot{\mathbf{c}}\rangle = \sum_{k=1}^{N}\left\langle P_k\mathbf{c}, v_k e^{\mathbf{i}\theta_k}\left(1 - \frac{u_k}{\omega_k}\right)\right\rangle.$$

We choose the control law

$$u_k = \omega_k + \omega_k \gamma \left\langle P_k \mathbf{c}, v_k e^{\mathbf{i}\theta_k} \right\rangle, \tag{B.28}$$

where $\gamma > 0$ is a positive constant. This yields

$$\dot{V} = -\gamma \sum_{k=1}^{N} \left\langle P_k \mathbf{c}, v_k e^{\mathbf{i}\theta_k} \right\rangle^2 \leq 0,$$

i.e., the Lie-derivative $\dot{V}$ is rendered negative semi-definite by the control law (B.26). Let $\mathcal{I}$ be the set of all $(\mathbf{r}, \boldsymbol{\theta})$, for which $\dot{V} = 0$. By LaSalle's Invariance Principle (Theorem A.5), all solutions converge to the largest invariant set contained in $\mathcal{I}$. Furthermore, $\dot{V} = 0$ is equivalent to $\langle P_k \mathbf{c}, v_k e^{\mathbf{i}\theta_k} \rangle = 0$ for $k = 1, ..., N$. Hence, on $\mathcal{I}$, $u_k = \omega_k$ and $\dot{c}_k = 0$ for $k = 1, ..., N$, i.e., $P_k \mathbf{c}$ is constant on $\mathcal{I}$. Since $\langle P_k \mathbf{c}, v_k e^{\mathbf{i}\theta_k} \rangle = 0$ can hold for all $k = 1, ..., N$ and $t \geq 0$ only if $P\mathbf{c} = \mathbf{0}$, the solutions converge to a configuration where $P\mathbf{c} = \mathbf{0}$, as desired. Note that

$$P_k \mathbf{c} = r_k + \mathbf{i}\rho_0 e^{\mathbf{i}\theta_k} - R - \mathbf{i}\rho_0 p_\theta,$$

with p_θ as defined in (5.10) and $R = 1/N \sum_{j=1}^{N} r_j$. Hence

$$\left\langle P_k \mathbf{c}, v_k e^{\mathbf{i}\theta_k} \right\rangle = \langle \tilde{r}_k, \dot{r}_k \rangle + \rho_0 \left\langle p_\theta, \mathbf{i} v_k e^{\mathbf{i}\theta_k} \right\rangle.$$

With potential (5.13) and gradient (B.22), the control law (B.28) can be compactly written as

$$\begin{aligned} u_k &= \omega_k + \omega_k \gamma \left\langle P_k \mathbf{c}, v_k e^{\mathbf{i}\theta_k} \right\rangle = \omega_k + \omega_k \gamma \langle \tilde{r}_k, \dot{r}_k \rangle + \omega_k \gamma \rho_0 \left\langle p_\theta, \mathbf{i} v_k e^{\mathbf{i}\theta_k} \right\rangle \\ &= \omega_k (1 + \gamma \langle \tilde{r}_k, \dot{r}_k \rangle) + \gamma v_k^2 \frac{\partial U_{p_\theta}}{\partial \theta_k}, \end{aligned}$$

as stated in (5.25). □

Bibliography

Y. Ameho and E. Prempain. Linear parameter varying controllers for the ADMIRE aircraft longitudinal dynamics. In *Proc. American Control Conf. (ACC)*, pages 1315–1320, 2011.

B. D. O. Anderson and J. B. Moore. *Optimal Control: Linear Quadratic Methods*. Prentice Hall, 1990.

M. Andreasson, H. Sandberg, D. V. Dimarogonas, and K. H. Johansson. Distributed integral action: Stability analysis and frequency control of power systems. In *Proc. 51st IEEE Conf. Decision and Control (CDC)*, pages 2077–2083, 2012.

M. Andreasson, D. V. Dimarogonas, H. Sandberg, and K. H. Johansson. Distributed control of networked dynamical systems: Static feedback, integral action and consensus. *IEEE Trans. Autom. Control*, 59(7):1750–1764, 2014a.

M. Andreasson, D. V. Dimarogonas, H. Sandberg, and K. H. Johansson. Distributed PI-control with applications to power systems frequency control. In *Proc. American Control Conf. (ACC)*, pages 3183–3188, 2014b.

P. J. Antsaklis. Intelligent control. *Wiley Encyclopedia of Electrical and Electronics Engineering*, 10:493–503, 1999.

P. Apkarian, P. Gahinet, and G. Becker. Self-scheduled $\mathcal{H}_\infty$ control of linear parameter-varying systems: A design example. *Automatica*, 31(9):1251–1261, 1995.

M. Arcak. Passivity as a design tool for group coordination. *IEEE Trans. Autom. Control*, 52(8): 1380–1390, 2007.

K. J. Åström and P. R. Kumar. Control: A perspective. *Automatica*, 50:3–43, 2014.

L. Bakule. Decentralized control: An overview. *Annu. Reviews in Control*, 32(1):87–98, 2008.

G. Balas, J. Bokor, and Z. Szabo. Invariant subspaces for LPV systems and their applications. *IEEE Trans. Autom. Control*, 48(11):2065–2069, 2003.

M. Bartels and H. Werner. Cooperative and consensus-based approaches to formation control of autonomous vehicles. In *Proc. 19h IFAC World Congr.*, pages 8079–8084, 2014.

G. Basile and G. Marro. *Controlled and Conditioned Invariants in Linear System Theory*. Prentice Hall, 1992.

F. Bauer. Normalized graph Laplacians for directed graphs. *Linear Algebra and Its Applications*, 436(11):4193–4222, 2012.

G. Becker, A. Packard, D. Philbrick, and G. Balas. Control of parametrically-dependent linear systems: A single quadratic Lyapunov approach. In *Proc. American Control Conf. (ACC)*, pages 2795–2799, 1993.

D. S. Bernstein. *Matrix Mathematics: Theory, Facts, and Formulas*. Princeton University Press, 2nd ed., 2009.

A. Bidram, F. L. Lewis, and A. Davoudi. Distributed control systems for small-scale power networks: Using multiagent cooperative control theory. *IEEE Control Syst. Mag.*, 34(6):56–77, 2014.

P.-A. Bliman and G. Ferrari-Trecate. Average consensus problems in networks of agents with delayed communications. *Automatica*, 44(8):1985–1995, 2008.

R. Blind. *Optimization of the Communication System for Networked Control Systems*. PhD thesis, Inst. Systems Theory and Automatic Control, University of Stuttgart, Germany, 2014.

V. D. Blondel and J. N. Tsitsiklis. NP-Hardness of some linear-control design-problems. *SIAM J. Control Optim.*, 35(6):2118–2127, 1997.

V. D. Blondel, J. M. Hendrickx, A. Olshevsky, and J. N. Tsitsiklis. Convergence in multiagent coordination, consensus, and flocking. In *Proc. 44th IEEE Conf. Decision and Control, and the European Control Conf. (CDC–ECC)*, pages 2996–3000, 2005.

S. Boyd, L. El Ghaoui, E. Feron, and V. Balakrishnan. *Linear Matrix Inequalities in System and Control Theory*. SIAM, 1994.

S. Boyd, N. Parikh, E. Chu, B. Peleato, and J. Eckstein. Distributed optimization and statistical learning via the alternating direction method of multipliers. *Foundations and Trends in Machine Learning*, 3(1):1–122, 2011.

C. Briat. *Linear Parameter-Varying and Time-Delay Systems*. Springer, 2015.

R. A. Brualdi and H. J. Ryser. *Combinatorial Matrix Theory*. Cambridge University Press, 1991.

R. A. Brualdi and B. L. Shader. Graphs and Matrices. In L. W. Beineke and R. J. Wilson, editors, *Topics in Algebraic Graph Theory*. Cambridge University Press, 2004.

E. Brynjolfsson and A. McAfee. *The Second Machine Age: Work, Progress, and Prosperity in a Time of Brilliant Technologies*. W. W. Norton & Company, 2014.

F. Bullo, J. Cortés, and S. Martínez. *Distributed Control of Robotic Networks: A Mathematical Approach to Motion Coordination Algorithms*. Princeton University Press, 2009.

D. A. Burbano Lombana and M. di Bernardo. Distributed PID control for consensus of homogeneous and heterogeneous networks. *IEEE Trans. Control Network Syst.*, 2(2):154–163, 2015.

M. Bürger and C. De Persis. Dynamic coupling design for nonlinear output agreement and time-varying flow control. *Automatica*, 51:210–222, 2015.

M. Bürger, G. Notarstefano, F. Bullo, and F. Allgöwer. A distributed simplex algorithm for degenerate linear programs and multi-agent assignments. *Automatica*, 48(9):2298–2304, 2012.

M. Bürger, G. Notarstefano, and F. Allgöwer. A polyhedral approximation framework for convex and robust distributed optimization. *IEEE Trans. Autom. Control*, 59(2):384–395, 2014.

K. Cai and H. Ishii. Quantized consensus and averaging on gossip digraphs. *IEEE Trans. Autom. Control*, 56(9):2087–2100, 2011.

Y. Cao and W. Ren. Optimal linear-consensus algorithms: an LQR perspective. *IEEE Trans. Syst. Man Cybern. B, Cybern.*, 40(3):819–30, 2010.

Y. Cao, W. Yu, W. Ren, and G. Chen. An overview of recent progress in the study of distributed multi-agent coordination. *IEEE Trans. Ind. Informat.*, 9(1):427–438, 2013.

Y.-Y. Cao, J. Lam, and Y.-X. Sun. Static output feedback stabilization: an ILMI approach. *Automatica*, 34(12):1641–1645, 1998.

B. W. Carabelli, A. Benzing, G. S. Seyboth, R. Blind, M. Bürger, F. Dürr, B. Koldehofe, K. Rothermel, and F. Allgöwer. Exact convex formulations of network-oriented optimal operator placement. In *Proc. 51st IEEE Conf. Decision and Control (CDC)*, pages 3777–3782, 2012.

R. Carli and E. Lovisari. Robust synchronization of networks of heterogeneous double-integrators. In *Proc. 51st IEEE Conf. Decision and Control (CDC)*, pages 260–265, 2012.

R. Carli, A. Chiuso, L. Schenato, and S. Zampieri. Distributed Kalman filtering based on consensus strategies. *IEEE J. Selected Areas in Communications*, 26(4):622–633, 2008.

R. Carli, A. Chiuso, L. Schenato, and S. Zampieri. Optimal synchronization for networks of noisy double integrators. *IEEE Trans. Autom. Control*, 56(5):1146–1152, 2011.

N. Ceccarelli, M. Di Marco, A. Garulli, and A. Giannitrapani. Collective circular motion of multi-vehicle systems. *Automatica*, 44(12):3025–3035, 2008.

F. Chen, Y. Cao, and W. Ren. Distributed average tracking of multiple time-varying reference signals with bounded derivatives. *IEEE Trans. Autom. Control*, 57(12):3169–3174, 2012.

F. Chen, G. Feng, L. Liu, and W. Ren. An extended proportional-integral control algorithm for distributed average tracking and its applications in euler-lagrange systems. In *Proc. American Control Conf. (ACC)*, pages 2581–2586, 2014.

Z. Chen and H.-T. Zhang. No-beacon collective circular motion of jointly connected multi-agents. *Automatica*, 47(9):1929–1937, 2011.

Z. Chen and H.-T. Zhang. A remark on collective circular motion of heterogeneous multi-agents. *Automatica*, 49(5):1236–1241, 2013.

M. Chilali and P. Gahinet. $\mathcal{H}_\infty$ design with pole placement constraints: An LMI approach. *IEEE Trans. Autom. Control*, 41(3):358–367, 1996.

M. Chilali, P. Gahinet, and P. Apkarian. Robust pole placement in LMI regions. *IEEE Trans. Autom. Control*, 44(12):2257–2270, 1999.

S.-J. Chung and J.-J. E. Slotine. Cooperative robot control and concurrent synchronization of lagrangian systems. *IEEE Trans. Robot.*, 25(3):686–700, 2009.

Daimler AG. *World premiere on U.S. highway: Daimler Trucks drives first autonomous truck on public roads.* press release (http://www.media.daimler.com), Stuttgart / Las Vegas, Nevada, May 5th 2015.

E. J. Davison. The robust decentralized control of a general servomechanism problem. *IEEE Trans. Autom. Control*, 21(1):14–24, 1976.

M. C. de Oliveira and R. E. Skelton. Stability tests for constrained linear systems. In S. O. R. Moheimani, editor, *Perspectives in Robust Control*, volume 268 of *Lecture Notes in Control and Information Sciences*, pages 241–257. Springer, 2001.

C. De Persis and P. Frasca. Robust self-triggered coordination with ternary controllers. *IEEE Trans. Autom. Control*, 58(12):3024–3038, 2013.

C. De Persis, H. Liu, and M. Cao. Robust decentralized output regulation for uncertain heterogeneous systems. In *Proc. American Control Conf. (ACC)*, pages 5214–5219, 2012.

M. Deghat, I. Shames, B. D. O. Anderson, and C. Yu. Localization and circumnavigation of a slowly moving target using bearing measurements. *IEEE Trans. Autom. Control*, 59(8): 2182–2188, 2014.

P. DeLellis, M. di Bernardo, and G. Russo. On QUAD, Lipschitz, and contracting vector fields for consensus and synchronization of networks. *IEEE Trans. Circuits Syst. I, Reg. Papers*, 58 (3):576–583, 2011.

D. V. Dimarogonas and K. J. Kyriakopoulos. On the rendezvous problem for multiple nonholonomic agents. *IEEE Trans. Autom. Control*, 52(5):916–922, 2007.

D. V. Dimarogonas, S. G. Loizou, K. J. Kyriakopoulos, and M. M. Zavlanos. A feedback stabilization and collision avoidance scheme for multiple independent non-point agents. *Automatica*, 42(2):229–243, 2006.

D. V. Dimarogonas, E. Frazzoli, and K. H. Johansson. Distributed event-triggered control for multi-agent systems. *IEEE Trans. Autom. Control*, 57(5):1291–1297, 2011.

F. Dörfler and F. Bullo. Synchronization and transient stability in power networks and non-uniform kuramoto oscillators. *SIAM J. Control Optim.*, 50(3):1616–1642, 2012.

F. Dörfler and F. Bullo. Synchronization in complex networks of phase oscillators: A survey. *Automatica*, 50(6):1539–1564, 2014.

A. Eichler, C. Hoffmann, and H. Werner. Robust control of decomposable LPV systems. *Automatica*, 50(12):3239–3245, 2014.

J. A. Fax and R. M. Murray. Information flow and cooperative control of vehicle formations. *IEEE Trans. Autom. Control*, 49(9):1465–1476, 2004.

M. Fiedler. Algebraic connectivity of graphs. *Czechoslovak Mathematical J.*, 23(2):298–305, 1973.

P. Finsler. Über das Vorkommen definiter und semidefiniter Formen in Scharen quadratischer Formen. *Commentarii Mathematici Helvetici*, 9:188–192, 1937.

B. A. Francis. The linear multivariable regulator problem. *SIAM J. Control Optim.*, 15(3): 486–505, 1977.

B. A. Francis and W. M. Wonham. The internal model principle for linear multivariable regulators. *Applied Mathematics & Optimization*, 2(2):170–194, 1975.

B. A. Francis and W. M. Wonham. The internal model principle of control theory. *Automatica*, 12(5):457–465, 1976.

P. Frasca, R. Carli, F. Fagnani, and S. Zampieri. Average consensus on networks with quantized communication. *Int. J. Robust and Nonlinear Control*, 19(16):1787–1816, 2009.

P. Frasca, H. Ishii, C. Ravazzi, and R. Tempo. Distributed randomized algorithms for opinion formation, centrality computation and power systems estimation: A tutorial overview. *European J. Control*, 24:2–13, 2015.

R. A. Freeman, P. Yang, and K. M. Lynch. Stability and convergence properties of dynamic average consensus estimators. In *Proc. 45th IEEE Conf. Decision and Control*, pages 398–403, 2006.

S. Galeani, S. Tarbouriech, M. C. Turner, and L. Zaccarian. A tutorial on modern anti-windup design. *European J. Control*, 15(3-4):418–440, 2009.

S. Ganguli, A. Marcos, and G. Balas. Reconfigurable LPV control design for Boeing 747-100/200 longitudinal axis. In *Proc. American Control Conf. (ACC)*, volume 5, pages 3612–3617, 2002.

F. Garin and L. Schenato. A survey on distributed estimation and control applications using linear consensus algorithms. In A. Bemporad, M. Heemels, and M. Johansson, editors, *Networked Control Systems*, volume 406 of *Lecture Notes in Control and Information Sciences*, pages 75–107. Springer, 2010.

C. D. Godsil and G. Royle. *Algebraic Graph Theory*. Springer, 2001.

Google. *Green lights for our self-driving vehicle prototypes*. blog post (http://googleblog.blogspot.de/2015/05/self-driving-vehicle-prototypes-on-road.html), May 15th 2015.

H. F. Grip, T. Yang, A. Saberi, and A. A. Stoorvogel. Output synchronization for heterogeneous networks of non-introspective agents. *Automatica*, 48(10):2444–2453, 2012.

H. F. Grip, A. Saberi, and A. A. Stoorvogel. On the existence of virtual exosystems for synchronized linear networks. *Automatica*, 49(10):3145–3148, 2013.

A. Helmersson. *Methods for Robust Gain Scheduling*. PhD thesis, Dept. Electrical Engineering Automatic Control, Linköping University, Sweden, 1995.

K. Hengster-Movric and F. L. Lewis. Cooperative optimal control for multi-agent systems on directed graph topologies. *IEEE Trans. Autom. Control*, 59(3):769–774, 2014.

M. Herceg, M. Kvasnica, C. N. Jones, and M. Morari. Multi-Parametric Toolbox 3.0. In *Proc. European Control Conf. (ECC)*, pages 502–510, 2013. (http://control.ee.ethz.ch/~mpt).

E. G. Hernández-Martínez and E. Aranda-Bricaire. Convergence and collision avoidance in formation control: A survey of the artificial potential functions approach. In F. Alkhateeb, editor, *Multi-Agent Systems - Modelling, Control, Programming, Simulations and Appliactions*, pages 103–126. InTech, 2011.

F. Heß and G. S. Seyboth. Individual pitch control with tower side-to-side damping. In *Proc. 10th German Wind Energy Conf. (DEWEK)*, 2010.

C. Hoffmann, A. Eichler, and H. Werner. Control of heterogeneous groups of LPV systems interconnected through directed and switching topologies. In *Proc. American Control Conf. (ACC)*, pages 5156–5161, 2014.

Y. Hong, J. Hu, and L. Gao. Tracking control for multi-agent consensus with an active leader and variable topology. *Automatica*, 42(7):1177–1182, 2006.

Y. Hong, X. Wang, and Z. P. Jiang. Distributed output regulation of leader-follower multi-agent systems. *Int. J. Robust and Nonlinear Control*, 23(1):48–66, 2013.

R. A. Horn and C. R. Johnson. *Martix Analysis*. Cambridge University Press, 1985.

J. Huang. *Nonlinear Output Regulation: Theory and Application*. SIAM, 2004.

J. Huang. Remarks on synchronized output regulation of linear networked systems. *IEEE Trans. Autom. Control*, 56(3):630–631, 2011.

A. Isidori. *Nonlinear Control Systems*. Springer, 1995.

A. Isidori, L. Marconi, and G. Casadei. Robust output synchronization of a network of heterogeneous nonlinear agents via nonlinear regulation theory. *IEEE Trans. Autom. Control*, 59(10): 2680–2691, 2014.

A. Jadbabaie, J. Lin, and A. S. Morse. Coordination of groups of mobile autonomous agents using nearest neighbor rules. *IEEE Trans. Autom. Control*, 48(6):988–1001, 2003.

E. W. Justh and P. S. Krishnaprasad. Equilibria and steering laws for planar formations. *Systems & Control Letters*, 52(1):25–38, 2004.

H. Khalil. *Nonlinear systems*. Prentice Hall, 3rd ed., 2002.

S. Khodaverdian and J. Adamy. Entkopplungsbasierte Synchronisierung heterogener linearer Multi-Agenten-Systeme. *at - Automatisierungstechnik*, 62(12):865–876, 2014.

M. R. Kianifar. Cooperative Control of Nonholonomic Vehicles under Communication Constraints. Diploma thesis, Inst. Systems Theory and Automatic Control, University of Stuttgart, Germany, 2015.

H. Kim, H. Shim, and J. H. Seo. Output consensus of heterogeneous uncertain linear multi-agent systems. *IEEE Trans. Autom. Control*, 56(1):200–206, 2011.

D. J. Klein and K. A. Morgansen. Controlled collective motion for trajectory tracking. In *Proc. American Control Conf. (ACC)*, pages 5269–5275, 2006.

H. W. Knobloch and H. Kwakernaak. *Lineare Kontrolltheorie*. Springer, 1985.

H. W. Knobloch, A. Isidori, and D. Flockerzi. *Topics in Control Theory*. Birkhäuser, 1993.

H. Köroğlu and C. W. Scherer. Scheduled control for robust attenuation of non-stationary sinusoidal disturbances with measurable frequencies. *Automatica*, 47(3):504–514, 2011.

Y. Kuramoto. *Chemical Oscillations, Waves, and Turbulence*. Springer, 2003.

G. Lafferriere, A. Williams, J. Caughman, and J. J. P. Veerman. Decentralized control of vehicle formations. *Systems & Control Letters*, 54(9):899–910, 2005.

E. Lalish, K. A. Morgansen, and T. Tsukamaki. Oscillatory control for constant-speed unicycle-type vehicles. In *Proc. 46th IEEE Conf. Decision and Control*, pages 5246–5251, 2007.

J. P. LaSalle. *The Stability of Dynamical Systems*. SIAM, 1976.

J. Lawton, R. W. Beard, and B. J. Young. A decentralized approach to formation maneuvers. *IEEE Trans. Robot. Autom.*, 19(6):933–941, 2003.

N. E. Leonard, D. Paley, F. Lekien, R. Sepulchre, D. M. Fratantoni, and R. E. Davis. Collective motion, sensor networks, and ocean sampling. *Proc. IEEE*, 95(1):48–74, 2007.

F. Lescher, J. Y. Zhao, and P. Borne. Robust gain scheduling controller for pitch regulated variable speed wind turbine. *Studies in Informatics and Control*, 14(4):299–315, 2005.

F. L. Lewis, H. Zhang, K. Hengster-Movric, and A. Das. *Cooperative Control of Multi-Agent Systems: Optimal and Adaptive Design Approaches*. Springer, 2014.

Z. Li, Z. Duan, G. Chen, and L. Huang. Consensus of multiagent systems and synchronization of complex networks: A unified viewpoint. *IEEE Trans. Circuits Syst. I, Reg. Papers*, 57(1): 213–224, 2010.

Z. Li, Z. Duan, and G. Chen. On $\mathcal{H}_\infty$ and $\mathcal{H}_2$ performance regions of multi-agent systems. *Automatica*, 47(4):797–803, 2011.

Z. Lin, B. A. Francis, and M. Maggiore. Necessary and sufficient graphical conditions for formation control of unicycles. *IEEE Trans. Autom. Control*, 50(1):121–127, 2005.

K. D. Listmann. *Synchronization of Networked Linear Systems – an LMI Approach*. PhD thesis, Control Methods and Robotics, TU Darmstadt, Germany, 2012.

K. D. Listmann, A. Wahrburg, J. Strubel, J. Adamy, and U. Konigorski. Partial-state synchronization of linear heterogeneous multi-agent systems. In *Proc. 50th IEEE Conf. Decision and Control (CDC)*, pages 3440–3445, 2011.

H. Liu, C. De Persis, and M. Cao. Robust decentralized output regulation with single or multiple reference signals for uncertain heterogeneous systems. *Int. J. Robust and Nonlinear Control*, 25(9):1399–1422, 2015.

J. Löfberg. YALMIP: a toolbox for modeling and optimization in MATLAB. In *Proc. CACSD Conf.*, pages 284–289, 2004. (http://users.isy.liu.se/johanl/yalmip).

E. Lovisari and U. T. Jönsson. A framework for robust synchronization in heterogeneous multi-agent networks. In *Proc. 50th IEEE Conf. Decision and Control (CDC)*, pages 7268–7274, 2011.

J. Lunze. Synchronization of heterogeneous agents. *IEEE Trans. Autom. Control*, 57(11): 2885–2890, 2012.

C. Ma and J. Zhang. Necessary and sufficient conditions for consensusability of linear multi-agent systems. *IEEE Trans. Autom. Control*, 55(5):1263–1268, 2010.

J. A. Marshall, M. E. Broucke, and B. A. Francis. Formations of vehicles in cyclic pursuit. *IEEE Trans. Autom. Control*, 49(11):1963–1974, 2004.

J. A. Marshall, M. E. Broucke, and B. A. Francis. Pursuit formations of unicycles. *Automatica*, 42(1):3–12, 2006.

P. Massioni and M. Verhaegen. Distributed control for identical dynamically coupled systems: A decomposition approach. *IEEE Trans. Autom. Control*, 54(1):124–135, 2009.

J. Mei, W. Ren, and G. Ma. Distributed coordinated tracking with a dynamic leader for multiple euler-lagrange systems. *IEEE Trans. Autom. Control*, 56(6):1415–1421, 2011.

J. Mei, W. Ren, and J. Chen. Consensus of second-order heterogeneous multi-agent systems under a directed graph. In *Proc. American Control Conf. (ACC)*, pages 802–807, 2014.

Z. Meng, T. Yang, D. V. Dimarogonas, and K. H. Johansson. Coordinated output regulation of multiple heterogeneous linear systems. In *Proc. 52nd IEEE Conf. Decision and Control (CDC)*, pages 2175–2180, 2013.

M. Mesbahi and M. Egerstedt. *Graph Theoretic Methods in Multiagent Networks*. Princeton University Press, 2010.

J. M. Montenbruck, G. S. Seyboth, and F. Allgöwer. Practical and robust synchronization of systems with additive linear uncertainties. In *9th IFAC Symp. Nonlinear Control Systems (NOLCOS)*, pages 743–748, 2013.

J. M. Montenbruck, M. Bürger, and F. Allgöwer. Practical synchronization with diffusive couplings. *Automatica*, 53:235–243, 2015a.

J. M. Montenbruck, G. S. Schmidt, G. S. Seyboth, and F. Allgöwer. On the necessity of diffusive couplings in linear synchronization problems with quadratic cost. *IEEE Trans. Autom. Control*, 2015b. doi: 10.1109/TAC.2015.2406971.

L. Moreau. Stability of multiagent systems with time-dependent communication links. *IEEE Trans. Autom. Control*, 50(2):169–182, 2005.

U. Münz, A. Papachristodoulou, and F. Allgöwer. Delay robustness in consensus problems. *Automatica*, 46(8):1252–1265, 2010.

R. M. Murray, K. J. Åström, S. Boyd, R. W. Brockett, and G. Stein. Future directions in control in an information-rich world. *IEEE Control Syst. Mag.*, 23(2):20–33, 2003.

T. Nagashio, T. Kida, T. Ohtani, and Y. Hamada. Design and implementation of robust symmetric attitude controller for ETS-VIII spacecraft. *Control Engineering Practice*, 18(12):1440–1451, 2010.

A. Nedic and A. Ozdaglar. Distributed subgradient methods for multi-agent optimization. *IEEE Trans. Autom. Control*, 54(1):48–61, 2009.

W. Ni and D. Cheng. Leader-following consensus of multi-agent systems under fixed and switching topologies. *Systems & Control Letters*, 59(3-4):209–217, 2010.

K.-K. Oh, M.-C. Park, and H.-S. Ahn. A survey of multi-agent formation control. *Automatica*, 53:424–440, 2015.

R. Olfati-Saber. Distributed Kalman filter with embedded consensus filters. In *Proc. 44th IEEE Conf. Decision and Control, and the European Control Conf. (CDC–ECC)*, pages 8179–8184, 2005.

R. Olfati-Saber. Flocking for multi-agent dynamic systems: Algorithms and theory. *IEEE Trans. Autom. Control*, 51(3):401–420, 2006.

R. Olfati-Saber and R. M. Murray. Consensus problems in networks of agents with switching topology and time-delays. *IEEE Trans. Autom. Control*, 49(9):1520–1533, 2004.

R. Olfati-Saber and J. S. Shamma. Consensus filters for sensor networks and distributed sensor fusion. In *Proc. 44th IEEE Conf. Decision and Control, and the European Control Conf. (CDC–ECC)*, pages 6698–6703, 2005.

R. Olfati-Saber, J. A. Fax, and R. M. Murray. Consensus and cooperation in networked multi-agent systems. *Proc. IEEE*, 95(1):215–233, 2007.

D. Paley, N. E. Leonard, R. Sepulchre, D. Grünbaum, and J. K. Parrish. Oscillator models and collective motion. *IEEE Control Syst. Mag.*, 27(4):89–105, 2007.

I. C. Paschalidis and M. Egerstedt. The inaugural issue of the IEEE Transactions on Control of Network Systems. *IEEE Trans. Control Network Syst.*, 1(1):1–3, 2014.

J. Ploeg, S. Shladover, H. Nijmeijer, and N. van de Wouw. Introduction to the Special Issue on the 2011 Grand Cooperative Driving Challenge. *IEEE Trans. Intell. Transp. Syst.*, 13(3):989–993, 2012.

J. Ploeg, D. P. Shukla, N. van de Wouw, and H. Nijmeijer. Controller synthesis for string stability of vehicle platoons. *IEEE Trans. Intell. Transp. Syst.*, 15(2):854–865, 2014.

Z. Qu. *Cooperative Control of Dynamical Systems: Applications to Autonomous Vehicles*. Springer, 2009.

Z. Qu. Cooperative control of networked nonlinear systems. In *Proc. 49th IEEE Conf. Decision and Control (CDC)*, pages 3200–3207, 2010.

M. Rabbat and R. Nowak. Distributed optimization in sensor networks. In *Proc. 3rd Int. Symp. Information Processing in Sensor Networks*, pages 20–27, 2004.

W. Ren. Consensus strategies for cooperative control of vehicle formations. *IET Control Theory & Applications*, 1(2):505–512, 2007.

W. Ren. Synchronization of coupled harmonic oscillators with local interaction. *Automatica*, 44 (12):3195–3200, 2008.

W. Ren and E. M. Atkins. Distributed multi-vehicle coordinated control via local information exchange. *Int. J. Robust and Nonlinear Control*, 17(10-11):1002–1033, 2007.

W. Ren and R. W. Beard. Consensus seeking in multiagent systems under dynamically changing interaction topologies. *IEEE Trans. Autom. Control*, 50(5):655–661, 2005.

W. Ren and R. W. Beard. *Distributed Consensus in Multi-vehicle Cooperative Control: Theory and Applications*. Springer, 2008.

W. Ren and Y. Cao. *Distributed Coordination of Multi-Agent Networks: Emergent Problems, Models, and Issues*. Springer, 2011.

W. Ren, R. W. Beard, and E. M. Atkins. A survey of consensus problems in multi-agent coordination. In *Proc. American Control Conf. (ACC)*, pages 1859–1864, 2005a.

W. Ren, R. W. Beard, and T. W. McLain. Coordination variables and consensus building in multiple vehicle systems. In V. Kumar, N. Leonard, and A. S. Morse, editors, *Cooperative Control*, volume 309 of *Lecture Notes in Control and Information Sciences*, pages 171–188. Springer, 2005b.

W. Ren, R. W. Beard, and E. M. Atkins. Information consensus in multivehicle cooperative control. *IEEE Control Syst. Mag.*, 27(2):71–82, 2007.

C. W. Reynolds. Flocks, herds and schools: A distributed behavioral model. *ACM SIGGRAPH Computer Graphics*, 21(4):25–34, 1987.

E. Roche, O. Sename, and D. Simon. LPV/$\mathcal{H}_\infty$ control of an autonomous underwater vehicle (AUV). In *Proc. European Control Conf. (ECC)*, 2009.

P. E. Ross. Robot, you can drive my car. *IEEE Spectrum*, 51(6):60–90, 2014.

A. Saberi, A. A. Stoorvogel, and P. Sannuti. *Control of Linear Systems with Regulation and Input Constraints*. Springer, 2000.

A. Sarlette, R. Sepulchre, and N. E. Leonard. Cooperative attitude synchronization in satellite swarms: a consensus approach. In *Proc. 17th IFAC Symp. Automatic Control in Aerospace*, pages 223–228, 2007.

A. Sarlette, R. Sepulchre, and N. E. Leonard. Autonomous rigid body attitude synchronization. *Automatica*, 45(2):572–577, 2009.

L. Scardovi and R. Sepulchre. Synchronization in networks of identical linear systems. *Automatica*, 45(11):2557–2562, 2009.

C. W. Scherer. LPV control and full block multipliers. *Automatica*, 37(3):361–375, 2001.

R. Sepulchre, D. Paley, and N. E. Leonard. Stabilization of planar collective motion: All-to-all communication. *IEEE Trans. Autom. Control*, 52(5):811–824, 2007.

R. Sepulchre, D. Paley, and N. E. Leonard. Stabilization of planar collective motion with limited communication. *IEEE Trans. Autom. Control*, 53(3):706–719, 2008.

A. Seuret, D. V. Dimarogonas, and K. H. Johansson. Consensus under communication delays. In *47th IEEE Conf. Decision and Control (CDC)*, pages 4922–4927, 2008.

G. S. Seyboth and F. Allgöwer. Clock synchronization over directed graphs. In *Proc. 52nd IEEE Conf. Decision and Control (CDC)*, pages 6105–6111, 2013.

G. S. Seyboth and F. Allgöwer. Synchronized model matching: a novel approach to cooperative control of heterogeneous multi-agent systems. In *Proc. 19h IFAC World Congr.*, pages 1985–1990, 2014.

G. S. Seyboth and F. Allgöwer. Output synchronization of linear multi-agent systems under constant disturbances via distributed integral action. In *Proc. American Control Conf. (ACC)*, pages 62–67, 2015.

G. S. Seyboth, D. V. Dimarogonas, and K. H. Johansson. Control of multi-agent systems via event-based communication. In *Proc. 18th IFAC World Congr.*, pages 10086–10091, 2011.

G. S. Seyboth, D. V. Dimarogonas, K. H. Johansson, and F. Allgöwer. Static diffusive couplings in heterogeneous linear networks. In *Proc. 3rd IFAC Workshop on Distributed Estimation and Control in Networked Systems*, pages 258–263, 2012a.

G. S. Seyboth, G. S. Schmidt, and F. Allgöwer. Cooperative control of linear parameter-varying systems. In *Proc. American Control Conf. (ACC)*, pages 2407–2412, 2012b.

G. S. Seyboth, G. S. Schmidt, and F. Allgöwer. Output synchronization of linear parameter-varying systems via dynamic couplings. In *Proc. 51st IEEE Conf. Decision and Control (CDC)*, pages 5128–5133, 2012c.

G. S. Seyboth, D. V. Dimarogonas, and K. H. Johansson. Event-based broadcasting for multi-agent average consensus. *Automatica*, 49(1):245–252, 2013.

G. S. Seyboth, W. Ren, and F. Allgöwer. Cooperative control of linear multi-agent systems via distributed output regulation and transient synchronization. *arXiv preprint arXiv:1406.0085*, 2014a.

G. S. Seyboth, J. Wu, J. Qin, C. Yu, and F. Allgöwer. Collective circular motion of unicycle type vehicles with nonidentical constant velocities. *IEEE Trans. Control Network Syst.*, 1(2): 167–176, 2014b.

G. S. Seyboth, D. V. Dimarogonas, K. H. Johansson, P. Frasca, and F. Allgöwer. On robust synchronization of heterogeneous linear multi-agent systems with static couplings. *Automatica*, 53:392–399, 2015.

G. S. Seyboth, W. Ren, and F. Allgöwer. Cooperative control of linear multi-agent systems via distributed output regulation and transient synchronization. *Automatica*, 68:132–139, 2016.

J. S. Shamma and M. Athans. Guaranteed properties of gain scheduled control for linear parameter-varying plants. *Automatica*, 27(3):559–564, 1991.

H. Shim, J. Lee, J.-S. Kim, and J. Back. Output regulation problem and solution for ltv minimum phase systems with time-varying exosystem. In *Proc. Int. Joint Conf. SICE-ICASE*, pages 1823–1827, 2006.

A. Sinha and D. Ghose. Generalization of nonlinear cyclic pursuit. *Automatica*, 43(11):1954–1960, 2007.

S. H. Strogatz. From Kuramoto to Crawford: exploring the onset of synchronization in populations of coupled oscillators. *Physica D: Nonlinear Phenomena*, 143(1-4):1–20, 2000.

Y. Su and J. Huang. Cooperative output regulation of linear multi-agent systems by output feedback. *Systems & Control Letters*, 61(12):1248–1253, 2012a.

Y. Su and J. Huang. Cooperative output regulation of linear multi-agent systems. *IEEE Trans. Autom. Control*, 57(4):1062–1066, 2012b.

Y. Su and J. Huang. Cooperative output regulation with application to multi-agent consensus under switching network. *IEEE Trans. Syst. Man Cybern. B, Cybern.*, 42(3):864–875, 2012c.

Y. Su and J. Huang. Cooperative robust output regulation of a class of heterogeneous linear uncertain multi-agent systems. *Int. J. Robust and Nonlinear Control*, 24(17):2819–2839, 2014.

Y. Su, Y. Hong, and J. Huang. A general result on the robust cooperative output regulation for linear uncertain multi-agent systems. *IEEE Trans. Autom. Control*, 58(5):1275–1279, 2013.

Z. Sun, G. S. Seyboth, and B. D. O. Anderson. Collective control of multiple unicycle agents with non-identical constant speeds: Formation stabilization. 2015a. under preparation.

Z. Sun, G. S. Seyboth, and B. D. O. Anderson. Collective control of multiple unicycle agents with non-identical constant speeds: Tracking control and performance limitation. In *Proc. IEEE Conf. Control Applications (CCA), Part of IEEE Multi-Conference on Systems and Control (MSC)*, pages 1361–1366, 2015b.

D. Swaroop and J. K. Hedrick. String stability of interconnected systems. *IEEE Trans. Autom. Control*, 41(3):349–357, 1996.

V. L. Syrmos, C. T. Abdallah, P. Dorato, and K. Grigoriadis. Static output feedback – a survey. *Automatica*, 33(2):125–137, 1997.

H. G. Tanner, A. Jadbabaie, and G. J. Pappas. Stable flocking of mobile agents, part i: Fixed topology. In *Proc. 42nd IEEE Conf. Decision and Control (CDC)*, pages 2010–2015, 2003a.

H. G. Tanner, A. Jadbabaie, and G. J. Pappas. Stable flocking of mobile agents, part ii: Dynamic topology. In *Proc. 42nd IEEE Conf. Decision and Control (CDC)*, pages 2016–2021, 2003b.

H. L. Trentelman, A. A. Stoorvogel, and M. Hautus. *Control Theory for Linear Systems.* Springer, 2001.

H. L. Trentelman, K. Takaba, and N. Monshizadeh. Robust synchronization of uncertain linear multi-agent systems. *IEEE Trans. Autom. Control*, 58(6):1511–1523, 2013.

J. N. Tsitsiklis. *Problems in Decentralized Decision Making and Computation.* PhD thesis, Dept. Electrical Engineering and Computer Science, Massachusetts Institute of Technology, Cambridge, USA, 1984.

J. N. Tsitsiklis, D. P. Bertsekas, and M. Athans. Distributed asynchronous deterministic and stochastic gradient optimization algorithms. *IEEE Trans. Autom. Control*, 31(9):803–812, 1986.

S. E. Tuna. LQR-based coupling gain for synchronization of linear systems. *arXiv preprint arXiv:0801.3390*, 2008.

V. A. Ugrinovskii. Gain-scheduled synchronization of uncertain parameter varying systems via relative $\mathcal{H}_\infty$ consensus. In *Proc. 50th IEEE Conf. Decision and Control (CDC)*, pages 4251–4256, 2011a.

V. A. Ugrinovskii. Distributed robust filtering with $\mathcal{H}_\infty$ consensus estimates. *Automatica*, 47(1): 1–13, 2011b.

V. A. Ugrinovskii. Gain-scheduled synchronization of parameter varying systems via relative $\mathcal{H}_\infty$ consensus with application to synchronization of uncertain bilinear systems. *Automatica*, 50 (11):2880–2887, 2014.

V. A. Ugrinovskii and C. Langbort. Distributed $\mathcal{H}_\infty$ consensus-based estimation of uncertain systems via dissipativity theory. *IET Control Theory & Applications*, 5(12):1458–1469, 2011.

D. van der Walle, B. Fidan, A. Sutton, C. Yu, and B. D. O. Anderson. Non-hierarchical UAV formation control for surveillance tasks. In *Proc. American Control Conf. (ACC)*, pages 777–782, 2008.

A. F. Vaz and E. J. Davison. The structured robust decentralized servomechanism problem for interconnected systems. *Automatica*, 25(2):267–272, 1989.

J. Veenman and C. W. Scherer. A synthesis framework for robust gain-scheduling controllers. *Automatica*, 50(11):2799–2812, 2014.

T. Vicsek, A. Czirók, E. Ben-Jacob, I. Cohen, and O. Shochet. Novel type of phase transition in a system of self-driven particles. *Physical Review Letters*, 75(6):1226, 1995.

J. Wang, Z. Duan, Y. Zhao, G. Qin, and Y. Yan. $\mathcal{H}_\infty$ and $\mathcal{H}_2$ control of multi-agent systems with transient performance improvement. *Int. J. Control*, 86(12):2131–2145, 2013.

J. Wang, Z. Li, Z. Duan, and G. Wen. Distributed $\mathcal{H}_\infty$ and $\mathcal{H}_2$ consensus control in directed networks. *IET Control Theory & Applications*, 8(3):193–201, 2014.

X. Wang, Y. Hong, J. Huang, and Z. P. Jiang. A distributed control approach to a robust output regulation problem for multi-agent linear systems. *IEEE Trans. Autom. Control*, 55(12): 2891–2895, 2010.

P. Wieland. *From Static to Dynamic Couplings in Consensus and Synchronization among Identical and Non-Identical Systems*. PhD thesis, Inst. Systems Theory and Automatic Control, University of Stuttgart, Germany, 2010.

P. Wieland and F. Allgöwer. An internal model principle for consensus in heterogeneous linear multi-agent systems. In *Proc. 1st IFAC Workshop on Estimation and Control of Networked Systems*, pages 7–12, 2009.

P. Wieland, J.-S. Kim, H. Scheu, and F. Allgöwer. On consensus in multi-agent systems with linear high-order agents. In *Proc. 17th IFAC World Congr.*, pages 1541–1546, 2008.

P. Wieland, J.-S. Kim, and F. Allgöwer. On topology and dynamics of consensus among linear high-order agents. *Int. J. Systems Science*, 42(10):1831–1842, 2011a.

P. Wieland, R. Sepulchre, and F. Allgöwer. An internal model principle is necessary and sufficient for linear output synchronization. *Automatica*, 47(5):1068–1074, 2011b.

P. Wieland, J. Wu, and F. Allgöwer. On synchronous steady states and internal models of diffusively coupled systems. *IEEE Trans. Autom. Control*, 58(10):2591–2602, 2013.

W. M. Wonham. Tracking and regulation in linear multivariable systems. *SIAM J. Control*, 11(3): 424–437, 1973.

W. M. Wonham. *Linear Multivariable Control: A Geometric Approach*. Springer, 1985.

C. W. Wu. *Synchronization in Complex Networks of Nonlinear Dynamical Systems*. World Scientific, 2007.

F. Wu, X. H. Yang, A. Packard, and G. Becker. Induced $\mathcal{L}_2$-norm control for lpv systems with bounded parameter variation rates. *Int. J. Robust and Nonlinear Control*, 6(9–10):983–998, 1996.

J. Wu and F. Allgöwer. A constructive approach to synchronization using relative information. In *Proc. 51st IEEE Conf. Decision and Control (CDC)*, pages 5960–5965, 2012.

J. Wu, V. A. Ugrinovskii, and F. Allgöwer. Cooperative estimation for synchronization of heterogeneous multi-agent systems using relative information. In *Proc. 19th IFAC World Congr.*, pages 4662–4667, 2014.

J. Xi, N. Cai, and Y. Zhong. Consensus problems for high-order linear time-invariant swarm systems. *Physica A: Statistical Mechanics and its Applications*, 389(24):5619–5627, 2010.

J. Xi, Z. Shi, and Y. Zhong. Output consensus analysis and design for high-order linear swarm systems: Partial stability method. *Automatica*, 48(9):2335–2343, 2012.

J. Xiang, W. Wei, and Y. Li. Synchronized output regulation of linear networked systems. *IEEE Trans. Autom. Control*, 54(6):1336–1341, 2009.

L. Xiao, S. Boyd, and S. Lall. A scheme for robust distributed sensor fusion based on average consensus. In *Proc. 4th Int. Symp. Information Processing in Sensor Networks (IPSN)*, pages 63–70, 2005.

G. Xie, H. Liu, L. Wang, and Y. Jia. Consensus in networked multi-agent systems via sampled control: Fixed topology case. In *Proc. American Control Conf. (ACC)*, pages 3902–3907, 2009.

T. Yang, A. Saberi, A. A. Stoorvogel, and H. F. Grip. Output synchronization for heterogeneous networks of introspective right-invertible agents. *Int. J. Robust and Nonlinear Control*, 24(13): 1821–1844, 2014.

J. Yu, S. M. LaValle, and D. Liberzon. Rendezvous without coordinates. *IEEE Trans. Autom. Control*, 57(2):421–434, 2012.

L. Yu and J. Wang. Robust cooperative control for multi-agent systems via distributed output regulation. *Systems & Control Letters*, 62(11):1049–1056, 2013.

T. Yucelen and M. Egerstedt. Control of multiagent systems under persistent disturbances. In *Proc. American Control Conf. (ACC)*, pages 5264–5269, 2012.

T. Yucelen and E. N. Johnson. Control of multivehicle systems in the presence of uncertain dynamics. *Int. J. Control*, 86(9):1540–1553, 2013.

A. Zecevic and D. D. Šiljak. *Control of Complex Systems: Structural Constraints and Uncertainty.* Springer, 2010.

D. Zelazo and M. Mesbahi. Graph-theoretic analysis and synthesis of relative sensing networks. *IEEE Trans. Autom. Control*, 56(5):971–982, 2011.

H. Zhang, F. L. Lewis, and A. Das. Optimal design for synchronization of cooperative systems: state feedback, observer and output feedback. *IEEE Trans. Autom. Control*, 56(8):1948–1952, 2011.

H. Zhang, F. L. Lewis, and Z. Qu. Lyapunov, adaptive, and optimal design techniques for cooperative systems on directed communication graphs. *IEEE Trans. Ind. Electron.*, 59(7): 3026–3041, 2012.

Y. Zhao, Z. Duan, G. Wen, and G. Chen. Distributed $\mathcal{H}_\infty$ consensus of multi-agent systems: a performance region-based approach. *Int. J. Control*, 85(3):332–341, 2012.

R. Zheng and D. Sun. Rendezvous of unicycles: A bearings-only and perimeter shortening approach. *Systems & Control Letters*, 62(5):401–407, 2013.

R. Zheng, Z. Lin, and G. Yan. Ring-coupled unicycles: Boundedness, convergence, and control. *Automatica*, 45(11):2699–2706, 2009.

K. Zhou and J. C. Doyle. *Essentials of Robust Control.* Prentice Hall, 1998.

J. Zhu, Y.-P. Tian, and J. Kuang. On the general consensus protocol of multi-agent systems with double-integrator dynamics. *Linear Algebra and its Applications*, 431(5-7):701–715, 2009.

L. Zhu and Z. Chen. Robust homogenization and consensus of nonlinear multi-agent systems. *Systems & Control Letters*, 65:50–55, 2014.

M. Zhu and S. Martínez. Discrete-time dynamic average consensus. *Automatica*, 46(2):322–329, 2010.